The EARTH WILL APPEAR *as the* GARDEN *of* EDEN

The EARTH WILL APPEAR *as the* GARDEN *of* EDEN

ESSAYS *on* MORMON ENVIRONMENTAL HISTORY

EDITED BY
JEDEDIAH S. ROGERS AND
MATTHEW C. GODFREY

THE UNIVERSITY OF UTAH PRESS
SALT LAKE CITY

The Defiance House Man colophon is a registered trademark of the University of Utah Press. It is based on a four-foot-tall Ancient Puebloan pictograph (late PIII) near Glen Canyon, Utah.

Library of Congress Cataloging-in-Publication Data

Names: Rogers, Jedediah S. (Jedediah Smart), 1978– editor. | Godfrey, Matthew C., editor.

Title: The earth will appear as the Garden of Eden : essays on Mormon environmental history / edited by Jedediah S. Rogers and Matthew C. Godfrey.

Description: Salt Lake City : The University of Utah Press, [2018] | Includes bibliographical references and index. |

Identifiers: LCCN 2018030659 (print) | LCCN 2018032522 (ebook) | ISBN 9781607816546 | ISBN 9781607816539 (pbk. : alk. paper)

Subjects: LCSH: Environmentalism—Religious aspects—Church of Jesus Christ of Latter-day Saints—History. | Environmentalism—Religious aspects—Mormon Church—History. | Zion (Mormon Church) | Ecotheology. | Environmentalism—Utah—History. | Environmentalism—West (U.S.)

Classification: LCC BX8643.Z55 (ebook) | LCC BX8643.Z55 E27 2018 (print) | DDC 261.8/80882893—dc23

LC record available at https://lccn.loc.gov/2018030659

"The Earth Will Appear as the Garden of Eden" is taken from the Mormon hymn "Now Let Us Rejoice" by William W. Phelps, which was first included in the LDS hymnbook in 1835.

Printed and bound in the U.S.A.

CONTENTS

INTRODUCTION

The Promise and Challenge of Mormon Environmental History

Jedediah S. Rogers and Matthew C. Godfrey

Despite its long-established presence in the academy, the perspective of environmental history has made few inroads into Mormon studies. This is true even as dozens of high-quality works on Mormonism appear annually; Mormon studies has perhaps never enjoyed more vitality. This is not to say that writers have ignored environmental themes, but the historians and scholars consciously examining Mormonism within the academic stream of environmental history have been few and are generally not recognized by the field.[1]

This is surprising given the prominence of environmental history in the academy and the subsequent way insights from the field have seeped into the literature. Like the new Mormon history, environmental history developed as a subfield in the milieu of the new social history—history "from the bottom up." Concerned as its practitioners were with the lives of ordinary people as well as the dynamics of gender, race, and ethnicity, environmental historians added "nature" as an active actor to the stories they told. The historical subfield developed alongside the American environmental movement of the 1960s and 1970s. Many of its practitioners, then and now, came to it through their own environmental awareness—a personal connection to the natural world and a commitment to protect it from despoliation. At the same time, historians who embraced this burgeoning field brought to their studies a methodological rigor that drew on the insights of historical geography, cultural ecology, and other disciplinary schools.

From the beginning, environmental historians examined "nature" as a category of analysis and situated it within the just as amorphous category of "culture." Early contending positions centered in one sense on the difference between seeing nature as an autonomous entity with its own set of values independent of humans or, unavoidably, as a product of culture—of human values and assumptions.[2] The trend now, as the historian Paul Sutter argues, is to complicate the category of "nature" by seeing "all environments as interweaving the natural and cultural in complex ways." Historians identify cultural traces in what was before considered "pure" nature and detect nature in what once appeared to be thoroughly human landscapes. Departing from earlier scholarship that saw the human-environmental relationship as a tale of declension, today's practitioners "have rejected the notions that environments transformed by human activity are sullied and fallen."[3]

However they define the nature-culture relationship or speak of nature as an "agent" similar to humans, environmental historians recognize nature as a vital force—one that acts on the human dominated world and that is unavoidably transformed by human activity. Environmental historians long recognized that while nature once largely set the terms of life, humans have continually modified, altered, and affected it. People have always stood in relation to nature. But the terms of that relationship have fundamentally changed with the Industrial Revolution and the attendant burgeoning population and capitalist economy. In the estimation of the historian Donald Worster, this made the nineteenth century distinct from the eighteenth, giving rise to environmental prophets like George Perkins Marsh, who lamented the destructive force of human society on the environment.[4] The paradigm shift from a static earth to a vulnerable one was not immediate, but in the twentieth and twenty-first centuries it has become a major way of representing the nation and the earth's environmental history. Some scholars today contend that we are entering a new geologic epoch—the Anthropocene—suggesting an unprecedented age of human dominance over the planet.[5]

In recent years, scholars have also increasingly investigated the impact of religion and religious movements on nature. Lynn White Jr., for example, argued in 1967 that Western culture's disrespect for the environment was a result of Judeo-Christian religious thought and culture.[6] Others saw religion as having a more salutary impact on the environmental movement; Mark Stoll asserted that many environmental leaders had religious backgrounds, although few attended church as adults.[7] Others, such as philosopher Roger S. Gottlieb, have explored the ways in which religious traditions have addressed the problem of humankind's destruction of its environment.[8] Entire journals have been devoted to studying the interplay between theology and the environment.[9] The connection between religion and nature is a burgeoning field, making the time ripe to examine Latter-day Saints and their past through the lens of environmental history.

Mormonism was born and matured during the past two centuries of industrialization, capitalism, and environmental change. From the beginnings of the church, its members interacted with nature in significant ways—whether perceiving it as a place to find God (as Joseph Smith did by going into a grove of trees to offer the prayer that put into motion his work as a spiritual leader), as a place to build communities that would usher in the Second Coming, or as a place brimming with natural resources they could use to further their economic well-being. One way to examine Mormon environmental history is to see it as a tale of environmental declension, both in perception (the theological shift away from salutary agrarianism) and practice (the transformation attendant with Mormonism's incorporation into American consumer culture). This sense of loss and decline is an important component of the environmental history of Mormonism, but there are multiple other ways of reading the environment in the Mormon past. We open with a reflection by Jedediah S. Rogers on the themes and tools of environmental history in Mormon historiography—how the terms *nature* and *environment* have variously been employed in traditional Mormon history, the catalog of works addressing Mormon themes, and the challenge and revisions environmental history offers to a study of the Mormon past. He shows how geographers and scholars from disciplines outside of history have traditionally taken the lead when examining environmental themes in Mormon country. Despite the relative paucity of historical works on Mormonism and the environment, a handful of texts in recent years represent the promise of the field.

The other essays are organized according to divisions generally characteristic of the field—i.e., representing ideas and perceptions about nature on one hand and practice and transformation on the other. The essays in Part I, "Theology and Ideology," offer contextualized histories of Mormon environmental thought, demonstrating the complexity of applying a nature-based theology in the nineteenth- and twentieth-century West. We begin with Sara Dant's reflections on the environmental ethic of Brigham Young, the influence such thinking had on Latter-day Saints, and how the Utah Mormons drifted away from Young's notions as they became more fully integrated into the nation and the economy. The next essay, by Thomas G. Alexander, argues that the Mormon de-emphasis of nineteenth-century environmental teachings during the twentieth century was an act of forgetting in the face of what he calls "secular entrepreneurship." The shift to secular entrepreneurship not only resulted in the abandonment of a pioneer environmental ethic but also major landscape transformations. Both essays demonstrate a common theme in the field of environmental history: that myths, beliefs, and ideas about nature are fluid and have on-the-ground consequences.

These essays also suggest a deep association of Mormons to a particular place: the valleys, mountains, and deserts of the Great Basin. In Part II, "Perception and Place," the essays bring more specificity to the Mormon connection with what Wallace Stegner called "Mormon Country" and Donald Meinig referred to as the Mormon cultural region, as well as expound on the connection between Mormonism and other environments. A strength of environmental history is its attention on place and place-making—on the richly textured examinations of people and groups interacting in particular locales. Consider the first two selections in this section: Matthew C. Godfrey's examination of environmental language in Jackson County, Missouri, and Brett Dowdle's investigation of foreign environments. Godfrey challenges traditional understandings of early Mormon identity and conflict by detailing Mormon thought and interaction in Missouri. Dowdle, meanwhile, explores the complex environments of Great Britain and Nauvoo, Illinois, showing not just how early Mormon missionaries and converts perceived and acted in these places but also how these places in turn acted on their worldviews and even theology.

The other essays in Part II address environmental concepts of Mormons living generally in the Great Basin. Richard Francaviglia explores how concepts of a spiritual geography influenced mapmaking among the Saints, providing specific examples of maps created in both the nineteenth and twentieth centuries in the Great Basin. Betsy Quammen describes the influence of Mormon culture on the creation of Zion National Park in southern Utah, including how rural Mormons came to embrace the park even though it threatened their economic livelihoods of ranching and farming.

The essays in Part III, "Agrarianism and Urbanism," speak to the varying orientations of Mormonism to a pastoral environment on one hand and to urban, industrial environments on the other. These essays are also good examples of historical practice, steeped as they are in the hard data of history beyond belief and perception. Jeffrey Nichols explores the early world of Mormon ranching and the deleterious impacts of livestock on the natural environment in prerailroad and capitalist Utah Territory. As much as Mormons crafted an image as agrarian farmers, his work reminds us that they depended on—and portions of their new homeland were suited for—livestock raising. Most significantly, these animals can help us rethink early Mormon narratives of the Great Basin and the Mormon people's exceptional relationship to it. Two essays on the Mormon relationship to water, irrigation, and agriculture appear in this section. Brian Frehner takes readers to Mormon settlements in the lower Colorado River basin to detail the challenges posed by the lack of water that at times undermined their attempts to build and sustain communities. Brian Q. Cannon, meanwhile, shows how Mormon discourse into the twentieth century continued to promote agriculture as both a necessary and divinely blessed endeavor, often with devastating consequences for those who attempted irrigation on the region's remaining submarginal farmland. The Mormon agrarian tradition in the twentieth century and beyond came to be tied up with cultural and environmental forces as well as federal efforts to establish land policy in the West.

The final essays examine the agrarian tradition in the urban context. Nathan Waite illustrates how Spencer W. Kimball, president of the church from 1973 to 1985, promoted the concept of gardening among the Mormon community, believing this to be a way to reconnect an increasingly urbanized population of Saints with the land, and thereby increasing their connection with God. In the last essay Rebecca Andersen argues that Mormons believed God had blessed the hills with sand and gravel so they could build up the Salt Lake Valley, but that extraction of the resource threatened Ensign Peak, one of the earliest natural features of significance in the valley. Hers is an insightful meditation on the juxtaposition of mine and mountain and the attendant suburban landscape that derives from landforms considered to have symbolic significance for Mormons.

The essays herein provide fence posts in a vast field: the possibilities of future scholarship are virtually unending. We conceived this volume to be a scholarly introduction to Mormon environmental history as well as a spur to encourage historians to consider the role of nature in the Mormon past. These essays may be seen both as works of Mormon history that draw on the tools of environmental history and as works of environmental history that focus on a specific religious movement. They introduce overarching environmental ideas to other historians and students of history, while also examining particular aspects of nature and Mormonism. A particularly salient area of study in this collection is the challenge of determining to what extent Mormon theology and belief matters in how people perceive and act in nature. The appendix includes a talk delivered at the University of Utah in 2013 by Marcus B. Nash, in which he provides a window into how Mormon beliefs and theology in the twenty-first century shape views on nature and the environment. This is notable given that Nash is a presiding elder in the LDS church hierarchy. As George Handley eloquently reminds us in the epilogue, environmental histories are moral stories in that they underscore the beliefs and values informing how people act in the world.

The field of environmental history has recently blossomed, to evoke a religious and environmental notion. Environmental historians have found a niche in history departments around the country, and some have offered a fresh rendering of general histories through an environmental lens, in the case of Mark Fiege's recent survey of U.S. history.[10] It is high time the insights from this vital field are carried into Mormon studies—and certainly in the process environmental history will benefit, too, from close inspection of a major religion's interactions with the natural world. We hope this volume leads to a greater understanding of Mormonism's connection to the natural world by critically challenging historical narratives and contemporary perspectives. It is, however, a starting point to move the conversation forward, enliven the study of Mormonism, and infuse the field of environmental history with rich perspectives from a religious community.

HISTORY, NATURE, AND MORMON HISTORIOGRAPHY

Jedediah S. Rogers

Earl Pomeroy, a prominent mid-twentieth-century historian, identified the dominant emphasis on the environment in western American history as evidence of the field's backwater status. He criticized western history for its environmental determinism—"that physical environment has dominated western life and has made the West rough and radical"—and the de-emphasis of "the spread and continuity of 'Eastern' institutions and ideas." Any student of western history will recognize the genesis of this emphasis—Frederick Jackson Turner's influential (but by midcentury much discredited) 1893 frontier thesis. In Turner's rendering, western history played out on the soil of a continually receding West. As settlers moved westward, encountering wilderness and savagery, they adapted to the environment and gradually transformed the landscape and advanced civilization. Pomeroy criticized this outmoded paradigm and the accompanying "environmental theme" of western history that had helped Turner in his generation to "integrate the story of the West into a more convincing pattern" and had "appealed to Americans in a nationalistic and ostensibly democratic era."[1]

Pomeroy weighed in on the classic debate that Turner broke open—the West's influence on American character and institutions. By midcentury western historians had mostly rejected Turner's ideas, even as they acknowledged his continued influence on the field. Other imaginative and theoretical historians, most notably Walter Prescott Webb and James C. Malin, had continued to point to the environment's prominence in western history—Webb in his work on technology and innovation in arid environments and Malin in his social science research on ecology and history.[2] Due in part to their influence on the writing of western history as well as to the perception of a place with wide open spaces and grand scenery, the environment continued to occupy a central place in western history narratives. The field's close association with environmental history—a subfield that grew out of the social ferment and activism of American environmentalism—also affected its prominence. The field of American environmental history was from inception tied to the West. The earliest environmental histories addressed the careers of conservation pioneers, the politics of environmental protection, and the genesis and development of environmental thought. Many of these had a western American orientation.

Pomeroy could rightly have addressed Mormon history in the same stroke as western history. Long associated with Utah and western history, traditional Mormon history exhibited an insularity that branded it as a field of study. Before the mid-twentieth century, Mormon history rather unimaginatively borrowed from concepts originally expressed by Turner and subsequently taken up in varying form by other western historians, though neither Turner nor his principal intellectual student, Ray Allen Billington, had much if anything to say about the Mormons.[3] Prior to the fracturing of the historical profession with the new social history in the 1960s, Mormon history was, largely, western history. After that time, Mormon history did not come to associate with environmental history as western history later did. For Mormon historians, the environment often implied those aspects "beyond" history's core scope, useful for understanding but not essential to the story. The term *nature* was a stale category, continually being manipulated, sometimes presenting obstacles but not possessing any compelling motive or force. Early works of Mormon history considered the environment insofar as it was a backdrop or an obstacle to overcome. In describing the community's religious history, some early Mormon historians associated "natural" or "environmental" explanations as faithless and dismissive of heaven's guiding hand.

Although the new Mormon history since midcentury has moved toward understanding the past as a series of natural rather than supernatural events, Mormon scholars have been relatively slow to apply the methods and insights of environmental history. One reason may be that the contemporary LDS Church's policies and doctrinal emphasis say relatively little about environmental issues of the day—urban pollution and blight, the whittling away of open spaces to development, extinction of species, preservation of wilderness, or the changing climate. Many Mormons, along with a broad coalition of other Christians, embrace conservative political discourse that make "environmentalism" a partisan issue. This may explain the relative slowness of some Mormons who write history to adopt environmental history, which has long been associated with environmental activism. A greater attention by Mormons to environmental issues may give lift to the study of environment in our histories. But here scholars can lead the way, too: by writing nature into Mormon history, by showing the deep connection of early Mormonism to an environmental ethic, by demonstrating the centrality of land and water to the Mormon story, scholars can lead a re-envisioning of our history and our connection to natural forces at play.

Since the nineteenth century, observers have equated pioneer-era Mormon country to a common American motif—the agrarian myth. In this form of the myth, farmers diverted water from streams that originated on snowy mountain peaks and carefully laid out their farms in square sections with straight-lined canals and laterals. In his travelogue *To the Rockies and Beyond*, first published in 1878, the journalist Robert E. Strahorn witnessed "the villages of thrifty Mormon farmers, usually almost buried in foliage, orchards and grain fields, and nestled under the sheltering walls of the Wasatch, with the glorious vista presented by the lake in front."[4] The newspaper editor William E. Smythe, writing in 1905, praised the Mormon irrigation system and cooperative efforts to reclaim the desert.[5] These nineteenth and early twentieth century accounts fed the idea of Mormons as masters of agriculture in a dry region and helped give form to a lasting powerful motif of the Mormons' physical settlement of the West. The image of a Mormon village at the base of the Wasatch Mountains powerfully represented order, permanence, and godliness.

The work of Strahorn and Smythe differed from nineteenth-century ethnographic studies that emphasized Mormon life and community over the physical appearance—and mastery—of the Mormon village. Beginning in the 1920s, Mormon village studies, many originating as master's theses, teased out environmental themes. Lowry Nelson's work represents a broad academic approach; beyond the usual parochial works on Mormonism and grounded in scholarly observation and research, his work demonstrates attention to the village beyond Mormonism and situates Mormon land settlement in the context of settlement elsewhere in the United States. Nelson helped define the genre of Mormon village studies that flourished through the 1950s. The studies by Nelson, Edward C. Banfield, Henri Mendras, Thomas O'Dea, Wilfrid C. Bailey, and others generally emphasized economy, lifestyle, and identity over environmental influence or impact. The physical aspects of the village are shown to reflect—and reinforce—the character of the Saints: righteous, orderly, insulated. There, it was argued, boys and girls developed a strong Mormon identity.[6]

No writer better made this point than the venerated novelist and historian Wallace Stegner, who wrote about Mormons in the early 1940s with affection. In *Mormon Country* Stegner begins with a meditation on the Lombardy poplar, the "Mormon tree," commonly lining Mormon canals, roads, and "boundaries between fields and farms." He calls attention to the tree not for its ecological presence on the landscape but as an emblem of "Mormon group life"—"symbolic, somehow, of the planter's walking with God and his solidarity with his neighbors." Stegner thought the tree an apt representation of a people who "look Heavenward, but their roots are in earth."[7]

Stegner was not the only one, though arguably the most eloquent, to point to the materiality of Mormon doctrine and culture. The idea is that Mormons are more earthly, their communities more bounded in earthly expression and perhaps in nature itself, than are other western groups—religious or otherwise. Stegner's *Mormon Country* makes the point by detailing how Mormons took the ideas of the "Heavenly City" and introduced it to their own homes and communities. He describes the distinctive culture that developed in a landscape that the Mormons made their own. Stegner's attention to place and to issues of materiality foreshadowed his later work as an eloquent writer on western issues, particularly conservation. But it would be a stretch to call this a work of environmental history; Stegner is interested in the place, and in the ensuing transformation wrought through Mormon industry and agriculture, but the reader will find no mention of ecology, hydrology, or the complex interplay between Mormon society and environmental change. His work is, rather, a blending of sociological observation, historical interludes, and thoughtful musings. Thus, though dated and rough in parts ("They were not pioneers of the itchy-footed and free-elbowed sort"[8]), it remains a classic and required reading in Mormon studies.

In essays and books published after *Mormon Country*, Stegner demonstrated his commitment to conservation.[9] From this ideological position, he and his better known contemporary Bernard DeVoto were sometimes critical of Mormon social and economic systems and practice that denuded the landscape and harmed the environment. Although they built communities that diverged from the boom-and-bust pattern common throughout the West and built a social system that departed from the capitalism of the American mainstream, by midcentury Mormons as a group had departed from many of the pioneer practices and tenets that made them distinctive. Both writers admired the Mormons' irrigation systems, communal energy, and industry—DeVoto wrote of "the conquest of the arid land that succeeded repeatedly in areas identical with areas where non-Mormon attempts have failed"—even as they loathed the contemporary conservative culture that bemoaned federal protection. These critiques would later be repeated by an ideological successor, Edward Abbey.[10]

Thus, even conservation-minded writers at midcentury characterized Mormon settlement of the West as uniquely successful. Whereas settlers' initial perceptions of the western landscape were typically of an unbounded, plentiful nature, only after lived experience did they adapt their impressions to the realities of a hardscrabble land.[11] Nineteenth-century Mormons created a counter narrative. DeVoto acknowledged that "jubilant Mormons told the world as early as 1849 that they had made the desert blossom as the rose" and had widely "incorporated the phrase in their daily talk."[12] Both pioneers' and travelers' accounts depicted the Mormon enterprise to transform a dry, pest- and predator-infested desert into a productive homeland. Early histories of Utah contributed to what Jared Farmer has called the "desertification of Zion" by validating these popular folktales. By the mid-twentieth century the remaking of "Utah's land of lakes as a desert" was complete.[13] In a 1949 centennial history of Utah, after describing the lifesaving qualities of irrigation, Wain Sutton argued that were it not "for a wise provision of nature—the mountains—few inhabitants would have been able to live in that desert region," while failing to acknowledge that a sizable Ute population were subsisting largely on trout from Utah Lake at the time of the Mormons' arrival in the Great Basin. Although irrigation demanded trial and error, the end result was a marvelous success, as reflected in the words of Herbert E. Bolton: "Irrigation was one of the signal contributions of the Mormons to the upbuilding of the Great West. Without it, starvation was as certain as death in old age so the Mormons built reservoirs in the mountains, ran ditches and great canals across the valleys, and poured the life giving waters of the Wasatch upon the thirsty soil of the sunbaked desert; causing it to bloom like the rose."[14]

After midcentury some histories diverged from the earlier historiography by de-emphasizing the miraculous landscape transformation. Joel Ricks, a professor of history at Utah State University, was an important voice. In *Forms and Methods of Early Mormon Settlement in Utah and the Surrounding Region, 1847 to 1877*, Ricks propounded that Great Basin settlement required "modifications" from earlier Mormon colonization in the West. The landscape and available resources dictated their economic enterprise. The Mormons' first fort was made of adobe, not wood. Agriculture required irrigation. The physical design of Salt Lake City, with its square-grid, ten-acre blocks extending from the temple site, copied the pattern established by Joseph Smith. Admittedly, Ricks primarily tells a tale of "plow[ing] the hard sod to plant their crops" and "erect[ing] a fort to shelter them from the weather and from the menace of the lurking Indian."[15] But readers will glean new insights from his examination of geography and environment in Mormon settlement. Consider the seemingly ordinary observation that "[l]ack of rainfall and the distance between streams made it necessary for Brigham Young to explore large areas of land for expansion." Could it be that the expansive state of Deseret—the Mormons' inland empire—was as much a product of environmental contingencies as geopolitical aspirations? Unfortunately, Ricks does not elaborate.[16] Like earlier writers, he, too, tied the success of Mormon settlement to the fundamental character of the settlers and the church, but he also examines the "physical features and the Indians [that] did much to influence the process of colonization."[17]

The classic and most complete work on Mormon pioneering is Leonard J. Arrington's *Great Basin Kingdom: An Economic History of the Latter-Day Saints*, published in 1958. Signaling the arrival of the new Mormon history, of which he became the most prominent practitioner, Arrington writes in his preface that his work provides a "naturalistic treatment of certain historic themes sacred to the memories of the Latter-day Saints." He was probably the best known of a new generation of Mormon scholars looking beyond

God to explain the arc of the past. This new crop attributed events not to a divine source but to natural explanations. Arrington offered a materialistic, rationalist approach to nineteenth-century Mormonism, reasoning that "[a] naturalistic discussion of 'the people and the times' and of the mind and experience of Latter-day prophets is therefore a perfectly valid aspect of religious history, and, indeed, makes more plausible the truths they attempted to convey." That Arrington found it necessary to pause and highlight this point is a statement on previous literature and sensibilities more than anything else. For Mormons who might be "troubled" by his approach, he rightly contends that "it is difficult, if not impossible, to distinguish what is objectively 'revealed' from what is subjectively 'contributed' by those receiving the revelation."[18]

Great Basin Kingdom addressed themes with an environmental, materialistic bent—land tenure, resource development, landscape architecture, and accompanying impacts to the land. Arrington's subsequent *Building the City of God*, cowritten with Feramorz Young Fox and Dean May, went beyond the popular trope that Mormons were more cooperative than other western groups to illustrate the institutionalized establishment of communal programs—the United Order, ZCMI, Zion's Central Board of Trade, and the welfare program of the twentieth century—in the LDS Church. Whereas *Great Basin Kingdom* detailed nineteenth-century economic and cooperative enterprises, this new volume carried the tradition into the twentieth century, offering a corrective to the idea that Mormonism in this era abandoned its communal ethos in favor of the American capitalistic system, though it may have overplayed the cooperative tradition in Mormonism.[19] By showing the temporal apparatus and orientation of the LDS Church, Arrington perpetuates the view that Mormons settled in a place where others would have failed. In terms of water use and the husbandry of other resources, the communal impulse never entirely subsumed the early Mormon tradition of private property.[20]

Some of the best and most sustained work on the environment within Mormon history is by geographers. The origins of environmental history reach back to that discipline; under the pioneering work of Carl Sauer, scholars in cultural geography and its subfield cultural landscape studies sought to understand how places are defined and transformed through human agency. Cultural geography is very often rooted in history, since landscapes and their meanings develop over time. An early important piece on Mormon cultural landscape studies is D. W. Meinig's 1965 essay on what he called the "Mormon culture region." Meinig identified regional patterns and "the processes which created them" in Mormon country. By bringing specificity to the spatial dimensions of Mormon settlement and western expansion, he concludes that "in building their Zion in the mountains, the Mormons have left their stamp upon districts within and along the North American cordillera from a corner of Alberta to a corner of Chihuahua."[21] Richard V. Francaviglia added to this notion in his classic work *The Mormon Landscape* (1978) in which he argued that Mormons "transformed a difficult environment into a distinctive cultural landscape." As a geographer, he was not as interested in the process of transformation as in the end result—the distinctive features that characterize Mormonism's built environment. When discussing "the creation of the Mormon landscape," his principal concern was the conscious act of designing a unique Mormon "look"—on the visual, not the ecological, elements.[22]

For environmental historians, the geographers' major contributions have been in revealing the contrast between the physical characteristics of a place and the ways settlers perceive and experience that place. In contradistinction to Meinig, Richard H. Jackson of Brigham Young University speaks of geography in terms of Mormons' spatial positioning and changing ideas of a place. Beginning with his 1970 dissertation, "Myth and Reality: Environmental Perception of the Mormon Pioneers, 1840–1865, an Historical Geosophy" (Clark University), Jackson richly detailed Mormon perceptions of the new Great Basin homeland. In a 1988 *Utah Historical Quarterly* essay Jackson argues that founders of Salt Lake City considered their new settlement a complement to the Great Salt Lake. He postulated that the name of the new settlement—Great Salt Lake City—reflected the earthy pragmatism of Brigham Young, resulting in "a pairing of physical and cultural geography unrivaled in America."[23] In another essay, Jackson placed the Mormon enterprise to "transform" their environment in the context of nineteenth-century efforts to plant trees and crops to fulfill the belief that "rain follows the plow" in the arid West. But rather than examine "environmental change," Jackson principally documented a number of firsts—Mormons' initial reactions to the Great Basin and early attempts at irrigation.[24]

In more recent works, Francaviglia's contribution is to show how landscape is variously imagined by peoples and groups. *Mapping and Imagination in the Great Basin* contrasts the bleak European American depictions of the region with the indigenous knowledge of the Shoshone, Paiute, and Washo of a land that gave sustenance.[25] Francaviglia demonstrates in *Mapmakers of New Zion* that Mormon mapmaking blended local knowledge of terrain with cultural—and, in this case, theological—notions of geography and place. But whereas the Mormons would develop a deep understanding of their new homeland to produce their own geographic representations, they originally came to know it through the work of others. This helps explain, as Francaviglia wrote, how in the mid-nineteenth century church leadership moved "westward so confidently into country they knew so little about."[26]

This work on landscape by geographers complements a couple of important books by architectural historians writing about Mormon city planning and the built environment as both concept and physical space. Thomas Carter's *Building Zion: The Material World of Mormon Settlement* is the best of them—a deep study into the land and community of central Utah's Sanpete Valley and the environmental and cultural design elements long associated with that area. It went beyond the landscape studies of Nelson and, later, Francaviglia and Jackson to make landscape a more dynamic process than a static category. Carter identified the dominant features of the Mormon landscape—wide streets, carefully laid out lots on a grid, low population density, temple as a sacred center—and explored the values, ideas, and beliefs manifest in that landscape. Although not a work of environmental history per se, Carter's interest in how diverse elements—ideology, economic system, material objects—all came together in the creation of a community and in the imprint of Mormon and American values on the landscape reflect the dominant mode of environmental history. His exploration of "the fundamental connection between . . . the conceptual Zion and the material Zion" goes to the heart of environmental history's project to blend the conceptual and the physical.[27]

Through increasingly sophisticated narratives and methodologies, we see the religious imperatives that powered Mormon communalism and industry and led to confounding irrigated and industrialized landscapes. Take Mark Fiege's groundbreaking *Irrigated Eden: The Making of an Agricultural Landscape in the American West.* In his framing, religious and cultural ideals may have endowed the agricultural landscape of southern Idaho—predominantly Mormon in the nineteenth century—as an Eden but also obscured the confounding ecosystem of the irrigationists' making. The canals were clogged with plants and silt, the fields infested with insects, the waterways reduced in flow. All this challenged farmers and the very myths that they created about their homeland and their place in it. The strength of this work is in its ability to explain the varying relationship between ideas about nature and environmental realities. Paradoxically, in this case, the attractiveness of the Garden of Eden myth may have "increased in proportion to the inability of Idahoans to ever attain it."[28]

Most writing about Mormonism and agriculture lacks the ecological complexity of Fiege's work, and more surely can be done to rethink Mormonism's agrarian, pastoral past. What seems to be clear from existing scholarship is Mormon spiritual life in the nineteenth century was as much if not more accurately represented by irrigation canals and well-watered fields as by Sunday worship. Mormons reconfigured geography to correspond to religious doctrine and organization imperatives and to reflect their community's core values. By the twentieth century, however, religious devotion in the Mormon community had oriented inward. The image of Mormon devotion was no longer primarily physical—of the pastoral scene, irrigated fields, and religious architecture.

From an environmental perspective, the history of twentieth- and twenty-first-century Mormonism in the West is generally cast in a declensionist narrative. Along the Wasatch Front the agrarian village pattern gave way to suburban sprawl and strip malls connected by multilane freeways. For all of the evident economic and technological gains, Mormon country has confronted ecological problems: physical scars on the land, polluted air, overpopulated communities. Ecological problems in Zion generally correspond to the Americanization of Mormonism and the transition from an isolated pioneer sect to a powerful global church.

Although over the last century Mormon country has become less distinctively Mormon, scholars have attempted to tease out the religious responsibilities and dimensions of ecological problems. When environmentalism flowered into a political movement in the last half of the twentieth century, religion came into the crosshairs of scholars and activists searching for causes of the world's ecological decline. In an influential essay published in 1967, Lynn White Jr. argued that Judeo-Christian religious thought and culture "not only established a dualism of man and nature but also insisted that it is God's will that man exploit nature for his proper ends." Some religions interpret the biblical creation story as God setting humans above nature, thus giving them dominion over the earth and its resources. This anthropocentrism created a hierarchy of humans holding "dominion" over other living things. Critics of these ideas argue that a mere summation of the idea from Old Testament language is insufficient to indict Christianity and several thousand years of history. Critics charge White with neglecting another central strain of Judeo-Christian environmental thought: the idea that humans are to be stewards, or caretakers, of the earth and other animals. White countered that this strain, popular in the twentieth century, was not dominant in the past two millennia.[29]

A handful of historians and writers have applied these ideas in a Mormon context. Dan Flores, John B. Wright, and Max Oelschlaeger have suggested that Mormon doctrine has contributed to a particularly virulent strain of anti-environmentalism in the church. Oelschlaeger argued that "the only denomination that has formally stated its opposition to ecology as part of the church's mission is the Church of Jesus Christ of Latter-day Saints."[30]

Dan Flores's essay "Zion in Eden: Phases of the Environmental History of Utah" explored the environmental consequences of Mormon thought and practice by using ecological studies to reflect on the overcutting of ponderosas, lodge-poles, and Douglas firs and detailed plant succession and the invasion of nonnative species in Mormon country. He contested Leonard Arrington's statement that Mormon policy on resources "seems to have protected Utah from the abuses and wastes which characterized many frontier communities in the West" and instead offers his own assertion: "Zion was yearly becoming less productive and more unstable." What Flores sometimes failed to do was directly connect ecological problems with Mormon religion and culture. The observations that "Mormons paid too little attention to the tenuous nature of the natural forces that held Zion together" and that "contemporary Utah seems less open to environmental sympathies than virtually any other part of the modern Rocky Mountain West" are insightful but fail to fully address the matter.[31]

On the other side are Mormon scholars pointing to an environmental ethic deeply embedded in Mormon scripture and discourse. A host of scholars and writers, beginning with the historian and theologian Hugh Nibley, demonstrated not just environmental sensibility but a holistic environmental ethic in the writings of Joseph Smith, Brigham Young, Parley P. Pratt, and other early church leaders. These authors pointed to scriptural utterances like that in the Pearl of Great Price of Enoch weeping for the earth's pain for "the sickness of my children" and "the filthiness which is gone forth out of me"; to surprising statements of Young reverencing crickets as "creatures of God" welcome to "nible [*sic*] away" on his crops; and to Spencer W. Kimball's more recent injunctions to beautify properties and care for the land as evidences of a once-dominant but still lingering thread of a salutary Mormon environmental ethic.[32] The theological emphasis on materiality, redemption, the fusion of secular and sacred, the temporal kingdom of God, and the imperative to make the land blossom as the rose bolstered Church teachings on the stewardship of resources, gardening, and the treatment of animals. Indeed, early Mormon theology and teaching emphasized a land ethic that if not ecocentric was at least in line with the view that humans are more participants in a larger ecological system than lords over the earth.

The literature on this front has grown in recent years. A literary compilation of environmental theology and history, considerably overlooked, is *New Genesis: A Mormon Reader on Land and Community*, a collection of essays edited by Terry Tempest Williams, William B. Smart, and Gibbs M. Smith designed "to begin the storytelling of sustainability with our own spiritual tradition."[33] Another offering is George B. Handley, Steven L. Peck, and Terry B. Ball's *Stewardship and the Creation: LDS Perspectives on the Environment*, which continues that storytelling.[34] These authors rightly point to the rich strain of environmental stewardship in the Mormon tradition, but they do so, in part, because they are heavily influenced by twentieth-century environmental sensibilities.[35] Moreover, however Mormon scripture and nineteenth-century discourse offered a beneficial way of acting on the land, these theologies are not presently emphasized in the LDS Church. Without question, modern Mormons' relationship to the environment is different than it once was. Early Mormons embraced a communalistic ideology largely at odds with the individualistic, capitalistic attitudes of today. Thus, rather than pointing to the presence of an environmental ethic in the Mormon tradition, these works highlight the twentieth- and twenty-first-century neglect of a rich nineteenth-century tradition.[36]

It might be tempting to point to a decline of a Mormon environmental ethic, but that would fail to account for a diversity of Mormon thought about nature. Admittedly, environmental history does not always do a good job of demonstrating diversity within groups. In its attention to the relationship between nature and culture, environmental history sometimes treats culture as a single homogenous entity.[37] This can spur an overly simplistic view of how groups understand and act in their environments. One may hear the charge that Mormons are anti-environmental, but that loaded term not only imprecisely characterizes Mormonism's relationship to the environment, it also does not account for divergent environmental thought within the religion. How widely shared were Smith and Young's views of the environment among nineteenth-century Mormons? What about diversity among Mormons globally? How do Europeans, Africans, South Americans, and Asians differ from Intermountain West Mormons?

We might also ask how an environmental ethic in Mormon theology and nineteenth-century discourse draws from or mirrors other religious discourse about the environment. In answering these questions, Mormon historians can draw on rich studies of religion and environmentalism from scholars such as Mark Stoll, Thomas Dunlap, Roger S. Gottlieb, and Stephen Ellingson. These scholars have demonstrated that the environmental movement has religious roots and mirrors the beliefs and behaviors of religious traditions, while also explaining that numerous religious traditions have and continue to address the destruction of nature caused by humankind.[38] Becoming familiar with the work of such scholars would help historians determine whether similar types of efforts are present in Mormon history.

Perhaps no place has inspired more commentary on the Mormon connection to land and environment than southern Utah and northern Arizona's plateau country, at the periphery of Mormon country. Edward Abbey, Terry Tempest Williams, Edward Geary, Charles Wilkinson, Stephen Trimble, and other acclaimed writers have written insightful commentaries on the region's land and peoples, and the religious influences that bear on them. Some of these writers have developed a cult following in their often passionate defense of the canyon country.[39]

A couple themes emerge among works that take a deep historical dive. For example, several works triangulate around environmental conflict. Trimble's *Bargaining for Eden: The Fight for the Last Open Spaces in*

America is an eloquent meditation on the clash and varying discontents of preservation versus development in the Capitol Reef region. Trimble examines "the combined weight of avarice, inattention, and denial" in the whittling away of open space in Utah and the West, yet he argues that the story of good guys and bad guys is not as black and white as we'd like to believe.[40] My own *Roads in the Wilderness: Conflict in Canyon Country* builds on Trimble's work to highlight endemic conflict over land access and use of wildlands. Like Trimble, I am interested in the persistence of certain religious beliefs and worldviews that inform a particular approach to economic development. I suggest that the polar ideologies at work in the region may be embodied by two personalities: the writer Edward Abbey and the local politician and developer Calvin Black. Black, the prototype of Bishop Love in Edward Abbey's novel *The Monkey Wrench Gang*, was a particularly fruitful subject in twentieth-century Mormon thought, as his views wove together the religious and the secular, presaging the views espoused by more recent "patriots" such as Cliven Bundy.[41]

The difficulty is teasing out the relative effects of religious tenets on the behavior of certain individuals and groups. Dominant views toward the environment in southern Utah are a blending of religious, political, and secular notions of land historically dominant in American culture. As Trimble has suggested, Mormons today espouse "the sanctity of industrious hard work" as zealously as they had once embraced communal ownership of resources. Mormon religious ideas now dovetail neatly into the dominant thinking of American secular, material, and capitalist attitudes. That said, we need more research on the role of religion and theology in our western wars over land, access, and water. More work needs to be done to understand the impact of contemporary conservative Mormon culture on not just the environment but also on environmental ideas, including the influence of conservative politics in Utah and the West, attitudes toward environmental initiatives, and the influence of conservative Mormon rhetoric over the pulpit.

One good start to these questions is a 2006 essay in the journal *Society and Natural Resources* by Joan M. Brehm and Brian W. Eisenhauer, which examines the influence of Mormon religious culture on environmental concern.[42] But the hard interdisciplinary work of crafting an environmental history of Mormon country, and the perhaps more difficult task of identifying the intricate connection between ecological health and religion, have yet to be fully undertaken. Study after study highlights the Mormon-induced built environment—villages, architecture, landscape design—but we need studies mirroring the georegional approaches of William Cronon's seminal history of Chicago to underscore the relationship between what is designed and built and the corresponding impact on land and resources.[43] We understand early Mormon thought about the husbandry of raw natural resources—the oft-cited proclamation of Brigham Young that decreed land, water, and timber as the property of God to be used by all—but we know less about how Mormons internalized those teachings and the environmental realities that underscored them. However Mormons may have taught an environmental ethic, did they actually practice one? Whatever Mormon theology says about environmental stewardship, the Mormon environmental legacy is complicated.

The need remains to see beyond high-ranking leader rhetoric to the broader interplay between nature and Mormon society and history. Rather than examine the writings of individuals who spoke about nature, we need richly textured histories of people acting within nature. We have Brigham Young's environmental teachings but know less on how he acted in nature. We know what early Mormons thought about land and water but less on how the rank-and-file understood and acted on their beliefs. One fine example is Thomas Alexander's profile of Mormon leader Sylvester Q. Cannon as the city engineer in Progressive Era Salt Lake City.[44] Though the twentieth-century environmental history of Mormonism has been completed in pieces, more sophisticated comparative and religious works are left undone.

Environmental history is richest when its tools of analysis are applied to a specific locale. Whatever might be said about Mormon theology or ideas about nature, all must be solidly grounded in a particular environment. For Mormons, the place principally identified with their history is what Meinig called the "Mormon cultural region," encompassing a large swath of the Intermountain West between Canada and Mexico. The promise of environmental history is in the practice of reconceptualizing place as neither monolithic in space nor time. Consider that, with its attention to physical forces beyond culture, environmental history is sometimes conceived on a broad scale. Geologic time might mark the parameters of a study. Prominent historians, beginning with Donald Worster, have made the case for framing our histories on the long durée to show the active power of natural forces.[45]

At the least this long view revises traditional narrative structures. For Mormon country, that history might begin with Lake Bonneville (35,000 years ago) or earlier with the uplift of the Rocky Mountains (80 million years ago), a geologic landform that has come to symbolize the Mormon faith. But while some environmental history may take this long view, the field is really built on narrower studies that use conceptually understandable time scales to frame their narratives. We need not have a solid understanding of geology, biology, and other physical sciences to benefit from an environmental lens of the past. What is needed is a dedication to the idea that far from being a static backdrop, the environment is a dynamic force that both acts and is acted upon.

Environmental history can also open up avenues of investigation little considered previously. Histories of biodiversity and ecological restoration are rarely examined, for example. Nearly the sole work in this area is Marcus Hall's comparative look at the Cuneo Alps in Italy and the Rocky Mountains in Utah. This sophisticated transnational comparative study looks at the response of land managers and locals in responding to soil depletion and erosion on Utah's slopes. The value of Hall's work is in his demonstrating the distinction between Europeans and Americans (the former more likely to see the influence of culture in landscape), but we need work pointing to distinctions between smaller groups—in our case, among and between Mormons. Nevertheless, his work on the subtle differences and shifting emphasis between "gardening" and "naturalizing"—or rehabilitating—wildlands is interesting given Mormonism's theological emphasis on making the earth as a garden.[46]

For Jon Coleman, in his wide-ranging study of wolves in America, Mormon thought and theology informed but did not signal Mormon interaction with predators. The treeless Great Plains afforded frequent interactions with wolves, which preyed on livestock and exhumed human bodies along the pioneer trail. Pioneers responded by waging war on wolves in ways that departed from Joseph Smith's previous injunctions to leave predators well alone. When ensconced in the Great Basin, settlers sought to rid the territory of "wasters and destroyers" only to find that the diminutive prairie wolf increased in number. Finding that they were not easily eradicated, and not nearly as threatening as their larger cousins, Mormons referred to prairie wolves as coyotes, a linguistic turn that signaled the "eradication of wolves" and "the end of the pioneer era." In this way Coleman showed how environmental change informs the very core of Mormon identity-making and narrative construction.[47]

Consider the relationship to and reliance on nature as a food source, an area that environmental historians have frequently discussed. We know Mormons consumed wild game and wild plants in their early settlement years but know little about how this reliance may have impacted ecology; we know they had harvest seasons, bringing communities together for feasts, but know less about the process of extracting edibles from the land; we know hunting—including a large-scale competition to kill "wasters and destroyers" in 1848–49—was common practice among Mormons, but we don't know much about that activity's impact on the land and its carrying capacity. One quite delightful book that provides a start, even if it doesn't answer all these questions, is Brock Cheney's *Plain but Wholesome: Foodways of the Mormon Pioneers.* Cheney unveils stories behind the most ordinary of tasks—cooking and eating—that naturally have an environmental connection. Granted, Cheney is not interested in how "what we eat and how we eat it" might reveal a connection to or disconnection from nature, as an environmental historian might approach it, but he does underscore a bit "about who we are and how we think."[48] The rest is left to an environmental historian dedicated to showing the ecological and cultural dimensions of changing Mormon "foodways" up to the present.

Perhaps the most insightful and groundbreaking work to suggest the promise of Mormon environmental history is Jared Farmer's *On Zion's Mount: Mormons, Indians, and the American Landscape.* In this history of Utah Lake and neighboring Mount Timpanogos, colloquially known as "Timp," Farmer shows the consequence of Mormon settlement on Indian fish culture and sustainability in Utah Valley. Farmer's narrative is one of environmental declension—a pristine lake turned municipal sewage waterbody—but this is presented against a larger, and frankly more interesting, backdrop of a group's relationship to a prominent mountain. Local scout and church-organized hikes left an imprint on the mountain; so, too, did Robert Redford's Sundance development against the eastern slope.[49] By showing how Timp came to be viewed as a landmark, as opposed to merely a natural landform, he is able to examine its cultural creation beyond its physical presence or even its environmental change.

Farmer mirrors a similar model in his first book, *Glen Canyon Dammed: Inventing Lake Powell and the Canyon Country*, where he examines not just the physical loss of Glen Canyon but the cultural and social relationships with, and constructed meanings ascribed to, the new reservoir.

The attention to Timp as a landmark belies the range of this sophisticated work. Classifying it is not easy to do. Telling a local story, Farmer shows the breadth—and possibilities—of Mormon environmental histories. His demonstration of how Mormon place-making both diminished Ute pathways and wrote them out of the dominant historical consciousness offers an example of how racial categories are imprinted on the environment. He illustrates how a history of one localized landscape is tied up in the larger American landscape. Farmer also provides a roadmap to examine the cultural creation of other landmarks in Mormon country. If few works carry the style and range of Farmer's, it is due to the limits of our thinking and not to the universe of stories waiting to be told.

We end with a book of some literary merit, *Home Waters: A Year of Recompense on the Provo River* by BYU professor George Handley. Part history, part memoir, this volume introduces us to the quiet mechanisms of a specific place by helping us think expansively about it—the gaze is on the land and its features, but also on the culture and worldviews that animate it. In the tradition of good environmental history, Handley demonstrates a dominant culture's impact on the land, as well as the land's ultimate influence on the culture. A Mormon writing about his home, he thus demonstrates through personal observations a larger point, that "probably no religious culture in American history has had such an intimate and sustained opportunity to determine a relationship to a homeland."[50] That intimate and sustained connection is clearly manifested in these pages.

If there is a weakness to this book it is that Handley falls into a trap common to environmental historians in his reference to a group/people—in this case, Mormons and Mormonism—in a nonspecific, gestural way. But this does not suggest Handley's work is simplistic in its renderings. He is deeply aware of his own (and his people's) contradictions—as sympathetic to the impulse to "develop" as his own commitment to "preserve." Handley's honesty about his own internal contradictions makes his history more accessible, more relatable, to people less committed to the environmental cause. We need more works that blend personal memoir and environmental history, for this combination produces a deep dive into the richness of place. Handley's, as well as Farmer's, work also points to the interdisciplinary nature of environmental history. Environmental histories frequently draw on ecology, biology, geography, anthropology, and a number of other fields, often cross-pollinating to obtain sources and craft their narratives. This is understandable given the messiness of ecological relationships. But it also suggests the centrality of nature and the environment not just to history but to other fields as well. Undoubtedly, for this reason a great deal of what we might consider Mormon environmental history has come from scholars of disciplines other than history.

Despite some good recent work in the field, Mormon historians continue to lag behind western American historians in producing sophisticated environmental histories. The resurgence of western history from the days of Pomeroy is due not in spite of, but in part because of, the field's marriage to environmental history. Among top-tiered western historians over the last generation—Donald Worster, Patricia Nelson Limerick, William Cronon, Richard White, Elliot West, among others—many were trained, and train their students, in environmental history. Coincidentally, some of these prominent historians have been ahead of traditional Mormon historians in writing about Mormon environmental history.[51] That Mormon historians have not generally addressed themes and methods of environmental history is bewildering given this and the fact that several scholars of religious studies have embraced environmental history as a useful lens through which to gain new understandings of religion and theology.[52] But just as western history and religious studies have gradually become more sophisticated in their analyses of nature and the environment, so, too, might Mormon history. Mormon environmental history is a process of evolution, affected by the development of the entire field.

In the end, environmental history has relevance to Mormon studies not only by helping us consider the centrality of nature in the histories we write. It also grounds us in a reality, found in Mormonism as well as other religious traditions, that while nature may exist independent of humans, we are all intimately connected to it. Environmental history reminds us in an academic way what many religious traditions have long taught—that we are connected with the earth, sky, and other living beings.[53]

PART I

Theology and Ideology

THE "LION OF THE LORD" AND THE LAND

Brigham Young's Environmental Ethic

Sara Dant

One of the most remarkable of the nation's overland pioneer migrations belonged to the Mormons of Utah. In 1847, this uniquely American faith group sought protection and salvation in the Great American Desert—testimony, indeed, to the powerful forces pushing the Saints westward. The Church of Jesus Christ of Latter-day Saints—also known as Mormon or LDS—arose in 1830 as the divine vision of a charismatic prophet named Joseph Smith during the foment of religious revivals in upstate New York that historians call the Second Great Awakening. In the nadir following Smith's murder in 1844, Brigham Young emerged to lead the fractured faithful, and he quickly determined that the safest refuge for the Mormons lay to the west, outside the boundaries of the United States near the Great Salt Lake. For many historians, journalists, bloggers, and students, the clearest expression of Young's position on the optimal relationship between the Saints and their new sacred homeland is his oft-quoted July 25, 1847, pronouncement that "there shall be no private ownership of the streams that come out of the canyons, nor the timber that grows on the hills. These belong to the people: all the people." It's powerful. It's prophetic. But unfortunately, it's also problematic. Yet in the end, my effort to document "the quote" necessitated a close examination of Brigham Young's thoughts and beliefs on human stewardship of the natural world that evinces a surprisingly modern perspective, one that John Muir and Aldo Leopold could have appreciated and embraced. So how and why, then, did Utah generally and faithful LDS in particular largely leave behind Young's divine covenant with nature? "The quote" quest not only leads to a deeper and more complete understanding of Young's environmental ethic but also helps place his ideas within the larger context of regional development, revealing, ultimately, why it failed to endure as Deseret developed.

For Young and the Mormons, Utah would be the place. John Frémont's 1844 California expedition had described "a region of great pastoral promise abounding with fine streams . . . [and] soil that would produce wheat" along the Wasatch Front, and church leaders believed that the isolation provided by this desert mountain terrain would prove their salvation.[1] Coursing through the Saints' early settlement efforts in this new Zion was Young's powerful environmental ethic inspired by scripture: "It is expedient that I, the Lord, should make every man accountable, as a steward over earthly blessings, which I have made and prepared for my creatures."[2]

The Saints' Young-led migration to the Great Basin was a feat of organizational skill and coordination; there would be no wandering in the desert for this Mormon Moses and his people. Their flight from Nauvoo, Illinois, to the Salt Lake Valley would be church-organized, church-financed, and church-led, establishing both Young and the LDS apostles as *the* authorities in the region. The French naturalist Jules Rémy marveled of Young, "He has set before him, as the object of his existence, the extension and the triumph of his doctrine; and this end he pursues with a tenacity that nothing can shake, and with that stubborn persistence and ardent ambition which make great priests and great statesmen."[3]

In the beginning, in their new Promised Land, the church was omnipresent and it placed a strong emphasis on cooperation, community, and environmental stewardship. As Young preached, "It is our privilege and our duty to search all things upon the face of the earth, and learn what there is for man to enjoy, what God has ordained for the benefit and happiness of mankind, and then make use of it without sinning against him."[4] This religious "land ethic" and insistence on the divinity of nature—"It does not matter whether I or anybody else owns it, if we only work to beautify it and make it glorious, it is all right"—suffused Mormon settlement of the American West and set it apart from the many individualistic and capitalistic pursuits that had dominated other Euro-American colonization efforts.[5]

So, let us begin with "the quote," which I first encountered while writing a history of the early uses of the Weber River. I found Young's dictum a perfect fit with the larger picture of public rights and access that I was describing. The reference most scholars use for "the quote" is Leonard Arrington's fine 1958 history of Utah, *Great Basin Kingdom*, where it appears as part of his discussion of Young's emerging land ethic and effort to provide direction to the Saints on the stewardship of their new colony.[6] "The quote" provides the perfect summation of the many sermons, pronouncements, and edicts of the "Lion of the Lord" and captures the essence of Young's focus on public access to vital natural resources that he sought to impress upon Mormon pioneers. But as far as I can tell, there is no evidence that corroborates Young's utterance of these exact words. Nevertheless, while the source of the exact quote may be dubious, its essential affirmation is spot-on.

My sleuthing of "the quote's" primary source began logically enough with Arrington's footnotes. His citation for Young's aphoristic commandment is the third volume of B. H. Roberts's six-volume *A Comprehensive History of The Church of Jesus Christ of Latter-day Saints.*[7] But "the quote" isn't there. Instead, there is only Roberts's passive voice statement that "subsequently it was announced there would be no private ownership in the water streams; that wood and timber would be regarded as community property."[8] Roberts does not say who announced this, when this announcement was made, or what was actually said. In fact, Young's oft-used quote does not appear anywhere in Roberts's *Comprehensive History.* Arrington used the same citation for "the quote" in his 1952 dissertation, which was the first of several iterations of the manuscript that ultimately became *Great Basin Kingdom.* The paragraphs immediately prior to "the quote" in *Great Basin Kingdom* cite the October 1848 *Journal History* of the LDS Church, but "the quote" does not appear there nor in any other of the many entries of *Journal History* that I checked. "The quote" is nowhere to be found in Joseph F. Smith's *Essentials in Church History*, the *Journal of Discourses*, Wilford Woodruff's extensive journals, or any other of B. H. Roberts's books, nor do any of the myriad texts that nevertheless use it to illustrate Young's land ethic and reprint it as the word of the prophet provide an accurate citation.[9]

In an effort to be as thorough as possible, I also contacted several prominent Mormon history and Utah history scholars, as well as the LDS

Church History Library. Their diligent efforts, which included searches in closed-to-the-public documents, also failed to locate "the quote."[10] In fact, and perhaps most notably, Arrington himself avoids "the quote" in his subsequent, award-winning biography, *Brigham Young: American Moses.* Moreover, in a curious admission at the beginning of the notes and references section of *American Moses,* Arrington writes that he hopes the manuscript "is fair to the original writer and to modern readers who want to draw their impressions directly from the original writing without having it 'improved' by well-meaning editorial intervention."[11] A veiled mea culpa perhaps?

Intriguingly, the best source for "the quote" appears to be Gordon B. Hinckley's 1947 *What of the Mormons?,* which predates Arrington's book. In chapter 15, the future president of the LDS Church states that on July 25, 1847, Young issued a statement of the policies that would prevail in the new colony: "There is to be no private ownership of streams of water; and wood and timber shall be regarded as common property."[12] Although this is not quite the verbiage that appears in Arrington's (and everyone else's) quotation of Brigham Young, it is the closest expression of "the quote" that I have actually found in any of the scores of documents I examined. But even this quotation is problematic because Hinckley does not cite any source for Young's declaration. Rather, like Arrington, he seems to have drawn from Roberts's discussion in *Comprehensive History.* What is clear from Hinckley's use of the quote, however, and countless other church histories that also utilize it, is that the Saints have long *believed* that Young stated this idea and have acted accordingly. Arrington perhaps just tightened up the pronouncement and gave it a punchier ending.

The question of whether Arrington may have committed the cardinal sin of historical fabrication is not one that can be posed directly, however, since the "Dean of Mormon History" passed away in 1999. But given my extensive research, I think it is not unreasonable to conclude that Arrington simply synthesized Young's basic ideas into perfectly pithy prose. For anyone who has spent much time reading Young's (sometimes meandering) writings, "the quote" sounds strikingly direct and strident, in both tone and meter far more twentieth-century than nineteenth. These two terse sentences effectively encapsulated the essence of Young's land ethic, rendering it far more accessible and powerful than any long litany of bits-and-pieces citations.

In all fairness, of course, the absence of evidence does not prove that Young never uttered these exact words, but unfortunately, Arrington's (and everyone else's) citation is not valid. Herein lies the rub: so long as Young's quote remains unverified, responsible scholars and writers must cease and desist in its use. But the *idea* of environmental stewardship conveyed by "the quote" *is* valid and has important historical implications. When Young and the Mormons settled in this Great Basin kingdom, they confronted long-term land and nature challenges—aridity and isolation in particular—that their earlier nomadism had enabled them to avoid. Their theocracy integrated environmental stewardship into daily life and practice as it laid out and assigned new communities, distributed land based on family size and ability to cultivate, and—even more importantly—presided over a communal irrigation system that initially was a notable departure from the "prior appropriation" allocation practices in other parts of the West. Political economics professor Richard Ely concludes that "individualism was out of the question under these conditions, and in Mormonism we find precisely the cohesive strength of religion needed at that juncture to secure economic success."[13] Water was the key. Young declared that the Saints would take the "fallen" deserts of northern Mexico (because the desert still belonged to Mexico at that time), and turn them into the Garden of Eden. Redemption of the land was the means to the Saints' salvation.[14]

For the Mormons, this winning of the West extended far beyond water allocation, however; it was about stewardship of the divine. Smith had instructed the Saints that "all things which come of the earth, in the season thereof, are made for the benefit and the use of man," and Young's abiding belief in the interconnectedness of all things, both spiritual and material, shaped his religious and temporal teachings and guided his frontier conservation ethic. In the LDS scriptural text Pearl of Great Price, Smith reiterated the idea from Genesis 2:9 that God had invested the trees and animals, like humans, with "living souls." Smith's successor Young believed his task and goal as the Lord's steward, then, was the promotion of, in Arrington's words, "the temporal *and* the spiritual welfare of his people," which Young effectively fused when he preached: "it is not our privilege to waste the Lord's substance."[15] The Saints' survival necessitated a reckoning with the reality of arid scarcity and a meting out of natural resources. Yet for Young, the religious nature of Mormon settlement meant that waste was nothing short of sinful: "It is all good, the air, the water, the gold and silver; the wheat, the fine flour, and the cattle upon a thousand hills are all good. . . . But that moment that men seek to build up themselves . . . and seek to hoard up riches, . . . it proves that their hearts are weaned from their God; and their riches will perish in their fingers, and they with them."[16] Instead, LDS doctrine clearly articulates that "the Lord, should make every man accountable, as a steward over earthly blessings."[17]

Because the Mormons arrived as such a large initial group, their interaction with the land and environment was instinctively collective rather than individualistic. Young proclaimed, "The earth is here, and the fullness thereof is here. It was made for man; and one man was not made to trample his fellow man under his feet" through the individual possession of it. "The Latter-day Saints will never accomplish their mission until this inequality shall cease on the earth."[18] Thus communal access to and use of natural resources, such as timber and water, would prevail, reinforcing the ideal embodied by "the quote." Yet Young also argued that anyone, Gentile or church member, who improved access to the canyons along the Wasatch Front, for example, ought to be able to profit from this endeavor and should be rewarded for making the canyons' natural resources, primarily timber, accessible to all: "Put these canyons into the hands of individuals who will make good roads into them, and let them take toll from the inhabitants that go there for wood, timber, and poles."[19] But the resources themselves—timber, water, etc.—were clearly still to be public resources. Even though early LDS leader Parley P. Pratt had his toll road and eponymous canyon, for example, he did not "own" all of its timber. Rather, his job was to regulate access to ensure the equitable distribution of the resource as a steward. In this way, the Saints sought to balance economy and environmental protection as best they knew how.

Scarcity necessitated supervised sharing. According to historian Dale Morgan, Young's policies "were remarkably well suited to the economic necessities of life in a desert region."[20] This goal of achieving utopian perfection enjoyed the singular advantage of starting from scratch, rather than laboring to reform an existing society. By 1852, with more than eight thousand Saints living in the Salt Lake Valley alone, territorial law granted county courts control over "all timber, water privileges, or any water course or creek, to grant mill sites, and exercise such powers as in their judgement shall best preserve the timber, and subserve the interest of the settlements."[21] Resource grants thus included rights of use and extraction, but not ownership of the land nor monopoly of its contents.[22]

The long-term success of Mormon settlement belies some initial ecological difficulties. For example, the nearly two thousand original migrants who struggled through their first winter in Deseret faced drought conditions and a plague of crop-destroying crickets the following spring. Although flocks of seagulls soon descended to feast on the insects, the chirping hordes continued to menace these newest agriculturalists for years to come (and still billow up in vexing clouds today). Historian Thomas Alexander demonstrates that the specter of starvation haunted the first decade of the Saints' settlements, bedeviling their efforts to "make the Earth like the Garden of Eden."[23] Various predators further compounded Mormon subsistence struggles by preying on their farm animals. Hungry pioneers soon depleted the region's big game populations at the same time as they introduced thousands of sheep, cattle, and other easy prey—nearly seventy thousand domesticated animals by 1870. The predictable livestock mortalities led to a pioneer war on predatory "wasters and destroyers." One Salt Lake Valley resident explained, "There is a general raid by the settlers on bears, wolves, foxes, crows, hawks, eagles, magpies and all ravenous birds and beasts." The death toll recorded between Christmas Day 1848 and March 5, 1849, for example, when these aggressive hunters convened to tally their "scalps," was shocking: 84 men had slaughtered "2 bears, 2 wolverine, 2 wild cats, 783 wolves, 409 foxes, 31 minks, 9 eagles, 530 magpies, hawks, and owls, and 1,026 ravens."[24] A similar if less well-documented "big wolf hunt" on numerous target predators occurred shortly after the settlement of Cache Valley in the 1850s and 1860s. While individual side hunts, which awarded competitive "points" for certain species, had been a feature of frontier settlement for nearly as long as there had been an American frontier, both Utah hunts were notable for their collective organization and their sizable scope and scale. Not coincidentally, the timing for them also coincided with a period of severe hunger in the earliest days of Mormon settlement. As historian Victor Sorensen humorously observes, "one way to 'keep the wolf from the door' was to kill and eat it."[25]

Additionally, one of the earliest ordinances of the new territorial legislature of 1850 was a two-dollar bounty on wolves, coyotes, and foxes. Pelts were also acceptable legal tender for tithing credits in the Salt Lake Stake for a time. This enmity directed at "noxious vermin" continued into the twentieth century, enabling entrepreneurial hunters to earn a one-hundred-dollar reward for every wolf killed near Kanab, for example. In September of 1912, the newspaper there dutifully reported that "W. E. Hamblin, S. L. Lewis, Sixtus and Jett Johnson, Frank and Jos. Hamblin have gone on the range to hunt wolves." This hatred for wild predators seems incongruous with the teachings of Young and the early prophets, yet the Saints embraced these exemptions to God's sacred creation in the animals they viewed as enemies of their efforts to sanctify Zion in the desert. Historian Michael Robinson writes, "wolves remained at the bottom of a moral hierarchy so obvious that it needed no articulation." Such wholesale slaughter brings to mind instead the poignant lesson of Greek philosopher Bion of Borysthenes: that animals killed for sport or profit do not die in sport, but in earnest.[26] Within a decade of their arrival, Salt Lake settlers had also ravaged local timber stands and their livestock had overgrazed the grasslands, which in turn led to damaging flash floods and debris flows. Woody sagebrush and exotic Russian thistles (tumbleweeds) capitalized on the disturbed soils and soon drove out native bluestem and gramma grasses.

This rapid depletion of natural resources represents a common thread running through much of Utah and the West's environmental history: the "tragedy of the commons" dilemma, a cautionary idea popularized by ecologist Garrett Hardin. Hardin argued that individuals acting in their own self-interest will ignore the best interests of larger society and deplete shared resources. To illustrate his point, he used the example of a local community grazing commons, "open to all," where each resident could sustainably pasture one cow. An individual herdsman could easily rationalize that the addition of one more cow to the pasture would have no appreciable negative effect on the commons, but would bring appreciably greater profit to the herdsman himself. So long as he is the only herdsman who thinks and acts this way, then the commons remains unharmed and stable. But the "tragedy" arises when each herdsman in the community reaches this same conclusion and each adds another cow to the commons. The individual's contribution does not measurably degrade the commons but the *collective* additions result in overgrazing. Even though the individual's intent is not malicious, the effect is nevertheless tragic. As Hardin writes, "Freedom in a commons brings ruin to all."[27]

While Hardin's example is oversimplified and abstract—not all people or groups utilizing community-managed resources careened headlong toward disaster—the "tragedy of the commons" predicament is most useful for explaining the exploitation of open-access resources such as forests, water, air, and grazing lands, rather than the more narrow, legal definition of "commons" Hardin outlined. Local, self-regulated commons—captured so effectively in "the quote"—often utilized successful resource management customs and practices, only becoming unstable when outside capitalists attempted to satisfy extra-local market demand with limited resources, leading to Hardin's "tragedy." For the Mormons of Deseret, the church *was* the government, and like future Progressives they believed that the community was a better and more impartial steward of the natural resources they all needed to survive in this new, arid Great Basin kingdom. Historian Milton R. Hunter concluded that "it is certain that cooperative ownership, wide distribution, and thorough utilization of water, became an established policy early in Utah colonial history."[28]

So how, specifically, did this early settlement effort fit into the "Lion of the Lord's" environmental ethic? Despite the "Great American Desert" label, the land chosen by Young and the Saints, nestled up against the Wasatch, contained fertile soils, a moderate climate, and perennial mountain streams. Yet the ecclesiastical nature of their endeavor—that the earth and its resources belong to God and that the Saints were only stewards—motivated Young to distribute land based on a family's ability to cultivate. No land was to be bought or sold, lest injustice, and thus blasphemy, develop and deprive future Saints of access to God's creation. Moreover, the Saints' assignment of small five- and ten-acre family farms, as opposed to the 160-acre national norm, ensured the equitable distribution of scarce water and land and reinforced community identity. In this era before the 1862 Homestead Act, Mormon settlement was village-based rather than individualistic. It was a practical solution. Not only was it "the only system of Mormon expansion that really worked," according to historian Charles Peterson, but it also facilitated the communal distribution of land and water Young advocated. "It was a sacrament upon the land—an edifice of worship."[29]

Heeding the call to come to Zion and "be gathered unto one place," more than seventy thousand Mormons trekked to Utah over the two decades following Young's arrival, bringing about significant changes in this arid land lying at the foot of the Wasatch Range. Although Young had brought the Mormon pioneers to the valley of the Great Salt Lake in 1847 in part to escape the United States and the persecution they had encountered at the hands of their fellow Americans to the east, the Saints were soon swept back into the national fold, and both religious devotees and Gentiles began the earnest search for economic possibility. Young's early employment as a carpenter, painter, and glazier provided him with a practical, real-world perspective that informed his leadership and land policy. He personally chose many of the Saints' settlement sites, oversaw the surveying and allotment of lands, and carefully considered the most propitious array of crafts and skills to assign to new settlements.[30] The Mormons' religious zeal may have distinguished them from other overland immigrants, but their basic outlook and effect on the natural environment mirrored that of pioneers generally. They, too, sought fertile soil for their nonnative crops, sufficient grasslands and meadows for their nonnative livestock, and suitable timber stands for building materials. The Saints also exposed local Indians to deadly waves of smallpox and measles, and while they grudgingly acknowledged the Native Americans' title to ancestral homelands, they often failed to respect it.[31]

"The Lord's Beavers," as environmental historian Donald Worster labeled the hardworking Saints, utilized strategies similar to earlier Indian practices of communal control and usufruct rights in their remarkable conquest of the desert they chose to make bloom. Native peoples had lived in the region for thousands of years, of course, and had altered the natural environment to suit their own needs. Paleolithic hunters had stalked mammoths perhaps into extinction, and ancestors of the Anasazi had extracted life from the harsh and arid environs through hunting and gathering until the rise of agriculture and irrigation allowed the Anasazi and Frémont cultures to thrive, while the Utes and Paiutes encountered by the Mormons used fire for hunting and to promote open grasslands enticing to game. But it was the *scale* of the Mormon transformation that was so striking. Although they brought zero experience in water management with them, by 1850 the Mormons were irrigating more than 16,000 acres. By 1890, these thrifty and persevering pioneers—who called their provisional state "Deseret," which means "honeybee" in the Book of Mormon—had adopted the beehive as the symbol of their work ethic and were watering more than 260,000 acres to support a population in excess of 200,000. Elder George A. Smith found this endeavor altogether fitting: "When the

Lord sees proper to break down the barriers that exist and cause the rain to descend upon the land, he can do it; but until then, he has very wisely provided that we shall take the streams in the mountains to irrigate the soil."[32] The church planned, financed, and managed the web of canals and dams that produced bountiful harvests, which, in turn, filled the pantries of enterprising Saints, albeit with nonnative plants and animals. Young and his followers fulfilled their promise to "make the desert bloom as a rose," but the environmental price for success here and in the rest of the American West would be high and facilitated the dilution of Young's land ethic.

Although the church endeavored to manage Deseret's natural resources for the good of the people, all of the people, Mormon economic stability would ultimately rest to a large extent on a growing trade with westward-bound migrants along the overland trails. Mormon prosperity resulted from their successful communal conquest and transformation of nature as well as their fortuitous settlement at an important waypoint along the Oregon and California trails. Thus, when a California Mormon named Samuel Brannan bellowed "gold" through the streets of San Francisco in 1848, spilling James Marshall's shiny discovery secret, the Saints stood poised to make a handsome profit by supplying the rush of forty-niners destined to flow through their territory.

As the LDS faithful shifted their production efforts to supply growing demand from extra-local consumers, they became inextricably entwined in the larger emerging market economy. It was ultimately unsustainable. This change in focus—from local ecclesiastically governed subsistence to national and even global secularly motivated commerce—fundamentally transformed Mormon lifeways and gradually drew Utahns away from Young's communal land ethic. The short-term effect of Young's land policies had produced numerous flourishing, stable settlements scattered across Deseret. But the long-term consequence was environmental erosion; it became impossible for Mormons to continue to live as they had initially lived in the American West. By the 1890s, the degradation that accompanied what historian Frederick Jackson Turner celebrated as the triumph of "civilization" over "savagery" was sobering: numerous species in peril, overgrazing and land erosion, and obvious limits to once abundant resources.[33]

Unlike a subsistence economy, a capitalist or market economy dictates not only the value of and means by which goods are traded but the very goods themselves—the merchantable commodities. The range of these commodities is notably far narrower than in a subsistence economy, where *any* item that sustains life has value, because merchantable commodities must possess a value mutually agreed upon by both seller and consumer. As anthropologist Marshall Sahlins argues, "wants may be 'easily satisfied' either by producing much or desiring little."[34] Commercial economies emphasize the "producing much" gambit. A commercial economy and the desires it creates also have significant environmental implications. The fundamental goal of subsistence societies like that of the early Mormon pioneers was survival, not the maximizing of production or profit. The ecological limits of their immediate surroundings and the demands they place on their local natural environment are finite and usually sustainable. Liebig's Law argues that the minimum number of resources available during the scarcest season further limits population sizes, as the Saints' first hungry years attest. Thus, subsistence economies usually live sustainably within their natural environment because population does not exceed the carrying capacity of the ecosystem.

A market economy bypasses these environmental checks and balances. The laws of supply and demand not of a local population but of a world market govern a capitalist system, and world markets generate potentially infinite demand for a narrow range of products. In addition, the profit incentive present in capitalism, and absent in subsistence, drives production increases in an attempt to sate what are ultimately insatiable demands. "Markets," writes Sahlins, erect "a shrine to the Unattainable: *Infinite Needs*" (emphasis in original).[35] Sometimes consumers desire certain commodities purely for the status that they convey. In the long run, this is unsustainable. These changes in social and cultural values are inextricably intertwined with changes in the environment, a biocultural process. When the Mormons abandoned their local subsistence agricultural economies to cater to the demands of the consumer market generated by the overland trails and the Transcontinental Railroad (completed in 1869), they overtaxed their soils, forests, and grazing lands, which soon became depleted and eroded, and set aside stewardship.

In many ways, the Puritan experience in New England more than a century prior to the Mormon migration provides useful insights.[36] In both cases, persecuted faithful fled their homeland in search of freedom, however carefully circumscribed and controlled. In 1630, Puritan leader John Winthrop implored his followers to cleave together as a Christian community, observing that "we shall be as a City upon a Hill, the eyes of all people are upon us."[37] Much like Young's call to "be gathered unto Zion," Winthrop's entreaty emphasized the divine covenant central to Puritan salvation and both groups regulated natural resource use and consumption. The Puritans' initial struggles to survive oceanic migration and the "starving time" of the first few years, combined with a constant influx of persecuted immigrants from England, initially ensured the community's cohesion and maintained religious zeal. Early Mormon settlement shared many of these same tribulations and resulted in a similar devotion to the words and guidance of the prophet. Smith and Young, both men of Puritan heritage, also stressed the importance of the sound stewardship of God's gift of the earth to humans. Yet in each case, as prosperity replaced suffering and practical reality replaced persecution, the utopian vision of "a City upon a Hill" gave way to more secular economic patterns. By the second and third generations, neither Puritans nor Mormons felt inextricably bound by the edicts of their leaders; as they embraced farming, trade, and other economic endeavors, these settlers began to replace their communal focus with more individualistic ambitions. For the Mormons, this meant that Young's land ethic, reflected in the 1847 quote and other earlier pronouncements, lost its prescriptive relevance and the Saints' original godly focus on environmental stewardship became more secularly oriented toward expansion and development. Quite simply, the farther Mormon settlement diffused across Deseret, the less effective Young and the Church's centralized control, scrutiny, and oversight became.[38] Local adaptations that made sense economically, if not environmentally, help explain why Young's initial teachings were not maintained over the long term.

Young was keenly aware that the lures of secular commerce were steadily eroding the Edenic mission of the Mormons. As he lectured in March of 1858, "we cannot avoid knowing that much of the conduct of this people has been directly in opposition to our becoming the kingdom of God in its purity on the earth." The Saints, he argued, "for the first time have the privilege of laying waste our improvements, and are not obliged to leave our inheritances to strangers to enjoy and revel in the fruits of our labors." Yet this turning away from stewardship, Young believed, meant forsaking their divine covenant, and he inveighed against this consumer impulse for making the people "more or less dependent upon our enemies [Gentiles] for many things that we could have produced or done without."[39] Historian Gregory

Umbach argues that Young's anxiety was well founded: "Mormons increasingly turned to an emerging national bourgeois sensibility expressed in a range of popular literature. Mormons' swelling reliance upon this broader net of cultural knowledge to decipher the goods around them washed away the symbolic salience of local points of reference that had previously given possessions meaning."[40] In other words, so long as Young's leadership held sway, Mormons generally did not regard water, for example, as a commodity or property, but held to the communal and spiritual understanding of natural resources. But increased federal oversight in the territory, such as the 1877 Desert Land Act that specified "bona fide prior appropriation" as the legal foundation for claiming western water, hastened the acceptance of water as a private property commodity.[41] The 1862 Homestead Act and federal land distribution accelerated a similar transition to more individualized and commodified land ownership. Furthermore, once the federal government established a land office in Salt Lake City in March of 1869, all legal land titles could only be conveyed to individuals, not a collective community.[42] This further shift away from Young's communal land ethic not only prioritized individual profit, but it also helped alter the human perception of the land itself from one of nearly infinite resources capable of ensuring survival to a simple commodity containing limited resources that could produce wealth in an exterior-defined market.

What Young's land ethic sought to promote, however, was equitable distribution to all the (local) people via a communal model, and historians like Donald Worster, Dan Flores, and Thomas Alexander have all used "the quote" to support this idea.[43] But Flores reinforces the shortcomings of Young's policies—that over time they did not translate into an ethic of conservation within a commercial economy, but instead became a misused justification for unbridled consumption that "overstrained the Wasatch environment."[44] Thus they formed only one side of the later Progressive equation. Young's goal for the Mormons was development (as a means of protection and power), and equitable distribution, unfortunately, did not guarantee thoughtful use over time, especially within the context of a capitalist system. The church's "theo-democracy," which exercised power over political and economic life in Deseret, maintained a strict "use it or lose it" perspective that ultimately produced Hardin's "tragedy of the commons."

Albert Potter's terrific 1902 report on proposed Utah forest reserves documents the massive overgrazing and overcutting that had occurred along the

Wasatch Front by 1900 and the desperate need for federal management of these vital natural resources: "it would be hard to find a seedling big enough to make a club to kill a snake."[45] While Potter's purpose was to assess the potential for various areas to be included in new national forest reserves, his detailed descriptions provide excellent insight into the state of the forests at the turn of the twentieth century and the continued erosion of Young's ethic. Potter's account describes the Wasatch, Gunnison, and Sevier Forest Reserves and the effects of mining, sawmills, and stockyards on the forests. In almost every canyon of the Weber and Provo Rivers, Potter describes the remnants of the previous years' tie and timber harvests—old sawmills, camps, and areas "cut out."[46] The evidence for the overuse of the mountain areas was extensive—top grade timber was a rare commodity in many mountain valleys by the turn of the century and sheep ranching had fully invaded the high country.[47] Flores echoes this assessment, noting that by 1881, "and continuing thereafter with mounting fury and frequency, the now deteriorated mountain watersheds, which geological evidence proved had not flooded since Lake Bonneville had receded 25,000 years before, began periodically to send tons of water, soil, and boulders rolling into the streets and irrigation works of the towns below them. . . . Even with Forest Service regulation [eleven national forests by 1910] . . . grazing and logging pressure continued to be too intense on the Wasatch. By the 1920's [*sic*] a widespread land collapse had begun."[48]

Which brings us full-circle back to "the quote." Clearly, "the quote's" lack of an authenticated source does not change its near pitch-perfect capture of the sentiments of Young and the early Mormons about land and natural resource stewardship.[49] It does, however, serve as a valuable and important lesson in primary source research and scholarship: verify your sources.[50] Moreover, this kind of interrogation of the past, and even of the best historians, leads us to a deeper and more complex understanding of historical events while also maintaining the validity and vibrancy of the profession. Young may not have said specifically that "there shall be no private ownership of the streams that come out of the canyons, nor the timber that grows on the hills. These belong to the people: all the people." But he certainly meant it. However, historians know that any argument must rest on actual historical evidence. "The quote," unfortunately, doesn't qualify.

Nevertheless, "the quote" quandary reveals that Young's land ethic was and is seminal, and it provides a vital link between Spanish and Mexican colonial resource management policies of the past, which included public ownership of waterways combined with priority rights for users as well as future Progressive-Era federal conservation efforts such as the U.S. Forest Service, which sought to manage timber to provide "the greatest good" for "the greatest number."[51] All three entities—Spanish, Saints, and Progressives—treated mountain resources as communal rather than private property. Uniquely in the West, public lands such as national parks, forests, and reserves endeavored to recreate this model in an effort to counter Hardin's predictions of environmental ruin through federally regulated natural resources. The U.S. Forest Service adhered to "the gospel of efficiency"—an almost religious dedication to the scientific management of the forests (as a crop) to ensure "that the water, wood, and forage of the reserves are conserved and wisely used for the benefit of the home builder first of all."[52] This "multiple use" philosophy, which sounds strikingly similar to "the quote," sought to avoid the "tragedy of the commons" by regulating cutting, mining, grazing, and recreation to ensure that the forests could be both used *and* saved, the hallmark of efficient conservation. Presidential hopeful (and Mormon) Mitt Romney must have missed that lesson when he expressed being confounded in the 2012 election about "why the government owns so much of this land" and "what the purpose is of the land."[53] Despite Romney's Sagebrush Rebellion rhetoric, the idea of sound stewardship of and public access to the state's natural resources endured (and endures), both in law and in practice. Young argued specifically for the protection of City Creek, for example, saying, "this creek should be adored."[54] And in his 1927 bulletin on "Mutual Irrigation Companies in Utah," irrigation economist Wells A. Hutchins wrote, "no monopoly in either land or water developed in the early days was due to the fact that the church leaders were constantly on guard against it."[55] Subsequent court cases sought to clarify further public and private rights and protections. In 1937, the state's high court waxed eloquent, and echoed Young, when it declared that Utah's "waters are the gift of Providence: they belong to all as nature placed them or made them available. . . . While it is flowing naturally in the channel of the stream or other source of supply, [water] must of necessity continue common by the law of nature . . . or property common to everybody. And while so flowing, being common property, everyone has equal rights therein or thereto, and may alike exercise the same privileges and prerogatives in respect thereto."[56]

Young's land ethic and encouragement of environmental stewardship also provide a powerful ideological connection between early Mormonism and modern environmentalism. His amalgamation of theology and ecology anticipated later nineteenth- and twentieth-century advocates like George Perkins Marsh, John Muir, and Aldo Leopold, who all championed a moral relationship between humans and nature. The Mormon genesis in the Second Great Awakening—the same spiritual impulse that gave rise to abolitionism and women's rights—certainly validates Worster's argument that environmental reform owes "much of its program, temperament, and drive to the influence of Protestantism."[57] Both Young and Muir, the product of Scottish Presbyterianism, tapped into the aesthetic spirituality of nature and saw salvation in wise environmental stewardship of the sublime. As Muir wrote in 1869, "Everything turns into religion, all the world seems a church and the mountains altars."[58] Both men would have agreed that while nature itself may not be divine, as the creation of God, it is nevertheless holy. This profound connectedness, which Rachel Carson would later call the "ecological web of life," creates a moral responsibility and obligation for sound stewardship.[59]

Young's ideas about the temporary nature of human tenure on God's earth also find modern expression in religious scholar Richard A. Baer, who writes that the avoidance of arrogance and anthropocentrism is essential since the earth "is a property that does not belong to us."[60] Baer continues, "Failure to fulfill our obligations as faithful trustees of the gift of God's creation will inevitably bring God's judgment upon us."[61] According to Worster, "Muir invented a new kind of frontier religion: one based on going to the wilderness to experience the loving presence of God." Almost a century earlier, Young had similarly evangelized that "the Lord has poured out his blessings on the atmosphere, on the water, and on the soil of this country."[62]

Today, environmental activist groups like Greenpeace manifest this greening of religion legacy when executive director Annie Leonard avers, "We 'bear witness' to environmental destruction in a peaceful, non-violent manner."[63] In the effort to provide for "all the people," the non-Mormon Reverend Vincent Rossi concludes, "the *deep* ecological task is . . . to awaken in human souls the sharpest possible awareness that belief in the existence of God absolutely demands the deepest respect and reverence for the rights of the Earth."[64] Young's environmental ethic sounded this same clarion call throughout the far corners of Deseret, advocating stewardship of the land and its resources and promoting an appreciation of and reverence for nature as nothing less than bearing one's religious testimony.

LOST MEMORY AND ENVIRONMENTALISM

Mormons on the Wasatch Front, 1847–1930

Thomas G. Alexander

SINCE LYNN WHITE JR. BLAMED environmental damage on the mid-nineteenth century union of science, technology, and Western Christianity, scholars have debated the relationship of religion to the environment. White argued that Western Christianity "not only established a dualism of man and nature but also insisted that it is God's will that man exploit nature for his proper ends. . . . Modern Western science was cast in a matrix of Christian theology." Thus, he said, "Christianity bears a huge burden of guilt." After Western Christians found that St. Francis preached that nature was sacred and that animals have souls, they suppressed his teachings. The "roots of our trouble are so largely religious, the remedy must also be essentially religious," White wrote.[1]

Arguments like White's stirred some scholars to blame certain Christians more than others, and other scholars have defended Christianity. Some have argued that Evangelicals tend to be less environmentally conscious than Catholics, liberal Protestants, and secularists. Some have blamed capitalism.[2] Still others like Wendell Berry, Susan Power Bratton, J. Baird Callicott, Paul Santmire, and Mark Stoll have defended religion by arguing that the Bible and Christian tradition support a salutary environmental ethic.[3] Historians and others have also argued over the role of natural change and change by human intervention.[4]

Such debates provide insights, but they do not grapple sufficiently with a persistent environmental problem that Western Christians have faced—reconciling two traditions or understanding what has happened to them over time. These traditions are: 1) the Christian teachings of stewardship and reverence for life and 2) secularized entrepreneurship. I have used the term secularized entrepreneurship as shorthand for any of the broad range of economic activities such as mining, commerce, manufacturing, banking, ranching, and farming. Even though the Puritans had originally based entrepreneurship on the religious concept of divine calling, by the late eighteenth century Americans had secularized entrepreneurship.[5]

White based his argument on broad generalizations, but in this essay I intend to test his thesis by examining in detail a specific Christian tradition—the Mormons. They meet White's conditions in time and space since they flourished in a Western tradition after both the industrial revolution and the conjoining of science and technology.

My investigation into the problem of the relationship between Christian traditions, science, and technology led some years ago to the publication of an article in *Western Historical Quarterly*.[6] Since publishing that article, I have wondered whether I really got the analysis of motivation right. In recent years a number of scholars have begun to examine the problem of collective memory and how people remember or fail to remember events and concepts from their common past. A consideration of this question led me to examine collective memory of environmental problems and change in the Mormon community. The following is the result of that reconsideration.

In examining the Mormon response to environmental challenges, I argue that the gradual loss of the collective memory of a salutary environmental theology and its replacement by a memory of progress associated with secular entrepreneurship facilitated, in part, the environmental destruction that plagued Utah's lands. I believe that the change was similar to the change that James Fentress and Chris Wickham observed in the loss of memory by working groups they studied. They observed that forgetting usually derives "simply from the different criteria of importance that working-class informants have for determining the personally and collectively relevant past, . . . sometimes the past entirely slips away."[7]

I argue that a similar phenomenon happened to the environmental teachings of nineteenth-century Mormon leaders. The collective memory loss of salutary environmental teachings and their replacement by the memory of progress under secular entrepreneurship rather than the marriage of Christianity, science, and technology as White argued played a central role in the loss of environmental consciousness in the Mormon community. This change actually began during the time these early leaders lived. This replacement seems to have taken place because of the economic progress and relative prosperity that began early in the Mormon occupation of Utah and accelerated after the arrival of the railroad.[8] Brigham Young died only eight years after the railroad arrived, and most of the other leaders who preached a salutary environmental theology either died before he did or shortly thereafter. I argue that in place of an environmental theology was an emphasis on restoration, atonement, and Zion building. Since 1900, some in the community have occasionally recovered the memory of the salutary environmental theology in the revelations and sermons of early leaders. They point to these revelations and sermons, in part, to support their arguments on the need to protect God's physical environment. This is not to say that environmental theology was emphasized more than fundamental doctrines like the restoration and the atonement—rather it was taught alongside other doctrines, though with less frequency.

By contrast, some authors have argued that Mormon theology and cultural attitudes lie at the root of what they perceive as Mormon environmental insensitivity. A problem with this writing seems to be that these authors simply did not study the early preaching of Mormon leaders to understand that they taught a salutary environmental ethic. John B. Wright argued that Mormon millennial theology and its cultural attitudes underpin Utah's environmental degradation. Max Oelschlaeger asserted that the LDS Church "stated its opposition to ecology as part of the church's mission." In an excellent study of changing conditions, Dan L. Flores blamed environmental damage on "stockmen possibly somewhat bewildered by the strangeness of the Mountain West, . . . [and with the lack of an] empirical understanding of how mountain land worked."[9] Charles Peterson argued that observers understood the environmental damage, but that the damage persisted because neither the Mormons nor the Forest Service could reduce the numbers of sheep grazed on fragile watersheds by the many small farmers.[10]

We can easily dispose of Wright's and Oelschlaeger's arguments. Those of Flores and Peterson are much more plausible. Wright linked what he perceived to be current Mormon actions and theology with what he perceived to be millennial theology. Oelschlaeger's arguments are outdated; since his book was first published, the LDS Church has issued a comprehensive official statement entitled "Environmental Stewardship and Conservation," which calls for environmental sensitivity and care for God's creations.[11]

Wright, on the other hand, showed a lack of understanding of Mormon theology. Contrary to Wright's argument, Mormonism's millennial theology does not require insensitivity to nature as the church made clear in its statement on environmental stewardship. Its statement on the millennium relates to conditions following Christ's return to the earth to rule for a thousand years, not to insensitivity about the environment.[12] Wright demonstrated confusion over both Mormon ideals about environmental stewardship and their conception of the millennium. Moreover, he would also have been unaware of the official statement on environmental stewardship mentioned above since his book was published before the LDS Church issued the statement. Flores and Peterson, on the other hand, seem to have understood the problems that ranchers and farmers faced in an environment quite different from their native Midwest or Europe.

Some scholars have made the mistake of comparing the ideals of a favored group with the actions of a group they criticize, generally to the detriment of the latter. The actual behavior of virtually every group of human beings falls short of their ideals. Robert F. Berkhofer Jr. has dealt cogently with this tactic, which some anti-Mormons use to the detriment of the Mormons.[13]

Recently, a number of authors have begun to recover and restore the earlier salutary environmental theology. Hugh Nibley was undoubtedly the first of these. Nibley recovered this theology as he researched Brigham Young's sermons and teachings. Richard Jackson argued that the Mormons expected the Lord to temper the climate and geography if they followed his commandments. Jeanne Kay and Craig Brown extended Jackson's argument. Several anthologies with articles by prominent Mormons, some of whom are or were high-ranking church leaders, argued that Mormons are environmentally conscious.[14] I have argued elsewhere that, in fact, a salutary Mormon environmental theology preached by nineteenth-century church leaders encompasses central tenets of Mormon doctrine—including creation, sin, redemption, the atonement, and millennialism.[15]

If we wish to investigate Mormon environmental theology we should explore those environmental teachings that for a time were forgotten or de-emphasized by the Mormon community. In recent years some in the Mormon community have begun to remember the teachings again. In part, the Mormon heritage resembles that of other Euro-Americans. Several features of that heritage seem paramount, and a number seem contradictory. Like many Christians the Mormons carried a theological disposition to live on earth as stewards with reverence for all things. At the same time, they shouldered the Euro-American collective practice of secular entrepreneurship. Some Mormons view entrepreneurship as a religious concept, but many—perhaps most—do not.

The Mormons' heritage, however, differed from some other Euro-American Christians in significant ways. Strongly communitarian, they sought to build the Kingdom of God on earth; and, assuming both the Christian and Western belief in linear progress that Lynn White rightly observed, they expected to use science and technology to refashion the semi-arid West both as a fit place for Christ's Second Coming and for their earthly home. To assist in accomplishing these goals, returning to the Puritan vision, they attempted, eventually without success, to subordinate the entrepreneurial tradition to the sacred.

Mormon environmental theology developed first in the revelations and contemplation of Joseph Smith. Smith was a radical materialist, but his materialism was also radically different from secular materialists. Smith taught that both visible substance and invisible spirits are material. Spiritual material, he taught, is simply more refined than physical material. Moreover, material is eternal. God did not create it. Most importantly, God did not create the earth and the universe from nothing (*ex nihilo*). Rather, he organized it from preexisting but unorganized material. In the case of the earth, He had Christ (under the name Jehovah) and Adam (under the name Michael) organize the material into the earth.

With views similar to the American Indians and modern Gaians—and heretical to those nineteenth-century Christians influenced by modern physics—Joseph Smith taught that animals and plants, like humans, had eternal spirits.[16] This meant that they had agency, or in Mormon terminology: "free agency." Both God and human beings had an obligation to respect that agency. In reporting revelations, Smith said that unlike the majority of people these nonhuman creatures lived "in their destined order or sphere of creation, in the enjoyment of eternal felicity," and that the earth, the "mother of" all humans, possessed a soul pained by "the wickedness of my children."[17]

In addition, he taught the spiritual unity of humans, the earth, and its nonhuman inhabitants under the fatherhood of God. In prophetic statements repeated by other church leaders, Joseph Smith taught the unity of the temporal and the spiritual. Speaking for the Lord, he said that "all things unto me are spiritual, and not at any time have I given unto you a law which was temporal."[18] From a theological perspective, then, the Latter-day Saints lived in an undifferentiated temporal and spiritual world, building God's kingdom on earth and in heaven under the leadership of divinely commissioned prophets. In the most profound sense, prophetic leaders expected the Mormons to weave entrepreneurship and stewardship into a seamless garment.

In interpreting Joseph Smith's teachings, some early church members drew upon a holistic concept of the relationship between the temporal and spiritual to regulate settlement and the utilization of resources. Beginning in the late 1840s other church leaders elaborated on Smith's teachings. President Brigham Young rebuked members of the pioneer company as they traveled to Utah for killing more animals than they could eat.[19] Orson Pratt, of the Quorum of the Twelve Apostles, taught that God had created "the spiritual part" of the earth and all earthly animals and plants in heaven "before their temporal existence," and that this creation sanctified them. Heber C. Kimball, of the First Presidency, urged the Saints to extend mercy "to the brute creation," since animals have spirits and God will resurrect them along with the earth and human beings. Only after the Saints had learned to live in harmony as stewards with one another and with the earth, Brigham Young said, could they expect to inherit it, presumably as exalted beings, from the Lord who owned it.[20]

As the Mormons settled Utah, church leaders elaborated on these teachings.[21] The church leadership assigned town lots and farms as inheritances. On July 28, 1847, Young said they could sell their inheritances under certain conditions. Young said when "improved," the owner might "sell the whole Lot or his inheritance in the country and go to some other place."[22] He vacillated on this view, but generally approved the sale of improved property.[23] Nevertheless, in an explicit reference to a sacralized entrepreneurial tradition, Young said that if stewards did not develop the land as good managers, the Lord required them to relinquish it to someone who would.

Young's views on property ownership seem consistent to a degree with John Locke's. Locke and Young, both religious men, essayed on how humans could convert God's property into private property. Locke believed humans first lived in a state of nature. God owned all property. God then gave the property to mankind in common. People could transform the commons into private property if they "mixed [their] . . . labour with, and joined to it something that is . . . [their] own."[24] Young believed that God had designated the priesthood as stewards over His property. They had the right to assign God's property to His people as inheritances. He knew that the federal government had to extinguish Native American title to the land, but for the time being said "the land was given to me by the Indians and I have [given] it to you to use [after paying for] . . . surveying and recording."[25]

County courts (now called commissions) issued certificates of ownership that provided a color of title until occupants could acquire legal titles from the federal government. Mormon pioneers understood that these titles carried no legal authority with the federal government. They knew they would have to obtain clear titles after Congress authorized them to do so.[26] Although the Mormons sent numerous petitions, Congress did not extend the land laws over Utah until 1868.

Before long the Mormons seem either to have forgotten or ignored the salutary environmental theology. As early as the 1850s, years before the railroad facilitated a radical transformation of Utah's economy, Mormons allowed the market rather than religious stewardship to control the value of lots. The price of property in Salt Lake City increased rapidly. By mid-1851 lots in Salt Lake's business district cost more than $1,000. In 1847 settlers paid a $1.50 surveying fee. The increase in value resulted from extensive development.[27] In 1854, Charles H. Oliphant paid $250 for a lot in the Salt Lake City Twelfth Ward that, just six years earlier, someone had bought for a $1.50 surveying fee. Anyone with sufficient cash could purchase, rent, or lease property that Mormons had transferred from the Lord to individuals. In a sermon in October 1865, Orson Hyde of the Quorum of the Twelve Apostles chastened the Saints for their "inordinate desire for wealth and extensive possessions." He deplored the destruction of overgrazing in the valleys.[28] Hyde lived in Spring City, a Sanpete County community. He had seen the overgrazing by small farmers, and commented on their misuse of the land. Overgrazing in the mountains followed thereafter as farmers took their sheep into the uplands.[29] This problem persisted until the Forest Service succeeded in regulating grazing.

Young preached on the biblical injunction to multiply and replenish the earth. He urged the Saints both to conserve native plants and animals and to increase the diversity of God's creations, since they were "all designed to be preserved to all eternity." In view of this belief, Young fostered the importation of large varieties of flora and fauna to the Intermountain West, while he urged the people to protect the species already there.[30] In some cases, introduced species competed with native species to such a degree that the native species declined and disappeared. The Mormons apparently did not realize the contradiction at the time.

The Saints carried a sacred obligation to build God's kingdom on earth as they exercised their stewardship over property in an environmentally responsible way. The Mormon view—as taught by Joseph Smith and later church leaders—bore little relationship to classic American agrarianism or to nineteenth-century capitalism. Mormons believed the earth and its animal, mineral, and vegetable inhabitants were living organisms with souls. They were neither possessions nor commodities. Everything, whether folks perceived it as living or not, occupied a place in God's domain, and each creation—the earth, the animals, the plants, the rocks, and human beings—relied on one another. The earth, a living creation of God, belonged to Him. In September 1850, Young spoke of the duty to protect the earth. This is "the Zion of our God," he said, "here the air and the water—the grass—you are not at liberty to foul the water, but let it run free for all. . . . The pasture belongs to our Father in Heaven—then do not destroy it—the timber is free for you—but you have not right to destroy."[31]

Young believed the earth did not then "dwell in the sphere" in which God created it, but He had "banished [it] from its more glorious state . . . for man's sake." He said, "The spirit constitutes the life of everything we see." This included rocks, mountains, grass, flowers, trees, and even "the different ores of the mineral kingdom. . . . There is a spirit in the earth; . . . the earth is a living creature and breathes as much as you and I do." Tides, he said, acted like the "beating of a man's heart."[32] The earth's breathing causes the tides and "forces the internal waters to . . . the highest mountains which often gush out forming lakes, and springs." He said that the spirits in plants like those blades of grass, no two of which are exactly alike, engender the variety we see in nature.[33] In 1855, he said humans could plant, water, and reap, but a "superior power" springs the seed "into life" and ripens it "for the sustenance of man and beast." If people "are covetous and greedy," their "nerves twitch and they have the jerks in their sleep."[34]

In May 1871 Young discoursed extensively on the earth and science. He said that Moses "obtained the history and traditions of the fathers, and from these picked out what he considered necessary." It does not "matter whether it is correct or not, . . . or whether he made it in six days or in as many millions of years, it is and will remain a matter of speculation in the minds of men unless he give[s] revelation on the subject." He continued, "We differ very much with Christendom in regard to the science of religion. Our religion embraces all truth and every fact in existence, no matter whether in heaven, earth, or hell. A fact is a fact, all truth issues forth from the Fountain of truth, and the sciences are facts as far as men have proved them." And again: "The Lord is one of the most scientific men that ever lived; you have no idea of the knowledge that he has with regard to the sciences."[35]

Although Young sought to convert wilderness into cities and farms, he also expected Mormons to preserve some wilderness to gladden the eye and heart. In September 1848, he ordered a mill removed from City Creek Canyon because it had polluted the water.[36] He told the "brethren to keep out of City Creek Canyon." He wanted "to leave . . . every little shrub on the creek," to restore it to its 1847 condition of being "beautifully shaded and cool. . . . This creek should be adorned."[37] Young, however, later constructed a road up the canyon and permitted people to log in the upper canyon.[38] In Big Cottonwood Canyon in 1856, he said: "Here are the stupendous works of the God of Nature." He urged Saints to "calmly meditate . . . upon the wonderful works of God, and his kind providence that has watched over us and provided for us, more especially during the last ten or fifteen years of our history."[39]

In 1852 the legislature assigned regulation of canyons to county courts. Unfortunately, many who traveled into the canyons used roads without paying, and violent conflicts often resulted from confrontations between builders and freeloaders.[40] Addressing this problem in an October 1852 General Conference, Young proposed: "We do not own the kanyons [*sic*] . . . but . . . let them go into the hands of individuals who will make them easy of access." The conference approved, and Young told the county courts to "put these kanyons into the hands of individuals who will make good roads . . . and let them take toll from [those] that go there for wood. . . . Note this is my order from the President of the Church."[41] Young's policy also militated against what Garrett Hardin called "the tragedy of the commons,"[42] which occurs when people can obtain a benefit from the community without paying. Although the toll policy did not prevent profligate use, it did prevent "free" use.

The county courts also issued permits for herd grounds. To receive permits, users had to file bonds and have lots "surveyed and located." The permits lasted one year, after which herders had to apply again.[43] In 1857 and 1858, the legislature granted herd grounds to individuals. In some cases, the grantees had to allow others to graze and log on the grants. In others, the grants made no such restrictions.[44] The legislature assigned parts of Cache Valley to Brigham Young; southern Weber Valley to Thomas J. Thurston, Jedediah M. Grant, and others; northern Weber Valley to John Stoker, William Smith, John Hess, and Abiah Wadsworth; and parts of Utah Valley to a party headed by George A. Smith.[45]

The regulation of timber and herd grounds elicited vigorous opposition from federal officials. Utah surveyor general David H. Burr complained to U.S. attorney general Jeremiah Black that such grants compelled "the settler . . . to pay the Grantees for all the timber and firewood they use." "County Courts," he wrote, "control all the timber, water privileges and so forth." He deplored that "minions of the Priesthood" controlled "herd grounds" in "neighboring Valleys."[46]

The Mormons generally ignored such complaints because they believed they had regulated use in the common interest. It is not clear if such regulation always served the public interest. After all, some of the large grants went to prominent individuals. The actual may not have coincided with the ideal.

Mormon settlers developed an orderly landscape of wide streets, large blocks, and detached houses with barns, corrals, large gardens, and fruit trees inside the towns. They cultivated farms, from ten to twenty acres, outside the towns. In Salt Lake City, Governor Young expected irrigation ditches to run down the sides of each street, not only to water the gardens and orchards, but to carry any sewage and refuse to the lower Jordan River, which he seems to have conceived as a natural sewage treatment facility. They dug ditches and laterals, plowed and planted the land, sent parties into the mountains to cut timber, and detailed volunteers to manufacture adobes, from which they constructed most of their early buildings.[47]

We do not know to what degree Young's environmental theology motivated this orderly development. It may have been a pragmatic solution to immediate problems. He may have been inspired by Joseph Smith's plan for the city of Zion, but the streets in Salt Lake City were wider and the blocks larger. Nevertheless, such controlled development was consistent with his environmental theology. If pragmatism underpinned his developmental orders, then pragmatism and environmental theology reinforced one another.

Following the patterns of the American Indians and Hispanics, the Mormons departed from previous practice not in the plants they cultivated, but in the development of irrigation technology and institutions.[48] In the nineteenth century, their projects consisted of small dams or weirs across creeks and rivers built by cooperative irrigation companies that also dug canals to divert water onto farms and town lots. The settlers generally appointed a watermaster to apportion the streams in rotating weekly turns. By 1900, although the Mormons had constructed some small reservoirs, most irrigation consisted of these ditchworks.

Of great importance to the Mormon story, the Wasatch valleys possess an abundance of water and rich soil. This abundance was concentrated in a relatively small area amid a vast expanse of arid lands. Estimates of the available water in the Bear, Weber, Jordan, and Provo River watersheds by Grove Karl Gilbert—published in John Wesley Powell's *Lands of the Arid Region*—confirmed the Mormons' initial assessment. Gilbert estimated that the Bear River and its tributaries carried more than enough water to irrigate all arable land in Cache Valley, and that the Bear and Malad Rivers could also irrigate the lower Bear and Malad drainage in the Salt Lake Valley. Furthermore, with the addition of surplus water from the Bear and Jordan Rivers, he estimated that the settlers could water the Wasatch Front land in the Weber and Ogden river valleys susceptible to cultivation. Tributaries of the Jordan River carried more than enough water, he said, to irrigate all of the land in the Salt Lake Valley, and unless settlers diverted excessive water for relatively inefficient agriculture in the Kamas and Heber valleys along the upper Provo River, the streams that flowed into Utah

Lake carried enough to irrigate the lands of Utah Valley. Thus, as the early observers anticipated, instead of a water shortage, the Mormons found the Wasatch Front richly endowed with the precious fluid. Investigations supervised by Elwood Mead a quarter century later supported Gilbert's analysis, though Mead included reservoirs in the equation.[49]

Unfortunately, in view of the paucity of timber in Utah's Wasatch Mountains, the admonition of Young and other leaders to care for the timber failed. Loggers worked inefficiently, they ignored Young's advice, and they cut a giant swath through accessible stands. Loggers soon found easily harvested timber in short supply. By the mid-1850s, Wilford Woodruff had difficulty finding accessible timber in the lower canyons near Salt Lake City. During his trip to the Salt Lake Valley and visits to the nearby canyons in 1860, Sir Richard Burton noted the summit of Emigration Canyon pass "well nigh cleared of timber." He noted that in other canyons timber was not plentiful. Surveying the Wasatch Mountains in 1902, Albert Potter found denuded slopes and extensive environmental damage throughout.[50]

Increasingly, people in the community who had ignored or forgotten Young's views separated the temporal and spiritual. They forgot or ignored the environmental theology, enthroning in its place progress through secularized entrepreneurship. The apostasy of prominent business leaders—like the Walker brothers during the late 1850s and the Godbeites during the late 1860s and early 1870s—sidetracked teachings about the sanctity of all life and environmental protection.[51] That is, prominent business leaders who left the church were not concerned about environmental matters. Their principal concern was with making profits in the prevailing economic system.

Examples from the logging industry illustrate the point. In one venture in Little Cottonwood Canyon, a speculator cut one million board feet of timber. Unable to sell the logs, he simply left them to rot on the ground. In other cases, logging in the upper valleys near the ridges left the canyon slopes vulnerable to avalanches and flooding. As ventures in mining and smelting expanded after 1869, responding to the market, loggers denuded the slopes of juniper and piñon to make charcoal.[52] That is not to say that problems with improvident use of resources did not occur before 1869. In some cases, overgrazing took place well before 1869.

In spite of such improvident use of resources, the application of improved technology elsewhere in the nation saved the forests of the Wasatch from complete destruction. The expansion of railroads throughout the Wasatch valleys by the late 1870s left Utah's forest industry vulnerable to competition from outside lumber operations. In the Mississippi Valley, the Great Lakes region, and the West Coast loggers had access to large stands, modern mills, and railroads.[53]

Pounded by competition from outside and suffering from a short supply, between 1880 and 1900 the relative volume and value of Utah's lumber harvest declined rapidly.[54] In 1870, Utah had more mills and a higher value of timber production than any of the Intermountain or West Coast states or territories except California and Oregon. By 1916, Utah mills supplied only about 10 percent of the 150 million board feet consumed annually in the state, and while the Forest Service and others tried to encourage increased timber production, little occurred—and, in general, the damage to Wasatch timber stands that Potter observed in 1902 had considerably healed.[55]

As the experience with timber management showed, on balance, part of the Latter-day Saint philosophy wore well in practice and part of it fit poorly. The separation of the entrepreneurial and stewardship traditions and the secularization of the former meant that immigrants seeking land (excepting those who could qualify under some federal land laws) found themselves at the mercy of the market.

By contrast, while new settlers had to buy their land, the Latter-day Saints followed Brigham Young's admonition to increase the diversity of plants and animals through the introduction of exotic species provided by nurseries that had opened in several towns. For example, Charles Oliphant purchased sprouts for root stock to graft large varieties of fruit trees. So anxious was Young to increase the variety of fruits that he called Oliphant to operate the nursery as a religious duty. Joseph Ellis Johnson, Luther Hemenway, William Rigby, and others also operated nurseries. Following the pattern, community leaders like George Q. Cannon, Brigham Young, Albert Carrington, and William Staines also grafted extensively and imported numerous exotic plants. Wilford Woodruff and others in the Deseret Agricultural and Manufacturing Society imported plants and animals from throughout the United States and western Europe.[56]

Although the efforts to increase the diversity of plants and animals on the irrigated farms in the Wasatch oasis proved ultimately successful, starvation and plant and animal destruction accompanied that success in the early years. To some degree the hunger and devastation resulted from Euro-American dietary habits. Crickets periodically attacked both native and exotic plants. Like John the Baptist, the American Indians had turned such plagues to an advantage and ate the crickets, but the Mormons were unable to overcome cultural attitudes and follow their example.[57]

Cricket and grasshopper invasions were not transient problems—though the devastating cricket invasion of 1848 is well known, recurring plagues reached Wasatch Front communities well into the 1870s. Young himself tried to eat the crickets but ultimately could not bring himself to do it.[58] Instead, during famines the Mormons hunted antelope, hawks, crows, and wolves, and scrounged up thistles, bark, roots, nettles, pigweed, redroot, and sego lily roots. As soon as familiar foods became available again, they reverted to their traditional diets.[59] Between 1847 and 1858, the Mormons experienced food shortages every year except 1852, 1857, and part of 1853.[60] They partially solved the problem in 1857 by constructing a canal from Big Cottonwood Canyon to southern Davis County.

In December 1849, in a desperate effort to survive in the face of starvation, Mormon settlers declared "a war of extermination against" wild animals. They organized parties that set out to kill predators and vermin, including wolves, wildcats, bears, catamounts (bobcats?), panthers (cougars?), skunks, and minks, as well as raptors and scavengers like eagles, hawks, owls, crows, and magpies—animals they called "wasters and destroyers." They also offered bounties for wolf and fox skins.[61]

By the mid- to late-1870s, the Mormon settlers had killed off enough crickets to change the Wasatch Front environment. These insects could no longer reproduce in large hordes. Under these circumstances, the railroad provided transportation facilities for the products of irrigated farms, grazing herds, and dry farms. Wheat production soared 109 percent between 1869 and 1879 and 512 percent by 1899. Oat production spurted ahead during the 1880s, potato output grew exponentially during the 1890s, and barley and hay yields took off and thrived during the 1920s.[62]

By replacing native vegetation with imported plants, Mormons altered the face of the land. After 1891, the European sugar beet became a mainstay of irrigated agriculture. From the Bear River and Cache Valleys on the north to Utah Valley on the south, families contracted with both the Utah-Idaho Sugar Company and the Amalgamated Sugar Company to supply beets to factories built from Lewiston to Spanish Fork.[63] In general, however, although these farmers produced for the market they did not establish large monocultural farms. Between 1870 and 1900, the median-sized farm increased only from thirty to fifty acres.[64] On the Wasatch Front, people continued to work family farms, allotting certain fields to sugar beets, raising a large variety of fruits and vegetables for home and market consumption, devoting upland fields to dryland wheat, and grazing milch cattle and workhorses in lowland pastures. In addition, many ran sheep or beef cattle on public lands in the mountains, driving them to the lowlands or the west desert in the winter.[65] Concurrently, transient Euro-American herders from outside Utah invaded the region to compete with local stock raisers in a free-for-all on the unregulated mountain lands. Brute strength governed access to the unappropriated public lands. Wool, mutton, beef, and hides brought cash to herders from the markets in midwestern centers like Omaha, Kansas City, and Chicago.

By taking advantage of inexpensive feed and new technology provided by a nationwide transportation network, Utah stock raisers rapidly increased the number of animals grazing on public lands. The number of sheep grazing in Utah increased more than 6,300 percent between 1870 and 1900, from just under 60,000 to more than 3.8 million. In 1900, Utah reached its peak sheep population. Afterward, numbers fluctuated between 1.6 million and 2.4 million. Cattle and calves increased over the same period by 860 percent, from just under 36,000 to nearly 344,000, and an additional 47 percent to nearly 506,000 by 1920.[66]

Because of the damage to mountain watersheds, Brigham Young's philosophy of increasing the variety of animals that worked well with plants in the valleys was a disaster for grazing in the mountains. In effect, the Mormon people reached one of Brigham Young's goals of multiplying and replenishing the earth by introducing a greater variety of plants and animals into Utah than had been there before. At the same time, they ignored his other goal by diminishing the native species. Moreover, some of the exotic plants were harmful. For instance, though wheat and sugar beets improved people's lives, cheat grass, orchard grass, and Russian thistle created fire hazards and sapped the soil of nutrients.

Another dramatic example of the change in the ecosystem was the air pollution spewing forth from smelters, various businesses, and home heating. Driven by secular markets to exploit available and unregulated mineral resources, the smelters flooded the valleys with polluted air. Lodes of silver, gold, lead, copper, and zinc in the Wasatch and Oquirrh Mountains created a demand for mills and smelters close to railroad lines. Most were built in the central Salt Lake Valley towns of Midvale, Murray, and Sandy, as well as communities like Bingham and Garfield at the eastern base of the Oquirrhs and Tooele at the west.[67]

As early as 1873, Mormon farmers in south-central Salt Lake began to complain about the smelter smokestacks that swirled sulphur dioxide and arsenic on their crops and livestock. Pressed by public opinion and the threat of lawsuits, the smelters paid reparations to the farmers. In 1904, four hundred of the farmers sued four of the smelter companies in federal district court. Judge John A. Marshall issued an injunction prohibiting the companies from smelting ore containing more than 10 percent sulphur or permitting arsenic to escape into the air. Two of the smelters closed, but the American Smelting and Refining Company and the United States Smelting Company remained open, with the court's permission, only after they contracted with the farmers to continue operating by installing pollution control equipment to remove the bulk of the sulfuric acid and all of the arsenic from the smoke. In addition, a large part of the smelting activity shifted to Garfield in the western Salt Lake Valley and to Tooele, west of the Oquirrhs, where prevailing winds carried the poisonous smoke beyond the farms.[68] At the same time, urban residents suffered from smoke spewed forth by industry, the railroads, and homes. By the mid-1910s, a letter to *Outlook* magazine complained that Salt Lake City had become the rival of Pittsburgh, Cincinnati, Chicago, and St. Louis as a smoke-plagued sinkhole.

While the valleys lay under banks of murky air and hungry livestock devoured mountain plants, during the 1880s and early 1890s a series of changes took place in Mormon society that further secularized entrepreneurship. Pressure by the federal government and evangelical Protestants eventually forced LDS church leaders to restrain themselves in political and economic matters, since most others in the United States considered this a secular rather than a sacred domain. Although the Mormon prophets continued to exercise a declining, but nevertheless direct, influence in politics and business well into the twentieth century, they increasingly turned their attention to teachings about individual morality, piety, and service to others.[69]

Moreover, although the LDS Church continued to promote various types of business enterprise, leaders began to separate such matters from moral sanctions. In the 1890s they stopped withholding tithing from business profits, and in 1922 they organized Zion's Securities Corporation to manage secular businesses.[70] Sometime in the late nineteenth century, probably by the mid-1880s, the Latter-day Saint social memory had forgotten the salutary environmental theology. I argue for the 1880s because in 1880 the Utah legislature passed a law assigning water rights based on priority of use and ordering the county courts to gauge streams and assign such rights, and in 1886 the Utah Supreme Court enthroned the prior appropriation doctrine of "first in time, first in right." Before this time, most Mormon communities allowed all of the water users to share in the scarce resource regardless of when they moved in.[71] In addition, people of other faiths and no faith poured into Utah, carrying with them a secularized Euro-American entrepreneurial tradition with the idea of progress. Engaged in mining, smelting, stock raising, and merchandising, they linked with the Latter-day Saints in further environmental destruction.

In response, beginning in the 1880s forces emerged in both American society and Utah that countered these destructive practices and beliefs. As environmental damage spread throughout the West, national leaders of the Progressive conservation movement began to worry. By the 1880s, the scientific community—led by people like Franklin Hough of the Department of Agriculture's Forestry Division, Charles S. Sargent of Harvard, and William H. Brewer of Yale—and the community of professional foresters (many European-trained)—led by scientists like Bernhard Fernow, Filibert Roth, and Gifford Pinchot—had come to believe (erroneously, as it turned out) that the United States faced imminent timber famine.[72] The fear of lumber shortages led to the passage of the Forest Reserve Act in 1891. Legislation provided for management of the reserves under the forestry bureau of the General Land Office in 1897 and their transfer as national forests to the newly created Forest Service in 1905.

Concurrently, a second generation of Mormons began to remember some of Joseph Smith's and Brigham Young's teachings about environmental stewardship, and some may have been influenced by the Progressive conservation movement then taking shape. At the direction of President Joseph F. Smith, who recognized the damage done to mountain watersheds by unrestricted logging and grazing, in a special general priesthood meeting on April 7, 1902, Mormons voted to support the withdrawal from the market of all public lands above Utah cities in order to protect them from damage.[73] Linking the wanton destruction of living things with personal morality while relying explicitly on the theological position that animals had eternal spirits, Smith condemned the needless destruction of animals. He opposed all killing for sport, stating that it was "wicked for men to thirst in their souls to kill almost everything which possesses animal life."[74]

Although many Mormon leaders from this generation shrank from trying to relocate entrepreneurship under religious control, some—imbued with the ideals of environmental stewardship—supported aspects of the Progressive conservation movement. These included Governor Heber M. Wells, Senator Reed Smoot, and especially stake president, city engineer, and later presiding bishop and apostle Sylvester Q. Cannon. The list also included women active in civic affairs, especially those associated with the Salt Lake City Council of Women—Leah Eudora Dunford Widtsoe, Susa Young Gates, and Emily L. Traub Merrill, who, in addition to their civic affiliations, were married to high church leaders. Significantly, unlike the second generation in the Progressive pattern chronicled by Robert Crunden, these men and women remained actively committed to Mormon Christianity rather than simply translating religious ideals into community service.[75]

At the same time, some of those who worked to heal the environmental damage more closely fit Crunden's model. George W. Snow, director of Salt Lake City's mechanical department, provides an example. Others with no connection to the LDS Church operated either from a sense of commitment to the community or from the ideals of the Progressive conservation movement. Among those was Utah governor and Protestant George H. Dern and women of other faiths like Elizabeth M. Cohen, Anna M. Beless, and Maude Smith Gorham. Non-Mormon members of organizations like the Utah Federation of Women's Clubs, the Salt Lake Council of Women, and the Ladies Literary Club worked shoulder to shoulder with their Mormon sisters.[76]

Beginning in the 1890s, these leaders tried to mend critical environmental damage. After Utah achieved statehood in 1896, Governor Wells withdrew from the market all state lands enclosed in national forest reserves. In 1905, the Mormon-dominated state legislature authorized

Governor John C. Cutler to establish a conservation commission to investigate environmental damage. Governor William Spry broadened the mandate of the conservation commission.

After his election to the Senate in 1903, Reed Smoot bucked the anger of legislators from surrounding western states like Weldon Heyburn of Idaho, who wanted to destroy the Forest Service. Supporting Theodore Roosevelt's and Gifford Pinchot's programs of utilitarian conservation, Smoot promoted the establishment of national forests to protect watersheds and regulate grazing and logging. He also sponsored the National Park Service Act in 1916, and he introduced legislation to establish the Zion and Bryce Canyon National Parks.[77]

Even the cooperation between these Mormons, Progressive conservationists, and the federal government did not immediately repair the destruction that two generations had inflicted on Utah's watersheds. Although numbers of sheep decreased after 1900, they did not begin to decline to the carrying capacity of the steep ranges until after 1950.[78] The resultant overgrazing led to serious rock-mud flooding in Sanpete County as early as 1888 and along the Wasatch Front during summer storms, especially in 1923 and 1930.[79]

In September 1930, following disastrous summer floods in Davis County, Governor Dern appointed a commission chaired by Sylvester Cannon to investigate the causes of the damage. Not surprisingly, the commission found that the floods resulted almost entirely from overgrazing. Beginning with the New Deal of the 1930s, the Forest Service began to restore overgrazed watersheds.

In the 1930s and 1940s, a number of Mormon farmers and community leaders recognized the problems caused by overgrazing as well. Some from Cache and Box Elder Counties purchased land in the Wellsville Mountains and transferred it to the Forest Service. Some of that land became part of a national wilderness area. After World War II, regional forester Chet Olsen, a member of the Latter-day Saint Ogden 29th Ward, fed Bernard DeVoto much of the information that fueled his environmental crusade.[80]

Turning to the problem of smoke pollution, in a pattern followed by others in the Progressive movement, people from various civic improvement organizations protested to the Salt Lake City Council about the damages to health and property.[81] George W. Snow, various women from the Federation of Women's Clubs, and some members of the Chamber of

Commerce—like Dern and Cannon—fought against air pollution. Eventually, bowing to public pressure, the Salt Lake City Council enacted a series of ordinances beginning in February 1914 that established a bureau of smoke inspection and dictated fines on residents and businesses that polluted the air.

Armed with legal authority, Snow and Cannon began fining polluters. Drawing on the services of consultants from the United States Bureau of Mines, the University of Utah College of Mines, and other institutions, they began pressuring businesses to install pollution control equipment during the early 1920s. By 1928, most businesses had complied with the city ordinances, and most remaining pollution issued from railroad locomotives and private residences. The introduction of diesel electric engines and natural gas for home heating reduced these sources after World War II, but industrial growth and the proliferation of automobiles has since expanded air pollution on the Wasatch Front. It remains one of the region's most serious problems.[82]

At the same time, water pollution led to epidemics of typhoid and cholera until underground water pipes and sewer systems replaced the open ditches that had served the Wasatch Front since its early years. Salt Lake City began constructing these improvements in 1884, but not until the early twentieth century did the city have an adequate water and sewer system, largely through the efforts of engineers like George Snow and Sylvester Cannon.[83]

Although these measures did not solve all the environmental problems that plagued the Wasatch Front, they did solve some. More to the point for this case study, they reveal a great deal about the effects of the heritage and memory carried by the Mormons in their environmental perceptions and actions. By the late nineteenth century Mormons seemed to have forgotten most of the environmentally creative teachings of Joseph Smith and Brigham Young about ecological stewardship, sacralized entrepreneurship, and the fellowship of all things both living and inanimate under the fatherhood of God. Soon after, however, the commitment of Mormons to the values of stewardship and in some cases the memory of these teachings, coupled with the Progressive sentiment in the community, facilitated the attack on the worst damage.

In recent years both religiously motivated and secular folks have worked to mitigate environmental damage. In his encyclical "Laudato Sí" and in talks during his September 2015 visit to the United States, Pope Francis emphasized the need to care for the environment. Elder Marcus Nash of the Quorum of the Seventy, in his 2013 Stegner lecture at the University of Utah, echoed teachings of the environmental theology when he called for an "approach to the environment [that] must be prudent, realistic, and balanced consistent with the needs of the Earth and of current and future generations."[84]

On balance, it seems clear that if we judge the Mormon occupation of the Wastach Front by the standards they set for themselves, the results are mixed. On the one hand, they created thriving irrigated farms and bustling cities in the valleys. On the other, they unleashed air pollution, watershed damage, and plant and animal extermination.[85]

Nevertheless, even with the secularization of entrepreneurship coupled with the idea of progress, the sense of community and religious values associated with the concept of stewardship and desire to build the Kingdom of God on earth helped in the long run to restore a more environmentally responsible community. During the Progressive Era, Mormons arose to work with non-Mormon Progressives. In recent years, a large number of environmentally conscious Mormons—including a number of general authorities—have remembered the earlier teachings of church leaders and have helped to promote environmentally sensitive policies and practices.

At least since the late nineteenth and early twentieth century, scholars like Max Weber and Émile Durkheim have debated the relationship between religion and other aspects of culture.[86] Scholars today recognize that the patterns of secularization that sociologists identified in European culture have not dominated in the United States.[87] As we have seen in the experience of the Latter-day Saints, however, at least in the realm of environmental consciousness, the Saints tended to reassign certain environmental practices within their culture from the religious to the secular realm and to forget their religious traditions.

Most important, by the late nineteenth century, the Mormon community had virtually forgotten the salutary environmental teachings of Joseph Smith, Brigham Young, and Orson Hyde. As Maurice Halbwachs, the father of the study of collective memory, pointed out, personal memory tends to fade unless it is periodically reinforced by contact with people who have shared the same experiences. Moreover, for people living in a particular time, there "is only one framework [for collective memory] that counts—that which is constituted by the commandments of our present society." In effect, "the mind reconstructs its memories under the pressure of society."[88]

If Halbwachs is correct in the case of collective memory—and I believe he is—repetition in the community and reinforcement in contemporary society are absolutely necessary for its retention. In the late nineteenth century Mormon community it is virtually impossible to find the repetition of environmental teachings of early prophets. In their place are sermons praising the progress made in building Zion and admonitions against sinful behavior. In recent years, however, a number of Latter-day Saints have restored the environmentally friendly teachings of Joseph Smith, Brigham Young, and Orson Hyde to the collective memory. In this connection, it seems clear that if the experience of the Latter-day Saints is any indication, the argument that the triumph of Western Christianity brought about our modern ecological crisis is seriously flawed. Lynn White is undoubtedly right that a combination of technology, science, and Christianity contributed to the damage. This occurred, however, in a Christian culture that had forgotten the doctrines of environmental stewardship, the sanctity of the earth, and the brotherhood of all creatures, while enthroning progress and secularized entrepreneurship. More recently, the recovery of cultural memory of environmentally salutary Christian theology, together with the work of secular environmentalists, has helped to correct some—but by no means all—of the degradation.

PART II

Perception and Place

THE NATURAL WORLD AND THE ESTABLISHMENT OF ZION, 1831–1833

Matthew C. Godfrey

THE FOUNDING OF THE CITY OF ZION in 1831 in Jackson County, Missouri, by members of what would become known as the Church of Jesus Christ of Latter-day Saints, has been the subject of much scrutiny by historians, geographers, and other scholars. Some have focused on the political characteristics of the city, while others have examined what its establishment meant from a millenarian perspective.[1] Still others have looked at the founding of the city through the notion of sacred space or through the ideology of place.[2] Yet few have really examined how Joseph Smith and church members at the time viewed the land itself and their relationship to it.[3] Doing so provides a glimpse into the ways that devoutly religious nineteenth-century Americans regarded the land and the environment, especially those who strongly believed that Jesus Christ's second advent was imminent. A close examination of correspondence, Joseph Smith's revelations, and church publications reveals that church members had specific views about the Jackson County area that influenced their theology, while also maintaining specific theological concepts that influenced how they perceived the land. In essence, they believed that land was owned by God, that wilderness needed to be improved to redeem it from its fallen state, that God used nature and the earth to both reward and punish the righteous, and that the physical possession of tracts of land demonstrated God's approval of them as a people. Many of these ideas had precedent in both the Bible and the Judeo-Christian culture from which church members came, but the Saints put their own twist on them. The Jackson County area was beautiful and even Edenic, they believed, but the civilizing hand of agriculture and industry still had to be applied to the area before the city of Zion could be built there and before the land was suitable for Jesus Christ's return.

The idea of building the city of Zion in the Americas was prevalent in Joseph Smith's theology from the beginning of his establishment as a prophet,[4] as was the central importance of the physical land where the city would be built. According to the Book of Mormon—which, according to Smith, contained a record of ancient inhabitants of the Americas and which he translated "by the gift and power of God"[5]—in the last days a New Jerusalem would be built in the Americas, to which the remnant of Joseph (from the Bible) would be gathered and to which Jesus Christ would return. This gathering would occur under the leadership of the Gentiles, understood by Joseph Smith and his followers to be those of European descent who had settled in the United States.[6] The New Jerusalem, according to one of Smith's revelations, would be "a land of peace a City of refuge a place of safety for the saints," where they would receive inheritances, or parcels of land, from God. It would be known as Zion.[7] "No man knoweth where the City shall be built," a September 1830 revelation declared, but it would be "among the Lamanites"—understood by the Saints at the time to be the American Indians.[8]

This same September 1830 revelation assigned Oliver Cowdery—who took with him Peter Whitmer Jr., Ziba Peterson, Parley P. Pratt, and Frederick G. Williams—to preach to Indian tribes living in Indian Territory west of Missouri.[9] Cowdery seemed to believe that part of his mission was to locate Zion—or at least to locate a spot for the temple that would be built there.[10] After these missionaries arrived in Indian Territory in early January 1831, they were ejected by a federal Indian agent because they did not have the necessary permits to preach, complicating the plan to establish Zion "among the Lamanites."[11] They then moved across Missouri's border into Jackson County, where they tried to convert white settlers.[12] Just a few months later, in June 1831, a revelation told Joseph Smith and Sidney Rigdon to travel to "the land of Missorie [*sic*]" where God would reveal to them the land of their inheritance.[13] After Smith arrived in Jackson County in July 1831, another revelation declared that "the land of Missorie" was "the Land which I [God], have appointed & consecrated for the gethering [*sic*] of the Saints," and Independence, the county seat of Jackson County, was the "centre place" of the city of Zion. The revelation commanded the Saints to begin purchasing lands in the area so that they could have "an everlasting inheritance."[14]

As church members targeted Jackson County as the location of Zion, they encountered a region located in the tallgrass prairies of North America.[15] U.S. government surveyors in 1826 classified approximately 48 percent of Jackson County itself as prairie, although some have estimated that 75 percent of the county "was open enough to have had substantial grass cover."[16] Several rivers and streams, including the Little Blue and Big Blue Rivers, traversed the region, most of which ran into the Missouri River cutting through the area. The southern part of the county contained grasslands devoid of trees and brush; land surrounding the Little Blue River was classified as wet prairie or marsh. Trees and brush encroached on the prairie in several places, but where the town of Independence was located, a "rich prairie"—grass with few trees or brush—existed.[17] One observer described the Jackson County landscape as "an undulating plane" consisting of "gently rolling" prairies interspersed with strands of timber.[18]

Before white settlement, American Indian groups such as the Osage, the Kansas, and the Missouri had utilized and claimed the land. The French also made several inroads into present-day Missouri in the 1700s through the fur trade. As they encountered the land, these groups relied on its resources for both sustenance and economic gain. The Osage, who were part of the Dhegian Sioux language family, "learned to exploit the seasonal diversity of their environment" by engaging in three principal hunts throughout the year and by cultivating corn, squash, pumpkins, and beans on the prairie lands where they settled. The French traded firearms and other goods to the Osage and other tribes for animal skins and furs. Although sparsely settled, western Missouri still had human imprints all over it.[19]

It was precisely the lack of white settlers in the area, though, that caused Mormons and other migrants to regard the region as a wilderness. The first white American settlement did not occur until 1808 when a treaty with the Osage, Kansas, and other tribes extinguished the Indians' title to nearly all of Missouri. However, a strip of land that covered most of Jackson County was not included in that treaty—that land would not be ceded to the United States until 1825. It was not until 1826 that the first concerted white American settlement of Jackson County began.[20]

By 1831, when Oliver Cowdery, Joseph Smith, and others first visited Jackson County, the area consisted of "a court-house built of brick, two or three merchant stores, and fifteen or twenty dwelling houses, built mostly of logs hewed on both sides."[21] Others described Independence, the county seat, as "full of promise," but as yet containing "nothing but a ragged congeries of five or six rough log huts, two or three clapboard houses, two or three so-called hotels, alias grogshops; [and] a few stores."[22] As such, Jackson County seemed to Smith and others to be largely an unsettled frontier landscape—but one that held promise because of its designation as the location of the city of Zion.

Before the land could be used for this holy purpose, it had to be redeemed through the application of the civilizing hand of agriculture and industry. This idea was prevalent in both the Judeo-Christian tradition and in nineteenth-century American culture, but it stretched back to colonial times, even to the first settlement of Massachusetts by the Puritans.[23] Many Americans considered wilderness a place of evil, "a moral vacuum" and "a cursed and chaotic wasteland."[24] To drive such evil away, pioneers and settlers had to bring the benefits of civilization, especially agriculture, as well as infrastructure resulting in villages, towns, and cities. By so doing, they could reclaim the wilderness from its fallen condition, fulfilling the divine injunction to "subdue" the earth.[25] Indeed, just as religious revivals that were part of the Second Great Awakening taught that an individual had "to remake himself or herself into a new person—to be 'born again,'"[26] so too did nineteenth-century Americans believe that wilderness had to be remade from its evil state and reborn into a pristine condition. In one case, Jesus Christ and the Holy Spirit were the facilitators; in the second, agriculture and industry provided the transformation.

When Joseph Smith arrived in Jackson County in July 1831 and surveyed the area, his thoughts epitomized this belief. According to a later history, Smith reportedly said upon surveying Jackson County and the Indian Territory just west of its borders, "When will the wilderness blossom as the rose, when will Zion be built up in her glory, and where will thy Temple stand unto which all nations shall come in the last days?"[27] This declaration echoed the ideas of Puritans migrating to America in the seventeenth century. Under the leadership of John Winthrop, the Puritans came to the American continent to build a city on a hill in the wilderness. This city would serve as "a garden from the wilds" and "an island of spiritual light in the surrounding darkness," as one historian put it—goals that Joseph Smith and others also had for the city of Zion.[28]

Smith's declaration may also have been influenced by passages in the Book of Mormon that depicted undeveloped land and unused resources as inferior to developed land. The book discusses a civilization known as the Nephites, who at one point began to expand into regions north of their main settlement. As they did so, they entered a land with "large bodies of water and many rivers," but with no timber. Because of the lack of timber, they constructed houses out of cement and protected "whatsoever tree should spring up upon the face of the land." The purpose of the protection was so "that in time they might have timber to build their houses, yea, their cities, and their temples, and their synagogues, and their sanctuaries, and all manner of their buildings."[29] Woodlands were not important in and of themselves, according to this passage; they were important only as material to be used to develop the land.

Other Mormons first viewing Jackson County also saw development of the area—or its blossoming as a rose—as a divine requirement before the city of Zion could be built. Some church members, in ways reminiscent of descriptions of New England by Puritan settlers in the 1600s,[30] spoke of Jackson County in glowing terms, almost as if it were already a Garden of Eden, but they always emphasized the necessity of developing the region to its full potential. Such declarations followed the example of boosters of settlement in undeveloped areas, who proclaimed that the natural advantages of certain regions meant that God himself had selected the sites for development.[31]

An example of this came in August 1831 when Sidney Rigdon was instructed to "write a discription [*sic*] of the Land of Zion" to be disseminated to the Saints.[32] The letter Rigdon subsequently wrote to church members extolled the virtues of nature that God had bestowed on Jackson County and pointed to the necessity of further development:

> This land being situated in the centre of the Continent on which we dwell with an exceeding fertile soil & cleared ready for the hand of the cultivator, bespeaks the goodness of God in providing so goodly a heritage & its climate suited to persons from every quarter of this continent . . . its productions nearly all the varieties of both grain and vegetables which

> are common to this country together with the means clothing, climate & surface all adapted to health, indeed, I may say that the whole properties of the country invite the Saints to come and partake in their blessings.[33]

A later Joseph Smith history provided an even more extravagant description, pointing to "the beautiful rolling prairies" that "lay spread around like a sea of meadows," as well as the "growth of flowers that seemed as gorgeous grand as the brilliance of stars in the heavens." Wild game also constituted "the rich abundance that graces the delightful regions of this goodly land." Yet the specter of development was present in this description as well. The "rich and fertile" soil in the region would allow production "in abundance" of wheat, corn, "and many other common agricultural commodities," while livestock would "nearly . . . raise themselves by grazing in the vast prairie range in summer, and feeding upon the bottoms in winter."[34]

Although she herself had not visited Missouri, church member Elizabeth Marsh, living in Kirtland, Ohio, described Jackson County to her relatives in much the same way, relying on what "the Brotheren" returning from Missouri had told her. Jackson County was "the very best of Land such as you or I never have seen for beuty & fertility," she declared. "The soil is good and the General prospect[s] are beutiful beyond discription." The land contained "plenty of wild plumbs, wild sweet grapes, mulberies, strawberies, raspberries, and Blackberies, hazlenuts hickery nuts &c," while "Common domesticated stock" were able to "take care of them selves year round on the praries." According to Marsh, "the spot where the temple is to be is now a forest," but it would not be long until "the forest shall be esteemed a fruitful field" and "would blosom as the rose."[35]

William W. Phelps, who began publishing the church newspaper the *Evening and the Morning Star* in Independence in June 1832, was even more direct about the necessity of improving the land before Zion could be established. After first encountering the land in the summer of 1831, Phelps wrote that "the prairies are beautiful beyond description."[36] Yet improvements were greatly needed. Zion, "at present," he noted, was "but a wilderness and desert" and settlers would experience many "disadvantages" when first arriving. However, Phelps declared, as the saints worked the land, Zion would "become like Eden or the garden of the Lord," just as the "prophets" had foretold.[37] Harking back to a Book of Mormon scripture that described land void of timber as "desolate," Phelps declared that most would not value this modern-day "land of Desolation" for anything other than a "hunting ground" because of its "want of timber and millseats." But, he continued, God had declared "it to be the land of Zion" and would bless it with "the precious things of heaven" and "the precious fruits brought forth by the sun."[38]

Non-Mormons who visited Jackson County in roughly the same time period also gave extravagant descriptions of the land—but, unlike the Mormons, they did not declare the need for the area to "blossom as the rose" before it could be useful. Instead, they saw the land as already blossoming. Charles Latrobe, an Englishman who had written travel books detailing his adventures in climbing the Alps,[39] encountered Independence in 1832 and waxed eloquent about "the autumnal Flora, covering these immense natural meadows, like a rich carpet." He also declared that "God has here, with prodigal hand, scattered the seeds of thousands of beautiful plants, each suited to its own season."[40] Noted author Washington Irving, Latrobe's traveling companion, told his sister that "the fertility of all this Western country is truly astonishing." He continued, "The soil is like that of a garden, and the luxuriance and beauty of the forests exceed any that I have seen."[41] Ten years later, author Rufus Sage spoke highly of the area as well, stating that "the soil is the richest and most productive of any I ever set my eyes upon" and that the prairie was "like a perfect paradise, covered with a virdure [*sic*] unknown to the east."[42]

Such descriptions seemed to equate Jackson County with the Garden of Eden, and, as one newspaper reported, it was not surprising that the Mormons had selected the area "as the 'promised land,' for it is decidedly the richest in the state."[43] But church members assigned to live in the area were disappointed in its lack of development, especially compared to the settled areas of the eastern United States from which they were moving. Edward Partridge, for example, seemed dismayed after a July 1831 revelation instructed him to move to Jackson County.[44] Writing from Independence to his wife Lydia, who was still living in Painesville, Ohio, Partridge said, "we have to suffer & shall for some time many privations here which you & I have not been much used to for years."[45] As Ezra Booth, who also traveled to Missouri from Ohio at this time, explained, Jackson County lacked the amenities that those from more settled regions in the east were used to. The Mormons "expected to find a country abounding with the necessaries and comforts of life," he stated. "But the prospect appeared somewhat gloomy, and will probably remain so for years to come."[46]

To overcome the "privations" of the wilderness, it was necessary to cultivate plants and crops like those that grew in the eastern United States. Accordingly, Partridge instructed his wife Lydia to bring a variety of seeds and plants with her to Missouri, including "all kind of garden seeds," peach stones, sprouts for currant bushes, apple seeds, and "a trifle of white clover seed."[47] Likewise, a General Conference held in Missouri in January 1832 counseled those coming to Jackson County to bring with them "one barrel buck wheat. [*sic*] and one of Clover seed" for planting.[48] As late as July 1833, two years after the designation of Jackson County as the location of the city of Zion, William W. Phelps still expressed surprise that many were migrating to Zion "without bringing garden seeds, and even seeds of all kinds" with them.[49]

Plants and crops were not the only things necessary to make the wilderness blossom. That same January 1832 conference recommended that "o[n]e blacksmith, two shoemakers one carpenter and joiner one mason one waggon and plow maker o[n]e tanner and currier one millwright one hatter one chair & cabinet maker, one silver Sm[it]h and one wheel wright" be sent to the area, thereby allowing industry to flourish.[50] It is difficult to believe that Independence did not already have at least some of these craftsmen, given that it was the jumping-off point of the Santa Fe Trail, but it appears that church leaders desired to establish a self-sustaining Mormon community and avoid relying on non-Mormons in the region. With these Mormon resources, the city of Zion could then be established and resemble more fully the communities in Ohio, New York, and other states from which church members had come.[51]

Joseph Smith even produced what he considered to be a divine layout for the city of Zion, although this plat was not sent to the Saints in Missouri until the summer of 1833, shortly before their expulsion. The plat showcased the orderly settlement that Smith foresaw in Independence, centered around twenty-four temples that would be used for both sacred and administrative purposes. From these temples, the city would be laid out according to a one-square-mile grid. Forty-nine city blocks would make up the square mile, containing between ten and sixteen acres each. Forty-six of the blocks were designated for residential use and divided into half-acre lots on which houses set at least twenty-five feet from the street would be constructed. Farmland would be utilized outside the city limits, separating urban living from agricultural production.[52] This orderly method of development drew on ideas from nineteenth-century urban planners, especially that city centers were to be places of gentility, populated by the city hall and large houses where the wealthy would reside.[53] Joseph Smith's plan for Zion both reflected this vision and sacralized it by making the temples—not city hall or wealthy residences—the center of Zion.

As Mormons gathered to Missouri, they learned from Joseph Smith's revelations that transforming the wilderness into an orderly settlement of agriculture and industry was conditioned on obedience to God's commandments. An August 1831 revelation specifically directed to those who had traveled to Missouri stated that "in the begining [*sic*]" God had "cursed the land." But in these last days, when Zion would be established, he had "blessed it in its time for the use of my saints that they may partake the fatness thereof."[54] Another August 1831 revelation highlighted more clearly the connection of obedience and productivity. If those in Missouri would keep the Sabbath day holy and offer to God their sacraments on that day, the revelation stated, he would bless them with "the fulness of the earth." That fullness included the use of "the beasts of the fields & the fowls of the air & that which climbeth upon trees & walketh upon the earth yea & the herb & the good things which cometh of the earth."[55]

Some members expected God to grant these natural blessings without any effort on their part. Although William W. Phelps stated in an article in the July 1833 issue of the *Evening and the Morning Star* that many in Jackson County were "beginning to enjoy some of the comforts of life" because of "fruitful soil, and a healthy climate," other elements in the article indicated that making the wilderness blossom as the rose was not as easy as church members expected. Some individuals had migrated to Jackson County without necessary means, he noted, "suppos[ing] that the Lord will open the windows of heaven, and rain down angel's food for them by the way." Such circumstances were depleting the church's resources, and God could not pour out blessings upon the Saints, Phelps asserted, "if all the means of the saints are exhausted." Phelps hastened to add that he did not "doubt in the least, that the Lord will fail to provide for his saints in these last days," but church members needed to do their part as well. Though the Lord had promised the Saints the fullness of the earth if they were obedient, church members needed to be diligent and hardworking.

If they refused to "use the means put into their hands to obtain the same in the manner provided by our Lord," they could not expect the Lord's blessings.[56]

Indeed, nature was not as pliable as some Mormons hoped. In January 1832, Edward Partridge reported that the Saints' settlement in Jackson County would not be able to sustain much immigration during that spring or summer. "Provisions are scarce & dear nearly double what they were one year ago," Partridge reported, and the church therefore did not have enough goods to support the migration of large numbers of people.[57] The high price of goods resulted from "excessive rains" and floods that damaged corn and wheat crops in the summer of 1831, making the fall harvest "very light" and increasing the price of grain.[58]

In addition, church leaders had hoped that Sidney Gilbert would be able to open a store in Independence in October 1831 to generate money to purchase both goods and land. Nature interfered with those plans as well.[59] Gilbert had gone to Kirtland, Ohio, to procure the "necessaries" for the store in August 1831,[60] but on the trip back to Independence he was delayed five weeks in Arrow Rock, Missouri, because of ice flows on the Missouri River. He did not arrive in Independence until January 1832. Because of the delays, the store itself did not open until February 1832.[61] Finally, Partridge noted that the type of land purchased by him for the city of Zion precluded much settlement in the spring. "It is mostly woodland," he declared, "& not in a situation to be improved this season even if it should be thought advisable to clear it faster."[62] Despite these setbacks, by November 1832 Phelps counted the church's population in Jackson County at 810, consisting of baptized members and children.[63] This increased to approximately 1,200 by July 1833. Phelps described these migrants as "generally so healthy, so industrious, so thriving," but he also noted that, because of the lack of "conveniences and even necessaries" in the land, their condition was "a matter of great joy, if not a miracle."[64]

According to Joseph Smith's revelations, one reason why the wilderness did not blossom as the rose as quickly as the Saints expected was because of church members' disobedience. In the spring and summer of 1832, correspondence between church leaders in Kirtland, Ohio, and church leaders in Independence convinced Joseph Smith that those in Missouri were conspiring against him. Smith told William W. Phelps that he had perceived "the displeasure of heaven" and "the frowns of the heavenly hosts upon zion [*sic*]" because of these leaders' recalcitrance.[65] Accordingly, in September 1832 a revelation condemned the "children of Zion" for their "vanity and unbelief," warning them that if they did not repent, a "scorge and a Judgment [*sic*]" would be "poured out upon" them. The "children of the kingdom" would not be allowed to "pollute my holy land," the revelation declared.[66]

Based on this revelation, Smith and other leaders told those in Missouri that if they did not repent, God "would seek another place" for his city. The Lord would not allow the sins of his people to defile the beautiful and productive land he had prepared for them.[67] When the Saints faced mobs in Jackson County, who ultimately ejected them in the fall of 1833,[68] some saw this as a result of God trying to keep the land undefiled. "It is a time of great anxiety," John Whitmer told Joseph Smith and other church leaders, "to behold the cleansing of this Church & also the land from wickedness & abominations."[69]

To disconnect church members from their Jackson County land was a particularly grievous punishment, for the physical land itself was a direct connection with God. One of the major reasons why church members were to settle in Missouri and build up the city of Zion was so that God could grant them inheritances, or parcels of land, as he did with the children of Israel in the Old Testament after they entered the Promised Land.[70] In January 1831, before Missouri had been identified as the land of promise, a revelation told the Saints that God would give them "a land flowing with milk & Honey [*sic*] upon which there shall be no curse." This land, the revelation continued, would be "the land of your enheritance [*sic*]" and would be possessed by the Saints both during their earthly lives and in the eternities to come.[71] After Jackson County was identified as the location of the city of Zion, Edward Partridge and Sidney Gilbert were told to buy up land in the county and the surrounding region so that Partridge could divide "unto the saints their inheritance."[72] They followed this instruction, purchasing over two thousand acres of land on behalf of the church.[73] Before members could receive their inheritances, however, they had to be willing to consecrate—or donate—their property to the church, thus providing church leaders with the means to purchase more land in Jackson County.

The significance of having an inheritance was emphasized in a November 27, 1832, letter that Joseph Smith wrote to William W. Phelps, which chastised those who were going to Zion but who were not receiving inheritances because they were not willing to consecrate property. Such individuals, Smith stated, would not "have there names enrolled with the people of God, neithe[r] is the geneology to be kept or to be had" in church records. Those whose names were not found on those records would "be cut assunder and their portion shall be appointed them among unbelievers where is wailing and gnashing of teeth." In contrast, those who were recorded in the book because they received their inheritances would "be exalted very high and shall be lifted up in triumph above all the kingdoms of the world."[74] In essence, if one had a physical parcel of land as an inheritance from God, one was tied directly to the church and to God, enabling one to receive all that God could give both in this life and in the hereafter. Without such an inheritance, individuals would be cut off from the Lord's presence and relegated to suffer his punishment. The actual land was thus a direct symbol of God's approval of the Saints. With it, they received blessings; without it, they were lost.

This was one reason why the law of consecration was so important to the establishment of the city of Zion. The principles of consecration were first given in a February 1831 revelation in Kirtland, Ohio, but consecration was really only practiced extensively in Missouri.[75] The necessity of living the law in Missouri was so that all who came to Zion would be able to have access to an inheritance—to land. Even though this land of promise was God's, the Saints still had to obey the laws of the land to possess it; they were told explicitly by revelation that land should be obtained by purchase, not by blood.[76] But not all church members had the means to purchase land. Thus, any surplus money or property was to be consecrated to the church "to administer to the poor and needy" and "for the purpose of purchaseing Land [*sic*] & building up of the New Jerusalem."[77]

Under the law of consecration, every church member coming to Zion would be able to possess a parcel of land—at least theoretically—as an inheritance, thereby tying themselves to God. Even though this land would not be deeded to the individual (the deeds for the vast majority of the land were in Partridge's name, who had purchased it), it would still be considered his or her stewardship, and, more importantly, it would be recorded in church records as an inheritance to be passed down to posterity, thereby allowing God to bless the individual and his family. As William W. Phelps put it, "each man receives a warranty deed securing to himself and his heirs, his inheritance in fee simple forever."[78] In some ways, this concept of consecrating property and receiving an inheritance was really not that far afield from the American capitalistic tradition of the necessity of individual land ownership to preserve a virtuous and righteous nation. In the case of the Saints, however, the possession of land would preserve an individual's connection with God.[79]

Because the physical land provided this direct connection to God and his blessings, it is no wonder that most church members driven from Jackson County by mobs in the fall of 1833 refused to sell their Jackson County land, even though non-Mormons who had not purchased it now possessed it. After mobs in Independence had destroyed the church's printing office, tarred and feathered Edward Partridge and Charles Allen in July 1833, and coerced church members to pledge to leave Jackson County by early 1834,[80] Joseph Smith told the Missouri Mormons to retain their land: "It is the will of the Lord," he said, "that not one foot of land perchased should be given to the enimies of God or sold to them but if any is sold let it be sold to the chirch [*sic*]."[81] He reiterated this point to Edward Partridge in December 1833, after the expulsion had occurred, stating that "the spot of ground upon which you were located is the place appointed of the Lord for your inheritance and it was right in the sight of God that you contend for it to the last."[82] Indeed, Smith declared five days later, "It is better that you should die in the ey[e]s of God, then that you should give up the Land of Zion, the inheritances which you have purchased with your monies."[83] Less than a week after writing this letter, Smith dictated a revelation declaring that it was God's will that church members should "hold claim upon" the land God had "appointed unto them though they should not be permited [*sic*] to dwell thereon."[84] The church made a concerted effort to regain their land in Jackson County, to no avail.[85] However, the idea that the city of Zion will still be built in Jackson County remains strong among members of the church, even to the present day.

Examining how Joseph Smith and early church members viewed land, especially in Jackson County, provides several insights into the mindset of church members in the early 1830s, thereby illuminating how individuals preparing for Jesus Christ's Second Coming viewed land and nature in early nineteenth-century America. In many ways, church members approached Jackson County with the same attitudes that other Americans had toward wilderness—it was both a place of great beauty and a place needing the touch of civilized hands to make it productive and beneficial.

Yet the Mormons also believed that agriculture and industry had to be applied to the land for it to fully reach its divine potential as the city of Zion. Their efforts to make the land blossom as the rose, they believed, would be rewarded by God pouring out the blessings of nature upon them. When such blessings did not come as quickly as the Saints expected, it indicated that God was not pleased with them because of discontent with Joseph Smith's leadership. According to Smith's revelations, this disobedience eventually resulted in their ejection from the land, which prevented them from holding the inheritances that tied them to God.

But even without a physical presence in Jackson County, the land itself still held great importance for church members. The ideas that the Saints held about land tying them to God, the necessity of bringing agriculture and industry to the wilderness, and God's utilization of nature to bless and punish them were passed down to other members, resulting, in large degree, in the environmental ethos proclaimed by the Saints and evidenced in the practices they followed after migrating to the Great Basin in the 1840s.[86] That legacy had its origins in the Saints' conceptions of land and nature in the early 1830s and their efforts to make Jackson County blossom as the rose.

Today, Jackson County continues to hold a place in the hearts of members of the church, precisely because it was the land designated as the city of Zion and the land where their inheritances lay. Many Latter-day Saints still believe that church members will someday go back to the county and recommence the establishment of Zion. If that happens, church members will find a land much different than the one Joseph Smith saw when he first came to Independence in July 1831. Instead of streams, timber, and rich prairies, they will see concrete, asphalt, strip malls, and highways. This was not what Smith saw when he envisioned Zion blossoming as the rose. Instead, he perceived a community of flourishing agricultural fields, interspersed with industry. Even if the area does not conform to this vision, it still holds the same sacred place in the hearts of the Latter-day Saints. It was once Zion and, to many, will be Zion again—both spiritually and spatially.

"WE SELDOM FIND EITHER GARDEN, COW, OR PIG"

Encountering Environments in Urban England and the American West

Brett D. Dowdle

On January 11, 1840, the *Oxford* sailed into Liverpool's harbor carrying 109 passengers, including Mormon elders Wilford Woodruff, John Taylor, and Theodore Turley. While both Taylor and Turley were British-born, Woodruff was an American making his first journey to England. After landing in Liverpool, the group "visited several noted places" including the markets, the custom house, and a monument to Lord Nelson, all of which Woodruff described as being "quite splendid."[1] Despite his genuine enjoyment over Britain's cultural offerings, however, Woodruff found himself dissatisfied with the country's high prices and the subpar sleeping accommodations in the inn they boarded at the first night.[2] By January 14, his focus had turned from a dingy inn to the more concerning sights of Preston's sixty thousand inhabitants and its streets "crouded with the poor both male and female going to and from the factories." This new environment and the image of urban poverty shocked Woodruff, who immediately drew comparisons between the circumstances of the poor and the Egyptian bondage of the children of Israel described in Exodus.[3]

Woodruff's observations of his new environment were the result of early Mormonism's expansion into a transatlantic denomination with congregations extending beyond the United States and Canada to the British Isles. But these American observations of industrial England were only one part of Mormonism's Atlantic exchange. The flow of American missionaries to the British Isles coupled with a flow of British converts to the United States and the Latter-day Saint Zion created a multidirectional exchange of people, ideas, and experiences. To a far lesser degree, these transatlantic experiences mirrored the earlier Columbian exchange.[4] In the case of both Mormon missionaries in England and British converts in the United States, this transatlantic exchange created opportunities for people to interact with and experience different environments. These exchanges included both positive and negative experiences and were highly influenced by the disparate nationalisms of the American and British Saints. Returning home from their missions, American missionaries like Brigham Young often incorporated environmental aspects of what they found in England, such as the idea of pleasure gardens, into their lives in Nauvoo. At the same time, not all experiences were positive. Negative interactions with new environments created vivid memories that participants recalled with pain even decades later. To understand these experiences, it is essential to contextualize both groups within their American and British nationalities.

The diaries and letters written by Mormon missionaries proselytizing in British cities beginning in 1837 provide valuable and occasionally overlooked insights into the way that Americans contextualized the environment of England's urban centers. These American missionaries discovered both positive and negative aspects of large industrial cities in England. While major centers like London offered museums and historic sites that exceeded the similar offerings in the United States, such cities also revealed levels of poverty that would have shocked even the most destitute of America's yeomen farmers, most of whom had spent most of their lives in the rural farming towns of the northeastern United States.

Originally united as citizens under a common flag, the United States and Britain had diverged widely from their common kinship. Following the American Revolution, the two nations continued to maintain a bitter rivalry and had come dangerously close to renewing hostilities on several occasions. Throughout the 1830s and the early 1840s, tensions were again rising between the two countries as American and British trappers battled over the rich landholdings in the American West.[5] In 1838, conflicts in Toronto began spilling over into the United States, convincing some

Americans like Kirtland resident Hepzibah Richards that the two nations would soon be going to war.[6] Similarly, John Smith expressed fears about Britain to his son George A., writing that persecution would soon "rage in England ere long worse than in Missouri and the Saints will have to flee out the best way they can."[7] It was in these circumstances that Mormon missionaries first made their way to England. Although less concerned about international struggles than Richards and Smith, they nevertheless carried with them longstanding American notions about the British Empire.

For many missionaries, the seabound journey to England posed a significant challenge—and gave missionaries their first encounter with an unfamiliar environment. Brigham Young described the trepidation of the voyage to his wife Mary Ann, writing that the night before his departure for England he went to bed having asked God for "a manifestation concerning [his] jorney [*sic*] across the water." Falling asleep, Young had a dream about crossing a large body of ice and water. In the dream, Young experienced some trepidation, but ultimately reached his destination "with out getting wet at all." Young's dream suggests that the passage across the sea was a matter of some concern. This is understandable, given the fact that this was his first experience with oceanic travel and given prevalent Mormon worries about water-bound travel.[8] Young's trepidation about the voyage proved justified. Owing to seasickness, he was almost unable to sit upright for several days of the journey. Rough seas made it impossible for passengers to go onto the deck "with out getting drenched with salt water[,] the waves continula [*sic*] coming over the deck." And at least for one of the shipmates, the voyage seemed to be the worst he had seen in fifteen years, making the arrival in England a welcome conclusion to the journey.[9]

Being used to the hot and humid summers of the eastern United States, many complained about England's constant rain and fog. Franklin D. Richards described "real English weather" as being "cloudy, rainy, and disagreeable."[10] Similarly, Woodruff complained that London's fog made the air "so dark and heavy" that it was impossible to see at midday, creating a gloomy feeling in the streets.[11] The damp conditions even affected the summer, causing George A. Smith to note that he was "obliged . . . to wear as much flannel" in July as he normally wore during an American winter.[12] While such weather conditions were little more than a nuisance, they served as a constant reminder to the missionaries of the differences between the United States and the British Isles.

Most noticeable, however, were the large populations of the British cities, which dwarfed all of the settlements that the rural American missionaries had come from. Woodruff noted that Manchester, with its population of 320,000 people, was geographically larger than America's largest metropolis, New York City.[13] Woodruff's mention of New York City is important. In many regards, the American metropolis seemed to have a relatively small effect upon the England-bound missionaries in 1839, not eliciting the same kinds of comments in letters and diaries as London, Manchester, and Liverpool would yield the following year. This is not to suggest that rural Mormons were unimpressed by New York City. But despite the numerous buildings and large population, the American metropolis lacked much of the cultural environment and sophistication of its British counterparts. Indeed, a visitor from Europe in 1838 dismissed American life as possessing neither "smartness" nor "music," including particularly the lack of opera and opera houses.[14]

While British visitors to America had criticized the country's lack of sophisticated sounds such as opera, rural Americans found the constant noise of industrial England extremely bothersome. Historian Richard Cullen Rath notes that for many early Americans, the "natural soundscape" was one of the defining aspects of the rural American environment. While "the sound of bells, guns, drums, trumpets, and musical instruments" all helped outline the auditory environment, it was the "unintentional sounds, not made by humans" that most dramatically defined the noises of the frontier.[15] Accordingly, Britain's crowded cities necessarily made for busy streets and a constant bustle that often continued throughout all hours of the night. The perpetual sounds of people, carriages, and horses occasionally made sleep difficult. Describing the scene in London, Woodruff wrote that he had difficulty sleeping at nights due to the "much passings of Drays & horses & singing & hollowing at all hours of the night in the streets."[16] Similarly, Brigham Young confided to his wife Mary that he wanted to "get out of so much noys [*sic*]."[17] Just as the noise in the streets had made it difficult for Woodruff to sleep in London, Young struggled to find adequate rest in Liverpool. Limited funds meant that the London missionaries lodged close to a local pub, which was frequented by men and women at all hours of the day. For Young, the lack of sleep, combined with the weight of directing the congregations and Mormonism's evangelizing efforts in England, led him to conclude that his "poor old Body" was wearing out at a rapid pace.[18]

In fact, England's climate actually proved beneficial to the health of the missionaries who made their departures from the malaria-infested shores of the Mississippi River in 1839. In his first letter home from England, Young reported a general improvement in his health, and elsewhere stated that he was in better health than he would have been if he had "remain[e]d in America."[19] Similarly Woodruff, who had been among those afflicted by malaria in the regions surrounding Nauvoo, reported that he had "been well" since his arrival in England.[20] This general state of good health seemed to continue throughout the 1840 mission of the Quorum of the Twelve Apostles to England. In his final letters home, Young noted the general health of the apostles, minus a minor illness from which Woodruff had quickly recovered.[21]

While most of the missionaries enjoyed general good health, some struggled with the damp environment and the polluted air. Soon after arriving in England, Woodruff wrote home to his wife and described his observations. Following a generally positive description, he wrote that there was one thing he found to be "disagreeable throughout England," which was that the "air is filld with <u>smoke</u>, <u>smut</u>, & gases" that made the air "vary heavy & bad for the lungs"[22] (emphasis in original). Other missionaries similarly struggled with their lungs in Britain. Writing to Brigham Young from Glasgow in January 1841, Reuben Hadlock noted that his health had improved significantly since moving his labors to Scotland, but that his "lungs remain[ed] some afflicted."[23]

While some of the lung ailments may have been attributable to the winter season, undiagnosed asthma, "the Damp air" in England, or other factors, Woodruff likely identified the most significant cause of problems when he attributed the poor air quality to the density of the population and to the constant "burning of coal."[24] Woodruff's willingness to point to coal as the cause of his ailment importantly recognized the constant—though at times nearly invisible—connections between the urban and natural environments.[25] Indeed, it was Britain's ready access to coal that was fueling the nation's industrialization, facilitating its important stature in the international world during the nineteenth century.[26] As Woodruff discovered, however, coal-powered industrialization was not without its serious limitations. Particularly when heavy concentrations were utilized in a relatively small area, coal introduced pollutants into the air that had the potential to create serious health challenges.

In discussing England's environment, it is important to recognize that it included more than just wet weather and a coal-filled atmosphere. Indeed, in many regards, these things were natural interlopers that crowded in upon an artificial environment of social, economic, cultural, and political factors. Indeed, geographer Ian Douglas has suggested that it is impossible to separate the physical environment of the city from the "historical, economic, and cultural factors" that accompanied those environs.[27] Drawing upon the raw materials of nature, industrial England had fashioned its own artificial landscape, more visibly marked by buildings, marketplaces, and railroads that commodified nature and perpetuated class distinctions rather than an environment characterized by hills, plains, and waterways.[28] Because England differed so dramatically in these factors from the United States, it was these things that were most frequently mentioned in the letters and diaries of the visiting Mormon missionaries in the 1840s.

Franklin D. Richards's first impression of England was that everything seemed "more staid and substantial than in America."[29] Whereas the inhabitants of the American West had day-to-day experiences with and in nature, those living in the British metropolises were more removed from the natural environment. Despite this distance, however, elements of the natural environment made their way into the metropolis. For instance, on his first day in England, George A. Smith toured Liverpool's marketplace with fellow apostle Heber C. Kimball. Smith found the market to be "filled with fruits and vegetables of great variety and beauty from all climates." Surveying the offerings, the rural-born Smith noted that he had never seen "anything to equal it." Still, when Kimball gave him the opportunity to select anything he wanted from the market, Smith's rural American identity manifested itself as he selected a large onion, recording that he "atc it with a craving appetite and shed many tears over it."[30] The peculiar selection demonstrated a kind of down-home sensibility that preferred the simple and familiar rather than the exotic.

The offerings of Liverpool's markets were unique. As Smith noted, farmers had cultivated the vast majority of the fruits and vegetables on display in different regions of the world, rather than in England. Furthermore, by this time, Liverpool likely boasted very little agricultural presence of its own. Nevertheless, this diverse display of agriculture had made its way into Liverpool's markets by way of Britain's vast trade networks, of which Liverpool was an important hub. Ironically, therefore, it was Liverpool's lack of an agrarian focus that allowed for a wider array of produce to be made available to customers.

During his first days in England, Young visited a well-kept garden in Liverpool and commented briefly on its beauty.[31] Similarly, several of the apostles visited Manchester's zoological gardens in July 1840 and came away impressed.[32] While most American homesteads boasted gardens, they tended to fill different purposes than those in England. Situated in the middle of urban centers, the manicured gardens in Liverpool and Manchester operated as places of leisure—extensions, as it were, of upper class architecture. On the other hand, American gardens largely filled a utilitarian purpose, often operating as extensions of the farm rather than places of resort. Only the wealthiest of antebellum Americans replicated the manicured look of British gardens.[33] Young, however, seemed to sense the value of such gardens and the leisure they offered. Writing to his wife, he expressed the hope that as she looked for a lot of ground in and around Nauvoo, she would find "a good place whare [*sic*] we can have water and a garden built . . . so when I doe com [*sic*] home I may have a place to rest."[34] Years later, after leading the Saints to Salt Lake City, Young consistently emphasized gardens and the beautification of the city, teaching that the cultivation of gardens was part of preparing the earth for Christ's return.[35] Accordingly, although Young later bemoaned the loss of family gardens and the sustenance they might have provided to working class families, he saw the value in the less-utilitarian gardens that he visited in England.[36]

Similar to the marketplaces and gardens, natural history museums helped bring the natural environment into the confines of major British cities. In Manchester, Woodruff and others visited the city's museum. Woodruff seems to have been particularly impressed with the museum's collection of animals. He wrote that they saw "every kind of Beasts Animals, fowles [*sic*] and birds fish and every creeping thing, serpents, and all things that live on land or in the sea." The visit also included opportunities to see exotic fruits, metals, and gem stones—most prominently a facsimile of a diamond that was reportedly worth £122 million.[37] Such exhibits seemed to place the allure of the natural environment within the grasp of England's urban population. At the same time, they presented an illusory view of the natural environment by extracting plant and animal life, as well as the earth's vast mineral deposits, from the realities of nature and placing them behind glass. Still, for these missionaries and other visitors, England's museums offered opportunities to glimpse and learn about plants, animals, and gems that they otherwise would never have seen. The impact of these cultural experiences upon the missionaries was palpable. Such a disparity between the two cultures led Woodruff to theorize that if "a Missourian from the western wiles of America" were to see many of the sights in England, "he would think he was in another world."[38] Although an obvious insult directed toward the hated state of Missouri, the comment likewise reflected the realities of the American frontier.

While the missionaries were complimentary of Britain's superior cultural opportunities, they were highly critical of many other aspects of British society. In visiting the country's urban centers, the missionaries could not help but be moved by the scenes of working-class poverty. The conditions of the working class were particularly shocking to yeomen Americans. Although they were well acquainted with the challenges posed by financial deprivation, the destitution of the British working class exceeded even the poverty they had faced in Ohio, Missouri, and Illinois. Describing the circumstances of the British poor, Willard Richards wrote that there were "hundreds of thousands short of food" in the country.[39] The same country that could fill its urban marketplaces with fruits and vegetables from around the globe was incapable of adequately feeding its population. As Richards and Brigham Young would later note, however, while industrialization allowed British consumers to access an increased variety of produce, it also created circumstances whereby the poorest segments of society became incapable of growing their own goods on their own patches of land.[40] Of necessity, a diversified diet seemed to favor the upper and middle classes, to the detriment of the laboring class. In such circumstances England's "beggars [became] almost as common as stones in the street." Even the ability to contribute to the systematic polluting of England's atmosphere seemed to come with a cost. Richards joked that even "the smoke passing through the chimney" and "the light passing through the window," came with attached prices so that the British people were "generally . . . taxed out of their last pound."[41] Such scenes served to strengthen the Americanism of Richards and the other visiting Mormon missionaries. Indeed, when asked by a group of British converts if he liked "this country as well as America," Lorenzo Barnes responded frankly, "No—Give me my native country before any other." For Barnes, the "hard times" and scenes of urban poverty in England solidified the superiority of America in his view.[42]

After visiting Preston, Young wrote to his wife, "When I look at the [difference] between poore People here and in America I rejoice that you and the children are there. . . . the Poor Peopel are rich in that contry for they have the priveledge of borrowing there they have the priviledge of [begging] and asking for somthing to eat if hungry but they have not that priveledge here let there circumstances be what they may they must not aske for food." Young went on to explain that if a person was caught begging for food or money in England, they would be "taken to the work house to rec[e]ive the due demerit of their crime."[43] Despite the criminalization of the practice, however, begging was commonplace among the British poor. According to Heber C. Kimball, it was not uncommon to see "whole families begging for bread" in the streets, or to have "small children . . . going from house to house" in search of food or funds.[44] For Young, Kimball, and their fellow missionaries, poverty could be most easily identified by the lack of sufficient food. Coming from agrarian backgrounds where each had owned farms and been able to cultivate the necessary resources for their families, the notion of having to beg for food was perhaps the most alarming issue that Mormon missionaries witnessed in England.

The scenes of British poverty were made all the more troubling by the factories and mines that facilitated that poverty. Visiting one of the nation's coal mines, Heber C. Kimball noted that the mineshaft was over five hundred feet deep, leading to a place where "about one hundred men and boys" worked through the day and night.[45] While Kimball's description of the mines was largely a benign statement of facts, Woodruff focused on the pollution they produced. He described the mines as places akin to "the Lake of fire & Brimstone" that caused "the whole face of the earth & heavens, air & horizon" to be "filled & Coverd with the composition of fire, cinders, Gas, sut [*sic*], and smoke." In summation, Woodruff concluded that the British mines were "more like Hell in comparrison [*sic*] than any place I have yet visited."[46] But while Kimball and Woodruff saw the mines and factories as unfortunate places to visit, they were workplaces for many of Mormonism's British converts. They were places where workers could be "almost worked to death" in the pursuit of the natural resources that fueled British industry. Given such conditions, some British laborers saw their experiences as being akin to the children of Israel who had labored "with hard bondage, in mortar, and in brick, and in all manner of service in the field" to build monuments for Egypt.[47]

In many ways, the social critiques of the Mormon missionaries mirrored the observations of others who visited England's industrial centers during the mid-nineteenth century. Following his visit to Manchester in the 1840s, German travel writer and historian Johann Khol wrote of the city: "It cannot be said that Manchester is either an ugly or a beautiful town, for it is both at once. Some quarters are dirty, mean, ugly, and miserable-looking to an extreme, [while] others are interesting, peculiar and beautiful in the highest degree."[48] Similar to Khol, many of the Mormon missionaries held mixed opinions of England's industrial centers. Although the scenes of urban poverty were deeply distressing, they were countered by the country's rich cultural offerings.

Friedrich Engels, a contemporary of Karl Marx and a critic of British poverty in the 1840s, particularly denounced England's industrial centers as places of squalor and appalling economic inequality. He wrote, "The slums are pretty equally arranged in all the great towns of England, the worst houses in the worst quarters of the towns. . . . The streets are generally unpaved, rough, dirty, filled with vegetable and animal refuse, without sewers or gutters, but supplied with foul, stagnant pools instead."[49] Engels highlighted the rank pollution that characterized Manchester's slums, suggesting that such pollution and the accompanying poverty were the end results of industrialization and laissez-faire capitalism. They were also, however, the results of economic policies that privileged extraction and profits over the environment.

Coming from the perspective of religious leaders, Mormon missionaries worried about the degenerative moral effects of poverty among the working classes. Writing to the First Presidency in Nauvoo, Brigham Young and Willard Richards provided a dismal description of the environment that contradicted the "histories of the people of England" with which many Americans were familiar. In their observation, dramatic changes in the British economy had "caused a mighty revolution in the affairs of the common people." It was clear that these changes had dramatically altered the environments and living conditions of the lowest classes. Young and Richards wrote, "A few years since, and almost every family had their garden, their cow on the common, their pig in the Stye [*sic*], which added greatly to the comforts of the household; but now we seldom find either garden, cow or Pig." Due to the consolidation of land and wealth in the hands of the upper and middle classes, flourishing gardens had given way to urban poverty, and the profession of farming to factory work. One so affected informed Young, "I had no place to keep my cow & I was obliged to sell her; I killed my pig to prevent its starving. . . . I have been obliged to go into the factory, with my wife & children, to get a morsel of bread." Concluding their assessment of the conditions of the poor, Young and Richards wrote to Smith, "Hunger & Rags are no curiosity here, & while things remain as they are we can expect but theft, robbery, murder which now fills the land—leaving out of the account, both as cause & effect the drunkards, gambling & swearing & debauchery—which are common on every hand." In short, Young and Richards felt that the lack of land and the inequality of wealth had created a condition of perpetual crime and debauchery in industrial England.[50]

Far from declaring the situation hopeless, however, Young asserted the necessity of British converts removing themselves from these conditions and immigrating to the United States. Without Joseph Smith's prior authorization, Young gathered together a company of poor converts and booked passage for them on an America-bound steamer. Explaining the decision, Young and Richards wrote to Smith, "our hearts are pained with the poverty & misery of this people & we have done all we could to help as many off as possible, to a land where they may get a morsel of bread." Hence, Young's decision to begin the British emigration grew out of his belief that the poor would find better circumstances in the United States than in England. Young specifically instructed the emigrants to "go to the western states," noting that in those regions they could "live among the farmers and wait orders from the Authorities of the Church," promising them that by doing so, "all will be well."[51] Such sentiments reflected not only the Latter-day Saint emphasis on the need for gathering, but also the prevailing romanticism with which Americans viewed the American wilderness.[52] For Young, there was a form of temporal salvation to be found in land ownership and farming.

Smith meanwhile suggested that America's farming frontier might not be the perfect answer to all of the problems of industrial poverty. Having been informed of Young's intentions to encourage British converts to gather to the American West, Smith responded that the British poor would need to "have certain preparations made for them" before they could successfully emigrate to Nauvoo. He suggested that Nauvoo and its surrounding regions had "advantages for manufacturing and commercial purposes" that could be used to establish "Cotton Factories, Founderies [*sic*], Potteries, &c." With the addition of British converts who had been trained to work in such facilities, he hoped to remake Nauvoo into an economic power on the Mississippi.[53] Smith clearly understood that not all people were made for farming on the American frontier. Further, not having seen the dire consequences of urban poverty firsthand, Smith seems to have focused on the possibilities presented by British-style industrialism in transforming Nauvoo into a dynamic center in the American West. Far from proposing yeomen agrarianism as the answer to urban British poverty, Smith seems to have believed that industrialism and economic growth in Nauvoo could be the answer to early Mormonism's transience and persecution in America.

Still, Smith's desires to encourage the industrialization of Nauvoo seem to have been made without either a full comprehension of the natural resources required or an understanding of the economic and environmental impacts that such efforts would have had upon the frontier town. Although situated advantageously along the Mississippi River, the Des Moines rapids near Warsaw would have created challenges for industrialization, as would the lack of any significant deposits of coal.[54] Furthermore, Nauvoo lacked both the capital and the large population necessary for industrialization. Finally, Nauvoo lacked the satellite communities that could create a constant flow of commodities to supply manufacturing centers.[55] On the other hand, as historian Kenneth Pomeranz has noted, Britain's natural advantages, including ready access to large deposits of coal and a large workforce of inexpensive labor, had enabled British industrialization.[56] Even if Smith had been able to build industrial factories near Nauvoo and populate them with a steady flow of emigrants from England, the factories would likely have perpetuated the very economic challenges that Young and the other apostles had decried in England. Furthermore, a steady use of coal would have introduced significant pollutants into Nauvoo and its surrounding areas, heightening the city's several health challenges. Accordingly, like other Americans of the period, Smith failed to understand the true costs of industrialization. Such plans represented the general lack of understanding that most nineteenth-century Americans and Europeans had regarding the heavy environmental costs of industrialization and economic development.

By encouraging an orderly gathering of the British Saints to Nauvoo, Smith accurately accessed the realities of 1840 Nauvoo and some of the challenges and potential "disappointment" that urban British converts might face in relocating to the frontier town.[57] Smith's concerns about potential disappointments were not unfounded. For some contemporary Englishmen, the prospect of living in America was a dreadful thought. Following an 1842 visit to Illinois and other parts of the United States, famed author Charles Dickens confided in a friend, "I don't like the country. I would not live here, on any consideration. . . . I think it impossible, utterly impossible, for any Englishman to live here and be happy."[58] Dickens wrote these dismal impressions after visiting the country's largest and most sophisticated cities. His feelings of disregard only deepened as he visited an Illinois town on the Mississippi River, which he found to be little more than "a dismal swamp on which the half-built houses rot away" leaving "a slimy monster hideous to behold; a hotbed of disease, an ugly sepulcher, a grave uncheered by any gleam of promise; a place without one quality, in earth or air or water, to commend it."[59]

Although situated in a different Mississippi River town, Dickens might well have been describing the Mormon settlement at Nauvoo during its early history. In 1839, the Latter-day Saints had settled on a small peninsula on the Illinois side of the river following their forced exodus from Missouri. At the time of their arrival, the peninsula boasted a small settlement of no more than one hundred residents who had ambitiously named the location Commerce, despite its general failure to live up to its aspirational name.[60] Similar to Dickens's description of Cairo, Illinois, the town was built on a malarial swamp that resulted in widespread illness throughout the community during much of the summer and early fall of 1839. By the time the first British converts began arriving in the city, the town's population had risen to approximately three thousand people, an impressive feat by frontier standards. But when compared with Britain's large industrial cities, Nauvoo was, by every estimation, a primitive location.[61]

Like their American counterparts, emigrating Saints underwent a strenuous and at times precarious sea voyage prior to their arrival in Nauvoo. And as with the American missionaries, British converts often found the environment of the high seas to be unsettling. In addition to crowded quarters, emigrants frequently described experiencing complications from seasickness. Joseph Fielding warned potential emigrants that strong winds frequently led to complications with seasickness, which was "very unpleasant, and sometimes painful," but assured them that "it does not continue long."[62] Others, however, were not as fortunate. Recalling his 1842 voyage from Liverpool to New Orleans, Edward Tolton remembered "enduring a siege of seasickness which continued . . . for four or five weeks." The experience was such that only the thoughts of "despondency and dread for the perils of the sea" prevented Tolton's immediate return to England.[63] William Ellis Jones similarly reported having been "sick nearly all the time [he] was on the sea," to the point that some "predicted that . . . I would have [a] watery grave."[64] Conditions were such that fears of sinking ships led some companies of Saints to hold daily prayer meetings in addition to their Sabbath worship services until arriving in either New York or New Orleans.[65]

Upon reaching the United States, emigrants immediately noticed the environmental differences between America and the British Isles. Having sailed into New York Harbor with his 1840 emigrating company, William Clayton offered an analysis of America's most prominent metropolis. He praised the city for its "very good harbor" and "elegant buildings." Clayton's analysis was not entirely positive, however, as he noted the city's lack of docks, its dirty streets, and the fact that the streets were "not so well flagged and paved as in England."[66]

The further west Clayton's company traveled, the more evident the distinctions between the two countries became. Being accustomed to cities with inns and other accommodations for travelers, the prospects of camping in the woods during a rainstorm, with "no houses near," proved taxing for the British emigrants.[67] Making its way through Ohio, the group stopped in Kirtland and determined that due to a lack of funds a portion of the company should winter in the town, where there was still a sizable Mormon community with whom they could associate and worship. The former church headquarters were less than impressive to several of the British converts. The conditions were stark enough that Clayton recorded, "Some was almost inclined to wish they had not left England"—this in spite of the serious challenges of urban British poverty.[68] By January 1841, at least some members of the Clayton company who had remained in Kirtland left Mormonism and determined to abandon their efforts to journey to Nauvoo.[69] That these emigrants desired to return to England seems to suggest that those in Clayton's company had likely not been numbered among the most destitute of the British Saints. Indeed, for those who desired to return, deprivation seemed to consist of the thought of living on the American frontier where food had to be cultivated rather than purchased. Distance from or control over aspects of the natural environment in some ways, then, became a marker of civilization.[70]

For the members of the group who continued onto Nauvoo that year, conditions were no better. It had taken the emigrants eleven weeks to make the five-thousand-mile journey from Liverpool to Nauvoo, a journey marked by deprivation and exposure to cold weather. Writing to Edward Martin of Preston, Clayton explained that the trip had included significant temperature changes, with conditions alternating between being "either to[o] hot or to[o] cold." In preparing fires, Clayton's company was reminded that, unlike the British, Americans burned firewood rather than coal.[71] While temperature changes were difficult, dietary changes proved to be one of the greatest struggles for the British Saints. As with all groups who visit new lands, Clayton's company had to learn to "live on food" that they "had not been used to." Still, despite occasional "hard times," Clayton assured his fellow British Saints that they had known "but little of the toils and difficulties of traveling."[72]

Following eleven weeks of travel, Clayton's company arrived in Nauvoo on November 24, 1840. The group quickly learned that Nauvoo was ill-equipped to accommodate the arriving emigrants. Still a relatively new community, the town lacked inns and boardinghouses. Indeed, Clayton's company arrived nearly two months before Joseph Smith's January 19, 1841, revelation instructed the Latter-day Saints to construct "a house for boarding" where "strangers may come from afar to lodge."[73] The company accordingly spent the first night in the region sleeping on the floor of one of the Nauvoo Saints' houses. Clayton's family ultimately shared a home with a few other families, making fourteen people in the same house, each sleeping on beds made of hay or occasionally oak leaves. In addition to the house being crowded, cooking often had to be done outdoors so as to avoid filling the home with smoke, a circumstance that created its own atmospheric challenges to breathing. These household inconveniences were made all the more challenging by the extreme cold and the lack of sufficient firewood. Reflecting on the situation, Clayton remarked that it seemed as though the new emigrants would "be froze to death."[74]

Despite the environmental and spiritual challenges that Clayton's party encountered, however, he remained optimistic about Nauvoo. According to Clayton, the change in climate had helped at least two British sisters who had suffered from chronic rheumatism in England. Clayton reported that one looked "considerably younger and more active" in Nauvoo than she had appeared to be in England.[75] Clayton similarly highlighted the natural beauty of the location, reporting, "We are pleased with the appearance of the country it is exceeding rich and beautiful."[76] Clayton doesn't explain exactly what he found to be most beautiful about the location, though his praise likely included at least some thoughts of his new closeness to the capital of Mormonism. Still, the prairie offered a kind of pastoral beauty that had not been readily available in Clayton's previous residence in Manchester. With regards to food, the frontier offered its own kind of culinary diversity, not dissimilar to what George A. Smith had encountered in the Liverpool marketplace. Clayton informed his fellow British Saints that the region offered "plenty of food of many kinds," and in greater quantities than was available to them in England. He reported that several in the company, including himself, had "grown very fat and healthy" in America.[77] In many regards, Clayton's letters to the Mormons in England served as a means of propaganda by which he hoped to encourage his fellow Saints to immigrate to the Illinois frontier. Clayton thus downplayed or omitted references to any of the natural challenges of the frontier town, while highlighting the opportunities presented there.

Not all British emigrants remained as positive about Nauvoo as Clayton. The "universal anxiety" of some to "get away to the land of promise" diminished in light of the environmental realities of the faith's frontier settlement on the Mississippi.[78] Thomas C. Sharp's *Warsaw Signal* stated that there was "great dissatisfaction" among the British converts who were discontented with "the sad state of things in the City of the Church." Encountering the frontier circumstances, many of the converts "determined to leave" the city and return to England, even sending letters to friends and kin warning them not to migrate.[79] The *Daily Missouri Republican* reported that several of the British converts had left Nauvoo and congregated in St. Louis, "hastening back to England, 'while their money holds out.'"[80] Such departures were common enough that Latter-day Saint letters back to England occasionally included warnings to would-be emigrants that some who made the journey would "shake out by the way" and return to England.[81] Mormon leaders worried that even some who remained would "send home an evil report" because of the challenges they faced on the frontier.[82] Such converts discovered that the American West could be a place of genuine concern as well as a place of romantic adventure.[83]

Environmental causes were often important components in the decision to leave Nauvoo. The *Preston Chronicle* detailed the story of the Thomas Margaret (Margarett) family, who had emigrated to Nauvoo in February 1841, believing it to be a "land of promise." The realities of the American frontier, however, did not match the family's expectations. For the Margaret family, rather than finding a land flowing with milk and honey, they found an area that contained "the most deplorable privations incidental to a new settlement." Whereas England was supplied with products from all over the world, Nauvoo was "badly supplied with the necessaries of life," indicating that they had probably not come from the poorest ranks of British society. Less than a week after moving to Nauvoo, the family had evidently exhausted their supply of flour, leaving the children "crying . . . for bread." By early September 1841, the family had returned to England where they took refuge in the house of a friend, "tired, jaded, and out of health."[84] While many factors likely influenced the Margaret family's return to England, the encounter with a world devoid of large markets and a steady influx of food and other commodities played a significant role in the decision.

Such stories posed problems for a community that aimed to draw upon new migrants for its continued growth and development. Accordingly, Mormon leader Sidney Rigdon tried to assure local papers that the British newcomers were generally "well satisfied" with the circumstances.[85] But while examples like the Margaret family may generally have been anomalous, it was clear that at least some of the emigrants were unimpressed with their new homes and the frontier setting of Nauvoo. Some of those who were dissatisfied with Nauvoo and its foreign environment quickly returned to their former homeland and native surroundings.

The potential disappointments, however, were far more significant than simply being dissatisfied over moving from an urban to a rural setting. Despite Smith's aspirations of creating manufactories in Nauvoo to capitalize on the abilities of British converts, the town never grew beyond its origins as a frontier farming community. As such, it provided relatively few occupations for many of the British Saints. Such deficiencies did not escape the notice of even the most committed of the British migrants. Soon after his arrival from England, Francis Moon authored a letter reminding the British Saints that Nauvoo had no factories, and that the town lacked both the funds and machinery to build them. For Moon, the lack of factories meant that not only jobs but also goods would be limited in the gathering place. Without factories, the Nauvoo Saints had to "do at home what would be done at factories," increasing both the cost and the time necessary to secure desired goods. Accordingly, Moon hoped that, in time, people with both the funds and property to build factories would come to Nauvoo, thereby providing jobs and goods for the citizenry. Beyond this, Moon believed that by industrially developing the rural area, Nauvoo would "feel more at home" to his fellow Englishmen.[86] Hence, for converts like Moon the natural environment was to be manufactured rather than lived in.

The lack of viable employment opportunities for British emigrants quickly became a concern for Mormonism's American leaders. On June 13, 1842, Smith attended a meeting at which he discussed "ways and means to furnish the poor with labor." He was particularly worried about the British Saints who were "unacquainted with any kind of labor except spinning weaving &c." In accordance with Moon's concerns, Smith informed the council that Nauvoo was not equipped to properly employ its growing populace of converts from industrial England. Without factories, the emigrants were "troubled to know what to do."[87] Making matters worse, British families frequently sent a single member of the family to Nauvoo to seek for financial assistance to help the remainder of their family emigrate. Such situations were problematic in Nauvoo, where payment for labor often came in the form of produce rather than cash. In 1842, currency was so scarce that it could not "be obtained by labor." Accordingly, Joseph and Hyrum Smith advised the British Saints not to emigrate to the Mormon capital until they could bring their entire family with them, as there was "no means here to be obtained to send back."[88]

Not all immigrants lived with unbearable situations, however. Despite the fact that many were compelled to live three or four families to a home, or even in pitched tents in the woods due to a housing shortage in Nauvoo, other families found good lodging at reasonable prices. Writing home to England, George and Ellen Douglas enthusiastically reported that they had found a house to rent "at 5 shillings a month." Along with the home, the family had obtained a steady supply of firewood, a well-stocked vegetable garden, "a flock of chickens," and a pig. Surveying their situation, the Douglases reported that they were "far better here than in old England." Contrary to the work problems that others experienced, George found work helping build the Nauvoo House, while other family members found jobs helping on farms.[89] In many regards, the Douglas family embodied the hopes and aspirations that had motivated Smith, Young, and the other apostles to encourage emigration to America. Indeed, the absence of personal property, including gardens and animals, had been among the most astounding features of urban England for the American missionaries. Accordingly, in the eyes of Mormon leaders, a well-supplied garden and a farm that included domesticated animals represented a significant success for the plan of British emigration.

In evaluating the responses of American and British Mormons to their new environments, we must consider not only the different conditions and experiences that they encountered, but also the national ideologies through which they interpreted these experiences. Despite the growth of a few urban centers, the United States remained largely rural and boasted an agrarian-minded populace throughout the Jacksonian Era. Ideologically, most Americans were motivated by Thomas Jefferson's hopes of creating an "Empire of liberty" with farming characterizing its citizenry.[90] Indeed, for many Americans, land and the frontier were at the very heart of Americanism, distinguishing the country from European nations.[91] Open lands and the farming opportunities they offered acted as a safety valve protecting the United States from the perceived dangers of urbanization. Accordingly, many nineteenth-century Americans viewed cities and factories as symbols of corruption, danger, and moral decline, rather than as evidences of civilization and progress.[92] It was with the ideological perspective of Americans that Mormon missionaries approached and interpreted the scenes of industrial England and urban poverty. Indeed, despite its expectations of a vast international growth that "no unhallowed hand" could stop from progressing, Mormonism in the late 1830s and early 1840s was a profoundly American faith.[93] Coupled with the theological concepts of communal cooperation and gathering to an American Zion in the West, Mormon missionaries viewed British cities through the lens of yeoman American farmers strongly believing in the ideology of agrarian virtue. This belief in agrarianism then helped expedite Latter-day Saint efforts to gather British Saints to an American Zion.

The British Industrial Revolution, however, had changed the way that the urban British populace interacted with the environment. Although Britain retained significant amounts of fertile and productive farmland, its population base had become steadily more urban. Factory work, urban housing, railroads, and streetcars had decreased the importance of skills like driving a team of cattle or oxen—skills that remained almost essential to survive on the American frontier. For the middle and upper classes, natural history museums had both expanded and diminished the interactions that people could have with the environment, expanding opportunities to see the exotic while decreasing real-world interactions with the animal life and plants they displayed. Within the mines and factories, interactions with nature revolved around the processes of extraction and production for profit. As many of the British converts prepared to encounter the American frontier, they found themselves ensconced in a variety of new environments, devoid of the comforts of urban life. For some British converts, frontier life and the lack of the commodities offered in Britain proved detrimental to their commitment to Mormonism.

Hence, while nineteenth-century British and American citizens maintained many similar views about the profitability of the land and its resources, the Industrial Revolution created a divergence in their opinions about the best paths to that profitability. Whereas nineteenth-century Americans saw agrarianism as the means of advancing the ideals of American democracy and civilization, British citizens believed that the civilizing path lay in the processes of industrialization and mechanization. Indeed, neither nation seemed cognizant of the long-term effects of their own national policies upon the surrounding environment. Entrenched in these competing national ideologies, both groups approached their new environments with a sense of wonder and skepticism, ultimately preferring their own national conceptions of the environment and civilization.

In evaluating the experience of American Mormon missionaries and British converts with differing environments, we find that nationality continued to play an important role in their various worldviews. Although both groups found levels of appreciation for various aspects of their new environments, they tended to prefer their own lands, climates, and regional demographics. The doctrines of evangelizing and gathering, however, induced both American and British Saints to encounter new environments. While these encounters were more permanent for British converts who gathered to Nauvoo, they nevertheless similarly influenced missionaries to leave their own homelands in search of new converts. By 1846, however, all Latter-day Saints who followed Brigham Young found themselves in new and unfamiliar environments, which would again both influence and be influenced by the Saints.

MAPPING DESERET

Vernacular Mormon Mapmaking and Spiritual Geography in the American West

Richard Francaviglia

In 1969, when I began researching records in the LDS Church Archives for my dissertation titled "The Mormon Landscape," I encountered a wealth of written narratives in the form of journals and diaries. Occasionally, I also found some fascinating examples of Mormon cartography—manuscript maps that were usually appended to letters or journals but sometimes were in their own separate file folders. As a geographer who had worked for two years at Rand McNally in San Francisco after graduating from high school but before starting college (1961–1963), these maps immediately caught my attention. Quite different from the scientifically accurate maps we made and sold at Rand McNally, many of these Mormon maps had a folkloric quality. They were hand drawn and often idiosyncratic, using techniques and producing results quite unlike those seen on professionally prepared maps.

I was grateful that these maps had been preserved because they revealed a unique perspective on the environment that the Mormons had settled. Like all maps, these consisted of combinations of graphics and words that provided a better picture of the landscape of an area than text alone could. A major premise of this essay is that the unique combination of graphic devices and text provided by maps makes them especially effective tools for environmental historians. In the context of Mormonism, these maps can provide a better understanding of the dynamic process by which the Latter-day Saints encountered, envisioned, and appropriated places. These Mormon maps helped me better understand what the writer Wallace Stegner called "Mormon Country" in 1942 and the geographer D. W. Meinig branded "The Mormon Culture Region" in 1965.[1]

These maps helped bring Mormon history to life. In addition to being idiosyncratic—each map varied in scale and technique—each of them revealed the personality of the individual who had drafted them. One of the things that made these Mormon folk maps different from those that I was familiar with at Rand McNally was that most Mormon cartographers had received no formal training. Instead of relying on professional surveying techniques adopted by governmental and corporate cartographers, they used intuition. As I further studied these maps in the Mormon archives, I noted that they reflected a different kind of investment in the places that were mapped. Whereas even an untrained, non-Mormon cartographer might draft a map for his own benefit, or the benefit of government or corporations, the Mormon mapmakers had a different agenda, one that reflected the fact that the Mormon experience was communal rather than individual in nature. True, each Mormon was an individual with concerns for his or her family, but Mormons were operating on behalf of their church. Religion rather than economics lay at the heart of why Mormons settled, and mapped, the West.

After conducting my field and archival research in 1969–1970, I concluded that the Mormon landscape vividly reflected the differences between the Mormon and non-Mormon experience.[2] Using the list of varied elements that I identified (including street width, architectural styles, and the like) anyone traveling throughout the West could tell which towns had been settled by Mormons, and which had been settled by non-Mormons (or "Gentiles" as they were called by the Latter-day Saints). Mormons built distinctive settlements for Christ's Second Coming,[3] and to facilitate this process they sometimes drew maps of the general areas they had been called upon to settle on behalf of the church. With each line inscribed on paper, the Mormons were making a visual record of the places they experienced through exploration. Step by step from the late 1840s and well into the 1870s, they simultaneously charted Zion and transformed their part of the American West into what geographer Lowell C. "Ben" Bennion later called "Mormondom's Deseret Homeland" (2001).[4] The term "Deseret" has long intrigued me as it supposedly translates into "honeybee" from the language of the Jaredites (a group in the Book of Mormon)—another subliminal cue that Mormons, like honeybees, were not only perpetually busy, but also collectively engaged in a task meant to sustain the community. Of course, the name Deseret is also phonetically similar to the word "desert," a coincidence perhaps but the Mormon homeland does indeed include a substantial amount of arid land.

The relationship between the natural environment and the Latter-day Saints' religion lies at the core of early Mormon mapping. The physical environment they encountered in the Intermountain West was new to them: in contrast to the well-watered East and Midwest, it consisted of arid and semi-arid basins juxtaposed with higher elevated lands that received more precipitation, and were thus better vegetated. In the Intermountain West, however, the desert-like lands in the basins had potential for agricultural development as waters from those higher lands often ran into them; hence the areas along the streams were often considered attractive and could be transformed into oases. At the same time, though, these promising areas along the river courses were subject to flooding. The art of community building in the Mormon West called for knowledge about this diverse and sometimes capricious environment. Much of this information could be conveyed by words, but Mormons soon realized that maps were an effective way to differentiate the landscape of their new home in the West.

This essay builds on the premise that mapping this new, challenging land was essential for Mormons to both claim and settle it. Mapping was part of the frenzied activity by frontier Mormons, though they rarely mentioned it in their journals and diaries. This omission may seem surprising, but by the time that Mormons were mapping and then building their homeland, maps had become so widespread in popular usage that reading—and even making—them was taken in stride. The Latter-day Saints had a job to do, and maps helped them do it. Of course, this mapping occurred at several levels, but normally the larger countryside was mapped first, and the individual town plats mapped shortly thereafter. A major exception to this rule was the platting of Salt Lake City, which occurred within the first few weeks of settlement there in 1847. Generally, though, area maps helped church leaders gain an understanding of the lay of the land, while a good eye and some elementary surveying helped them determine where the settlements would actually be built and how they were to be laid out.

In keeping with the original plan of the city of Zion as developed by Joseph Smith and put into map form by Frederick G. Williams as early as 1833, these communities were invariably configured into a compass-oriented grid plan—though the actual plan was modified by church leaders in the West to include larger lots. In reality, the process of Mormon settlement was sometimes hit and miss as a number of original Mormon towns proved to be located on land that was poorly situated (usually in flood plains) and had to be relocated shortly thereafter. As I noted in *The Mapmakers of New Zion*, the better the geographic information the more successful the settlement: in fact, some of the failures in Mormon settlements, for example the Little Colorado River of Arizona, coincided with—and may even have been caused by—a lack of mapping.[5]

It took me many years to begin thinking about what made Mormon cartography distinctive, but it lies in the deep bond that Mormons feel for their western American homeland, which developed in both space and time. The Mormons' phenomenal sense of direction and penchant for order is inherently cartographic.[6] Reflecting the conditions facing the Mormons at the time of encounter with new places that they transformed into home almost overnight, these maps served several purposes when they were first made. These earliest maps were essential to the process of exploration and initial colonization, which lasted from about 1847 to the 1870s and which increased the Latter-day Saints' geographical knowledge of the area they claimed. At a slightly later date, maps were also drafted to facilitate and enhance communication (for example, in establishing postal routes) throughout the Mormon West. These later maps characterize the post-exploration period, which lasted from the 1870s to around the turn of the century. From the outset, I should also note that Mormon mapmaking was not accomplished in a vacuum; the Latter-day Saints and federal authorities actively cooperated in making maps on a number of occasions. Albert Carrington's role in the expedition of Howard Stansbury is a case in point. Stansbury himself mentioned Carrington as an important contributor to the expedition's success.[7]

Maps were (and are) inseparable from the process of exploration and colonization. As hinted at above, they comprise more than simple graphic illustrations but are actually complex texts—in the case at hand, texts that employ words and graphics can sustain the Mormon encounter with, and appropriation of, the American West. Another premise of this essay is that these maps can be deconstructed to enlighten us about the attitudes of those who made them—in this case, Mormon pioneer cartographers whose religious beliefs motivated, sustained, and ultimately celebrated exploration on behalf of the church. As will become apparent, that process may be practical on the one hand but it also has deeper aesthetic significance. Naturally, wherever possible, I call upon additional narratives of the kind written in Mormon journals and diaries to further demonstrate the connection between what is mapped and what is believed.

In a sense, all mapping of the Intermountain West by Mormons begins with Salt Lake City, for it was to be the initial place of the "gathering" of Saints from around the world. With considerable precision, the city's earliest maps employed the orthogonal, compass-oriented grid, in part to honor the spirit of Joseph Smith's city of Zion plan and also to provide a template for future Mormon land division. It bears repeating here that the maps listed in the LDS archives as being by a particular cartographer—for example, Thomas Bullock, August 1847—often represent a "team effort."[8] In compiling these maps, Bullock likely consulted earlier sketch and survey maps, including original early surveys by Henry G. Sherwood and others done as early as August 2, 1847. From these, Bullock prepared several maps of the city that would become iconic, including one that he later drafted for Sir Richard F. Burton's popular book titled *The City of the Saints: Among the Mormons and across the Rocky Mountains* (1860).

The earliest maps of Salt Lake City depicted the site not as it actually appeared (a gently sloping plain at the base of the Wasatch Mountains) but rather as the Saints envisioned it becoming—an orderly, large city. As they ventured forth from that city in varied directions, the Mormons envisioned other cities taking root in wild country that they first sketched, which is to say mapped, and then sent those rudimentary maps back to church leaders in Salt Lake City. In a manner of speaking, these embryonic maps were thus not only depictions of places in the present, but they also served as templates of the future wherein Mormon communities would essentially blanket the area. Before that could happen, though, the Mormons needed to sketch out the armature of the countryside, which first appeared as a maze of mountain ranges and river courses.

Mormon colonization was a complex process by which the grandeur of nature would be enhanced, as it were, by Mormon settlements intended to collectively form a new Zion. Although the concept of sustainability is a phenomenon of the twentieth century, the Saints clearly envisioned self-sufficiency that depended on careful use of natural resources. Again using Salt Lake City as an example, although Brigham Young envisioned it growing rapidly southward, one of the first things he commanded the Saints to do after their initial arrival was to set aside a large parcel of land as pasture where development would be prohibited. The regulation of such matters required maps, as did the necessary exploitation of resources such as the establishment of coal and iron mining in parts of the area (for example near Cedar City). Through such settlement based on varied resources the Mormons hoped to develop either a separate and autonomous religious kingdom or a self-sufficient religious state within the expanding United States.[9]

This settlement expansion scenario was part of an intense religious drama unfolding—a drama in which the Latter-day Saints envisioned themselves playing a central role as modern-day Israelites helping to restore the gospel of Jesus, which Joseph Smith claimed had been lost through neglect and willful disobedience. In 1849, Apostle Orson Pratt wrote an explicit missive about the role that the Intermountain West was playing in the grand scheme of religious history. Pratt's missive was grandly titled "New Jerusalem; Or, the Fulfillment of Modern Prophecy," and it was published in Liverpool, England. As Brett D. Dowdle notes elsewhere in this volume, England was an important focus of the Mormon's proselytizing efforts at this time; however, although the call to Mormonism was envisioned as global in nature, in the Mormon mindset, the new converts were part of a worldwide migration to Utah.

In 1849, Orson Pratt noted that the Mormons were settling "one of the most wild, romantic and retired countries on the great western hemisphere." Whereas just two years earlier this wilderness was inhabited only by American Indians, Pratt now confirmed that it was becoming home to a new people. Pratt stated that these individuals had a name, a collective mission, and a place set aside for them by God. In one succinct sentence, he noted that "the people called Zion, who were to bring good tidings, were required to 'Get up into the high mountains.'" To Pratt and other Mormon leaders, the Saints' migration here was no accident but rather foreordained. As Pratt put it, "This prophecy the Saints are now fulfilling: they are moving by thousands from various parts of the globe into the '*high places of the earth*,' among the Rocky Mountains, where they are forming a prosperous settlement, elevated over four thousand feet above the level of the sea."[10] That settlement was Salt Lake City, but within a decade other Mormon communities would dot the entire region from the Rockies to the Sierra Nevada Mountains.

The Mormons' ambitious mapping of the interior West coincided with, and facilitated, their aggressive exploration of the region that would briefly be known as Deseret. Both Mormons and non-Mormons recognized that name in the mid-nineteenth century as the Saints were a highly visible minority bent on establishing their own state or kingdom. The name Deseret was, in a sense, synonymous with a new Zion in the mid-nineteenth-century American West. As such, it was easily polarized, or polarizing. Those sympathetic to Mormons saw it as positive, but the many who opposed the Saints viewed it as divisive if not downright subversive. Although Deseret was formally proposed with a constitution and soon operated de facto from the late 1840s and well into the 1850s under a series of ordinances, it would prove to be short-lived.[11] Its structure of government was theocratic and hence viewed by many as being at odds with American values separating church and state. Many American legislators regarded Mormonism with suspicion and officially kept Utah under territorial status rather than permitting it to become its own state during this time period.

For Mormons, though, Deseret became a cherished concept if not a political reality—a religious homeland where a persecuted people could build the kingdom of God on earth. Although Salt Lake City became the region's major city, the area claimed by the Mormons was huge and extended well beyond the Great Basin. Much of the Colorado Plateau as well as portions of present-day Colorado, New Mexico, and Arizona were included within its boundaries, which were defined by watersheds. An especially intriguing aspect of Deseret is that although it is represented on numerous maps by non-Mormons between 1847 and 1858, there is no known Mormon map that actually shows it by name. In the *Mapmakers of New Zion* I speculate that this is because Mormons tended to use maps by federal authorities to make their case that they should have their own state.

When the Mormons petitioned the U.S. Congress, instead of using a map to show the extent of the proposed territory or state of Deseret, they used a narrative description that beautifully captured both their religious

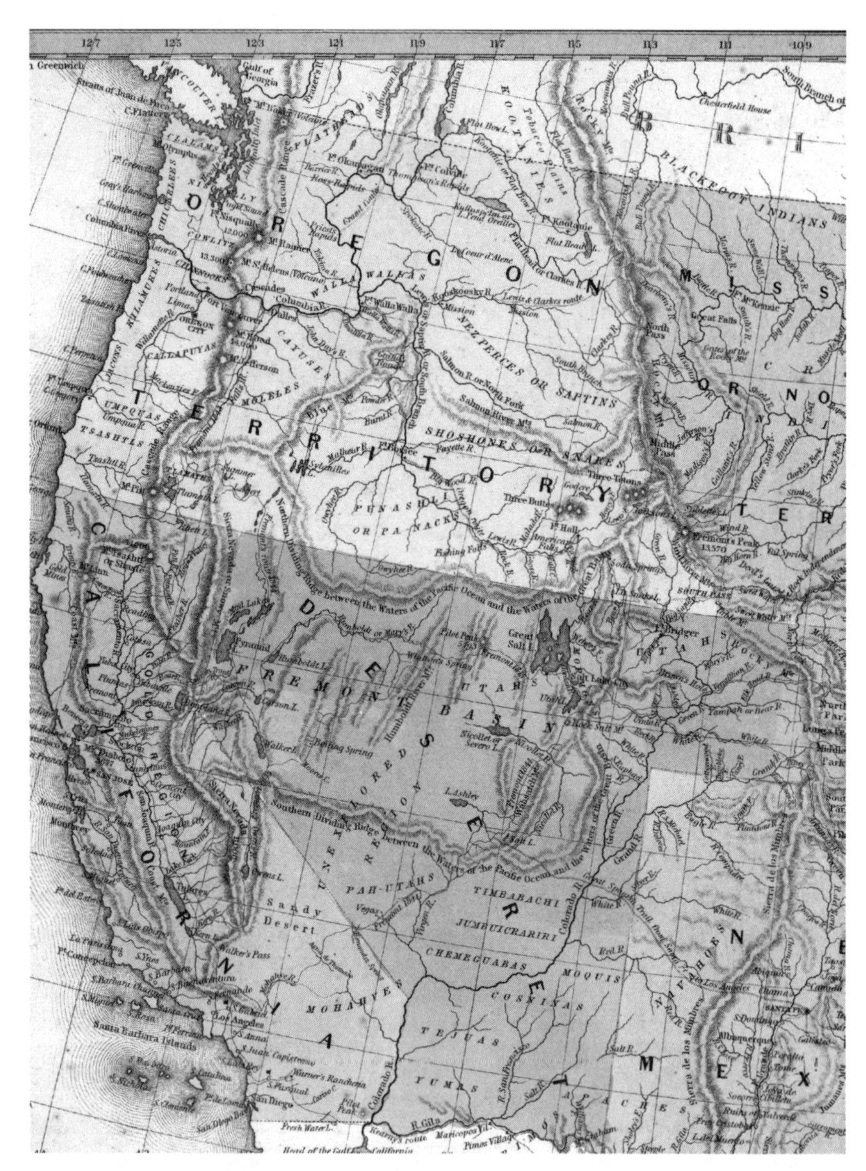

FIGURE 1. There is no known Mormon map of the proposed state of Deseret, but it was commonly depicted on non-Mormon maps in the mid-nineteenth century. As seen in this Map of the United States of America (1850) by J. H. Young, the name "D E S E R E T" is used for a portion of the "unexplored region" between the Rockies and the Sierra Nevada Mountains.

passion and their insistence on playing a part in shaping the geography of the American West. Their petition began with the following: "We, the people, grateful to the Supreme Being for the blessings hitherto enjoyed, and feeling our dependence on him for a continuation of those blessings, do ordain and establish a free and independent government by the name of the state of Deseret." It provided a detailed geographic description of Deseret's borders, concluding with the note that these points could be determined "as set forth in a map drawn by Charles Preuss, and published by order of the senate of the United States in 1848."[12] The narrative was certainly explicit enough to enable a person with that map in hand to draw the outline on it, but alas, other maps were often used as a template. On some, the proposed state covered fully half of the American West, but on others it was often depicted as far smaller than the huge state that the Mormons had envisioned (see figure 1).

The Mormons mapped many portions of this broad region they hoped to control. Step by step, each map helped them fill in the blanks, as it were. Like Mormon colonization itself, mapmaking here was governed by two seemingly opposing forces—centripetal and centrifugal. On the one hand, the Saints flocked to Salt Lake City, the major destination of new arrivals. Quickly thereafter, many of those same Saints were called upon to decentralize or scatter to far-flung areas of the West as they first explored and then colonized the region. Given the centralization of church decision-making in Salt Lake City, though, many of their maps ultimately found their way to church headquarters and the church historian's office there.

In this new and sometimes harsh environment, their experiences seemed to match counterparts in the Bible. In Mormon thought, the Saints were not settling just any land but the one that God had foreordained for them. They were awed by the natural environment but also tasked with transforming it into the place of the Second Coming. They interpreted experiences such as the salvation of their crops from locusts by seagulls as divine intervention and in their retelling these stories take on an aura all their own.[13] A historical geographer or environmental historian might consider their task a practical one—damming streams, laying out towns, dividing off farmlands—but it was difficult for Mormons to separate the practical from the religious and, moreover, from the beautiful. To Mormons, a job well done is not simply a job well done but also a thing of beauty. More to the point, this beauty is the result of following God's commands. As I travelled throughout the West interviewing Mormons, I found this to be a universally shared belief.

Small wonder that the Mormons came to look upon their handiwork, and the resulting landscape of Zion, with such pride. On numerous occasions, Brigham Young admonished the Saints in rural Utah villages to plant gardens and trees for both sustenance and spirituality, seamlessly connecting the two endeavors. In prose that typifies his colonization strategy and religious belief, Young urged the Saints to "render the earth so pleasant that when you look upon your labors you may do so with pleasure, and that angels may delight to come and visit your beautiful locations."[14] Other Mormon leaders easily conflated the Saints' new homeland with biblical prose, as when Apostle G. A. Smith noted: "I feel anxious that you should begin to beautify Parowan, and make it like the garden of Eden."[15] By this Smith meant "blossom like the rose," a Biblical metaphor that had been used in the Midwest but that took on new meaning in the arid and semi-arid West, where it became shorthand for transforming the desert through orchestrated irrigation projects.[16] Although it is difficult for lay persons and scholars alike today to comprehend the immediacy of God to Mormon pioneers, their journals abound in high praise to, as they often put it, "the Almighty."

As with all settlement, exploration came first. In contrast to federally sponsored explorers who might find a landscape to be scientifically interesting, aesthetically beautiful, or even awe inspiring on occasion, Mormons saw it as downright inspirational because it appeared to be linked to passages in scripture. Their perception transcended the popular "manifest destiny" wherein God decrees a nation's right to move westward. Whereas other settlers claimed the land individually, in Mormon society each settler had a role or roles to play on behalf of the church. One of those roles included mapmaking. More to the point, western Mormon explorers found themselves face to face with a land they believed God had given to them to live out according to both the Old Testament and the New Testament. Of the former, they made innumerable references to the desert blossoming as the rose (Isaiah) and of the latter they saw themselves living the final drama described in the book of Revelation, which explicitly states that the perfect city will be quadrangular in form. Understandably, this locale resonated as biblical to Mormons. As John Davis astutely observed of the early Mormons: "the singular landscape features surrounding them—such as the Great Salt Lake, with its evocation of the Dead Sea, or even the ever present desert, which inspired such town names as Moab, Utah—only

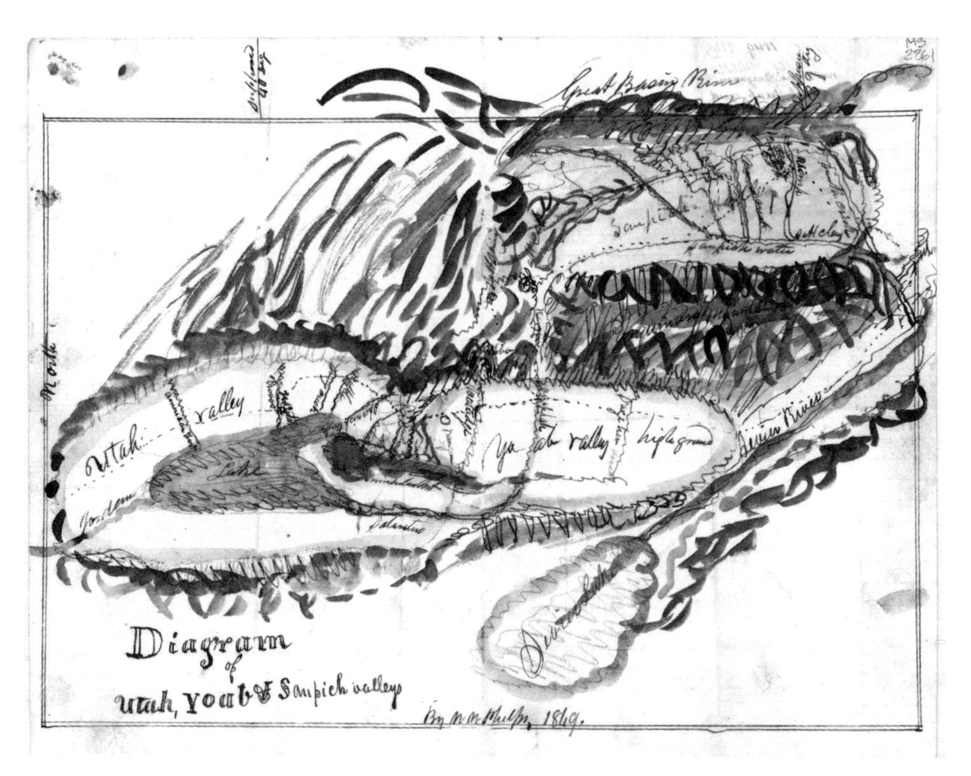

FIGURE 2. W. W. Phelps's sketch map titled Diagram of Utah, Yoab [Juab] and Sanpich [Sanpete] Valleys (1849) is oriented toward the east and depicts the landscape lying at the base of the Wasatch Mountains. Although the work of an untrained mapmaker, this evocative map succeeded in giving church leaders an idea of the topography and hydrology from the rim of the Great Basin (top) to Sevier Lake in the desert interior.

reinforced the connection and aided in the creation of their own 'sacred' space."[17] In short order, the Mormons named the river adjacent to the site of Salt Lake City after the real Jordan River in Israel. Like that fabled river in the Holy Land, their Jordan River of the West connected a fresh water lake (Utah Lake) with a huge salt sea (Great Salt Lake).

It was into such an evocative landscape that Mormon explorers ventured from the mid to late nineteenth century. Although very few of these explorers were formally trained cartographers, the maps they drew reveal much about not only the new land but also about the Mormons' newfound faith. One of the cartographic treasures in the church archives is W. W. Phelps's 1849 map of the countryside surrounding the biblically named Mt. Nebo in central Utah (see figure 2). This map beautifully articulates the corrugated topography consisting of tall mountains alternating with narrow valleys oriented pretty much north-south. As expedition leader, the New Jersey–born Phelps (1792–1872) was entrusted with conveying geographic information to Brigham Young to the best of his ability. That is exactly what Phelps did as he reconnoitered the rugged country bordering the Wasatch Mountains south of the Great Salt Lake. From his position on the flanks of Mt. Nebo, Phelps had a commanding view of the country that lay to the west, including the freshwater Utah Lake and the Sevier drainage system. Interestingly, Phelps oriented the map unconventionally: east is at the top, while many other Mormon mapmakers used the more conventional north as the top of their maps.

Phelps was modest about his mapmaking skills, for he called this map a "diagram." This, on the one hand, may reflect Phelps's recognition that his map was fairly primitive in design and content, at least when compared to professional maps such as those drawn by Charles Preuss on numerous expeditions led by John Charles Frémont. After all, at just this time, Phelps was likely aware that far more professional and scientifically accurate maps were being drafted by explorers working on behalf of the federal government. However, that term diagram hints at something inherent in Phelps's mission. He was not merely mapping the landscape but also systematically assessing its parts. Although Phelps's use of the term diagram suggests that it may be something different than a map, as a narrative/graphic device that depicts geographic features in relation to each other it does indeed qualify as one.

Rather than being a product of scientific observation, which characterizes maps drawn by federal exploration teams, Phelps's diagrammatic map has a very different—and far more intuitive—quality. Phelps's map does two things simultaneously: it both indicates geographic features as well as personifies them. As a geographer who has drawn maps of portions of Utah, the first thing that strikes me about Phelps's map is not its simplicity but rather its sense of drama. Unafraid to convey the "feel" of the land he was experiencing, Phelps emphasizes the dominance of Mt. Nebo by placing it at the top of the map; that desire on Phelps's part may explain why his map is oriented east rather than the conventional north. Using bold swaths to indicate the slope, Phelps depicts the topography as cascading downward from Mt. Nebo to the land lying to the west. The effect is as dizzying as it is arresting. Evidently not bound by increasingly scientific nineteenth-century cartographic conventions, Phelps renders the surrounding topography almost as if he were painting a watercolor rather than drawing a map. Watercolor is a key word here, for this stunning map is rendered in three colors—light sage green for the higher valleys, oxide red for the lower slopes of the mountains, and dark umber for the steeper slopes. I should also note that the content of the countryside itself—the varied geographical features—is rendered in a manner that might almost be called anatomical. Phelps's diagrammatic map shows the relationship of different physiographic units much like a medical illustrator might reveal the placement of organs—hence the map's almost visceral quality and impact. This map is indeed artistic in that it "reads" more like a watercolor painting than the line drawing we normally associate with a map. Its composition and technique emphasizes the mass rather than the outline of geographic features.

As I traveled throughout the Intermountain West in the late 1960s, I came to realize how important mountains were in helping Mormons situate themselves—literally and theologically. In Phelps's time those elongated mountain masses helped separate the Saints from their detractors farther east, but they were also part of their growing belief that mountains provided a direct connection to God. Compared side by side with LDS journals, maps like Phelps's gave spatial form to an enunciated vision of creating an American Zion that was cradled by mountains. As noted in an 1857 entry in a record book in the young Utah community of Nephi, Elder Hyde spoke to the gathered and said "Look upon the written word has I do on thease mountain peaks which act for land marks & o the spirit of God that gives general intelligence" (spelling retained from original). Leaving little doubt of the spiritual significance of such topography, the entry quickly added: "In the days of the children of Israil [*sic*] Moses went up into the Mountain to wait on the Lord."[18] This seamless equation of new landscape and its biblical counterparts is a reminder that Mormons were on a mission as a chosen people in a chosen land.[19] Small wonder that Mt. Nebo loomed above Nephi as its original namesake does above the real Holy Land.

Despite, or perhaps because of, this map's drama, it was easily comprehensible. Expedition members reported that Chief Wakara, the Ute Indian so familiar with the region, easily recognized each feature on this map. Wakara, whom the Mormons would have considered a Lamanite (a member of the lost tribes of Israel), described each feature Phelps showed much as a geographer would. This Indian's skill in visualizing the landscape was not lost on the exploring party, which equated him with a kindred professional in their own culture. The Mormons not only valued the Native Americans' input, but also hoped to convert them to Mormonism, and thus help them regain the faith that they had lost in the wilderness of the Americas. Wakara did indeed convert, though he later had a falling out with church authorities. For his part, Phelps also found the time to discuss the angels who had appeared to earlier Saints, and to discuss "translating Hieroglyphics." That term hieroglyphics was a metaphor for something grander, as Joseph Smith had interpreted ancient Egyptian papyri as the Book of Abraham. Now, as they traversed the mountains and canyons in the West, the Mormons' sense of adventure mounted, and their belief that ancient prophecies were being revealed became palpable. Many of them discussed encounters with Lamanites as if those indigenous peoples had stepped right out of the pages of the Book of Mormon. This was part of a much broader, and deeper, engagement with American Orientalism.[20]

I shall next turn to the fledgling community of "the Vegas," as Las Vegas was then often called. Although one tends to associate the city with modern gambling, it was an important point in the developing Mormon empire in the mid-1850s, at which time Mormons were actively laying out farming villages and even developing iron mines in nearby southwestern Utah. In 1855, the Vegas was a spot on the Old Spanish Trail but was basically only an outpost situated at a dependable spring until the Mormons arrived. As they sought to build a theocratic empire not unlike "the good land" in Deuteronomy 8—which is described as a place rich in diverse resources such as water, vegetation, and metal ores—the Mormons were aware that water and forage as well as strategic mineral resources were needed for them to create a self-sufficient empire, and the Vegas seemed to possess all of these. Its spring-watered pastureland stood out as emerald green patches in a buff-colored desert landscape punctuated by nearly bare, rocky hills. In addition to recognizing the site's potential as farming and ranching land, the early Mormons were also aware of reports about galena (lead sulfide) deposits located in the hills north of the Vegas. Furthermore, the native inhabitants were also of real interest as potential converts to the Mormon religion.

Seen in this light, the first known map of Las Vegas, a sketch by Scottish-born Mormon pioneer and cartographer Thomas D. Brown (1807–1874),

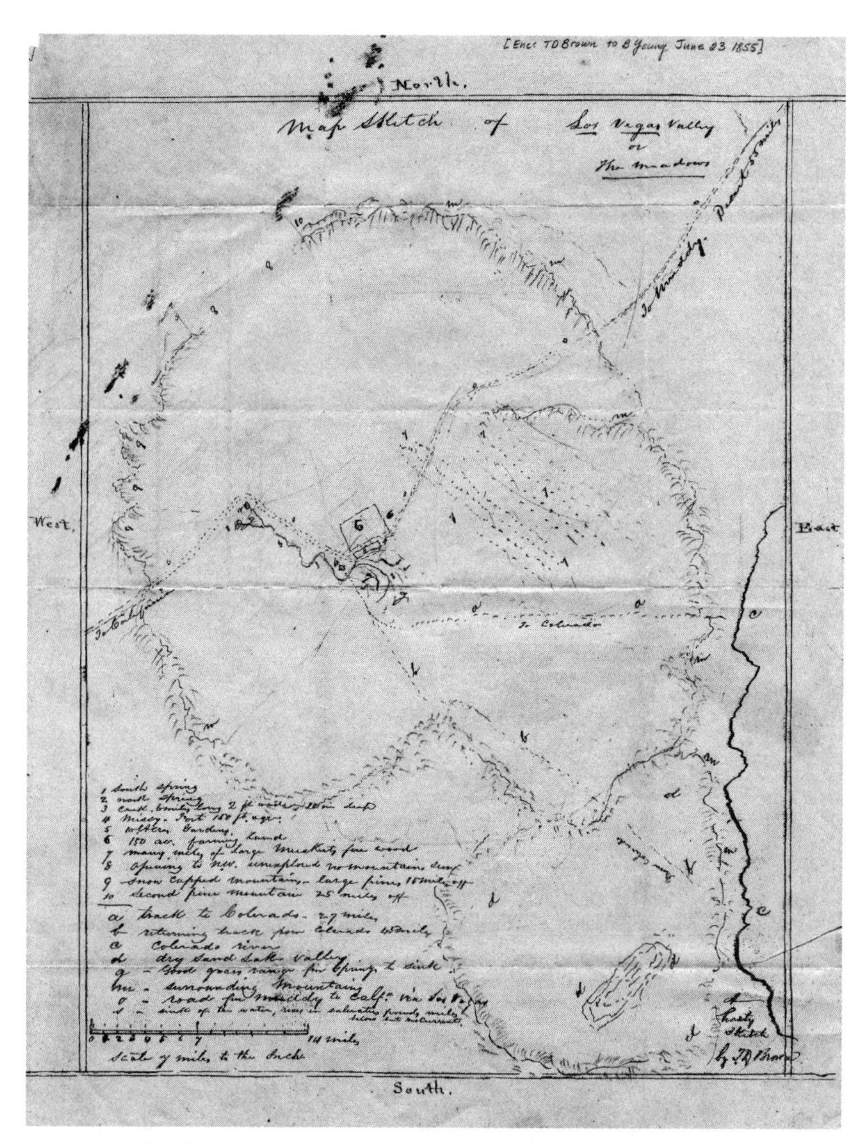

FIGURE 3. Made in the summer of 1855 and oriented with north at the top (as we have come to expect), Thomas Brown's Map Sketch of the Las Vegas Valley or the Meadows is the first known map of the city that began as a stop along the Old Spanish Trail. Brown modestly calls this map "a hasty sketch," but it enabled church leaders to better understand the site and its position near the Colorado River, which meanders along the right-hand margin.

is highly instructive (see figure 3). Although Brown's map is dated 1855, his activity in the southern edge of Mormon country began a year earlier. In the spring of 1854, Brown had been appointed by church leaders to explore the landscape and native inhabitants as part of the Southern Indian Mission. Keeping a detailed journal (as did many Mormons), Brown traversed the rugged country along today's Utah-Arizona border, believing he was gazing into the pages of Book of Mormon. At one point, he wondered aloud as to the origins of the natives here, speculating that they might be remnants of the lost tribes who, according to some Mormons, had first landed in South America."[21] To Brown and many other early Mormons, the indigenous people were part of the environment that would be colonized on behalf of the church. In addition to discussing the origins of the natives, he took notes on their handiwork. In the process, Brown drew a simple diagrammatic map of one of the dams that helped sustain the Indian communities in this arid land.

When Brown encountered the Vegas about a year later, his mapping had become more ambitious and accomplished. Given its well-watered site, the area was home to Paiute Indians as well as a prospective Mormon

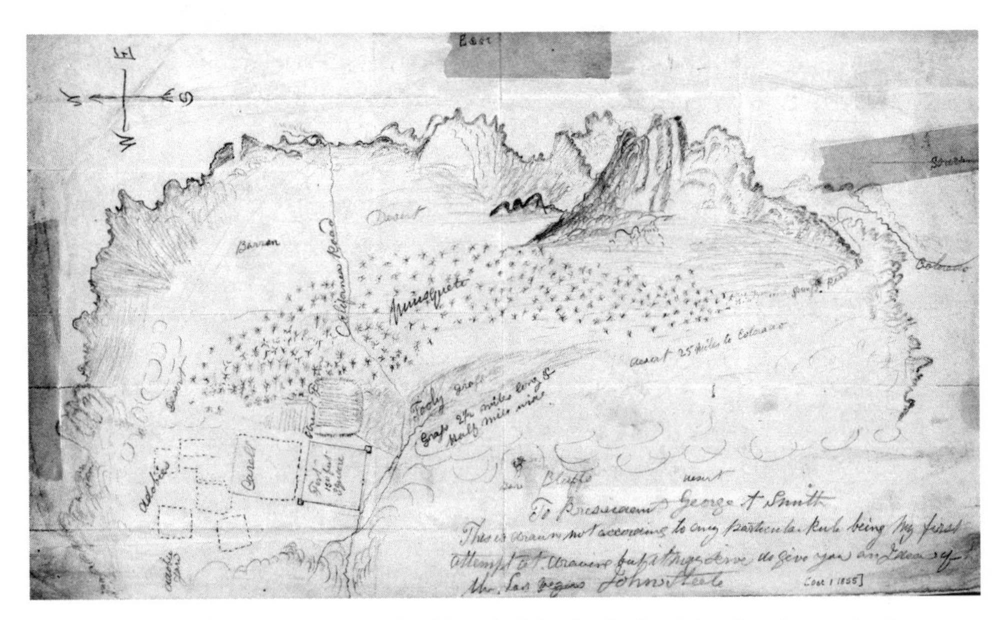

FIGURE 4. This stunning map of Las Vegas by John Steele, dated October 1855, evidently built upon Brown's map (figure 3) and was intended to give church leader George A. Smith "an idea" of the site by showing it in greater detail. Steele's eastward-oriented map also places Las Vegas in context at the foot of rugged mountains that lie adjacent to the Colorado River (near the right-hand margin).

settlement. With modesty reminiscent of Phelps, Brown characterized his map of the Vegas "a hasty sketch," but it is much more. On it, Brown depicted irrigated areas and buildings of importance. Like other Mormon maps of the period, this one had strategic value, for it represented an exchange of information in the field to church authorities headquartered in Salt Lake City.

Another map of Las Vegas prepared with much the same purpose in mind was drafted by John Steele (1821–1904) that same year (see figure 4). Born in Ireland and growing up to become part soothsayer and part practicing doctor, Steele was also a talented if home-trained mapmaker. In one insightful passage Steele noted that he was "a pret[t]y fare [*sic*] hand at whatever I undertook."[22] That adaptability characterizes many Mormon mapmakers. Steele represented that classic combination of scientific curiosity and religious faith that was common in the mid-nineteenth century. The Mormons seem to exhibit this combination of romanticism and pragmatism with great ease and frequency, but Steele may have epitomized it. By 1870, Steele could call himself a "Dr. & surveyor"—a designation that hints at his breadth of interests; however, he was far more than that, including a mystic. An article titled "John Steele: Medicine Man, Magician, and Mormon Patriarch" astutely notes that "because John Steele saw science and theology as united, he could subscribe to *Raphael's Prophetic Almanac* at the same time he was soliciting subscriptions for *Scientific American*."[23] The point to recall here is that Steele's map embodied that same combination of traits.

We have seen that unique combination of spiritual mysticism and practical mapmaking before—Phelps possessed some of it—but Steele exemplifies its near perfect fusion. Steele's mapmaking was part of the Mormon tradition but stands out as highly original. Prepared a few months after Brown's map, Steele's map deserves some interpretation. First, because it represents much the same area in a similar style, it is likely that Steele patterned his map after Brown's. They are both oriented similarly and use similar topographic features as map boundaries. Although both maps are technically competent and accurate enough to serve their purpose admirably, Steele's map represents a truly remarkable combination of scientific discipline and imaginative interpretation. Note how competently he renders the topography using a technique far more refined than Phelps's, and more enthusiastic than Brown's. By so doing, Steele effectively creates a sense of drama.

Steele's use of perspective, a partial bird's eye view supplemented by shading, enables him to bring the place to life in a remarkable way in this truly seminal, promotional map of a western city. It not only documents what is there but also conveys the "feel" of the locale. Steele himself had this in mind, for this map is a redrafting of one he made in the field. Although that earlier map is rather more prosaic, the one under discussion here was drawn specifically to enable Apostle George A. Smith to "see" the site of Las Vegas, as Steele noted in a letter to Smith dated October 1, 1855. On the map itself, Steele notes that it can give the user an "idea" of the place. This, in retrospect, was the perfect word, for "idea" could mean many things, including not only a representation of something but also a formulated thought or opinion. In other words, the map could both inform as well as motivate the map user. As I noted in *The Mapmakers of New Zion*, Steele's map was one of three drawn at about the same time. Two of these maps might look similar at first glance, but there are real differences between them. One—clearly the first—was evidently drawn as a draft that served as the basis for the second, much more confidently drawn, map. The third map, a postcard-sized plat of only a portion of the other two maps, was meant to serve as a blueprint for the land tenure. This third map is far less impressive, though it too served a purpose when it was enclosed in a letter to George A. Smith. In its careful articulation of the pattern of lots, it confirmed that the Mormons were branding their characteristic rectangular order onto the wild landscape, and thus making it their own.

Steele's maps of Las Vegas resonate today, for we have become used to seeing topography exaggerated in the vertical scale to convey the lay of the land in rugged areas. As I became impressed with Steele's map, I wondered if others might be similarly impressed. That question was answered when I showed it to my two grandsons without saying a word as they were helping me prepare a PowerPoint presentation on Mormon maps. Upon seeing the Steele map, fifteen-year-old Alex exclaimed "Wow!" When his similarly computer savvy seventeen-year-old brother Nick, who heard Alex's comment, came over to the computer, he exclaimed "Whoa! That's a wild map!" I had thought Steele was decades ahead of his time, but now believe that he may have been a century and a half ahead. Like many of their peers, these teenagers are expert video game players and admired how Steele's map had caught their attention and conveyed a sense of place. As Alex observed,

Steele's map "draws you into the place." Nick agreed, adding that "you could almost picture yourself moving through that setting." I could not help but notice that the Steele map was similar to the stylized locales the grandchildren navigated on their adventure-themed video games, where one has to scurry across ravines, up and down hills, and through patches of vegetation as they chased, or were chased by, nefarious characters.

By 1857, the Saints found themselves threatened by menacing federal forces and compelled to flee farther into the western interior. Their practice of polygamy was an issue, as was their belief in theocratic democracy. At this time, a budding Mormon surveyor named James H. Martineau

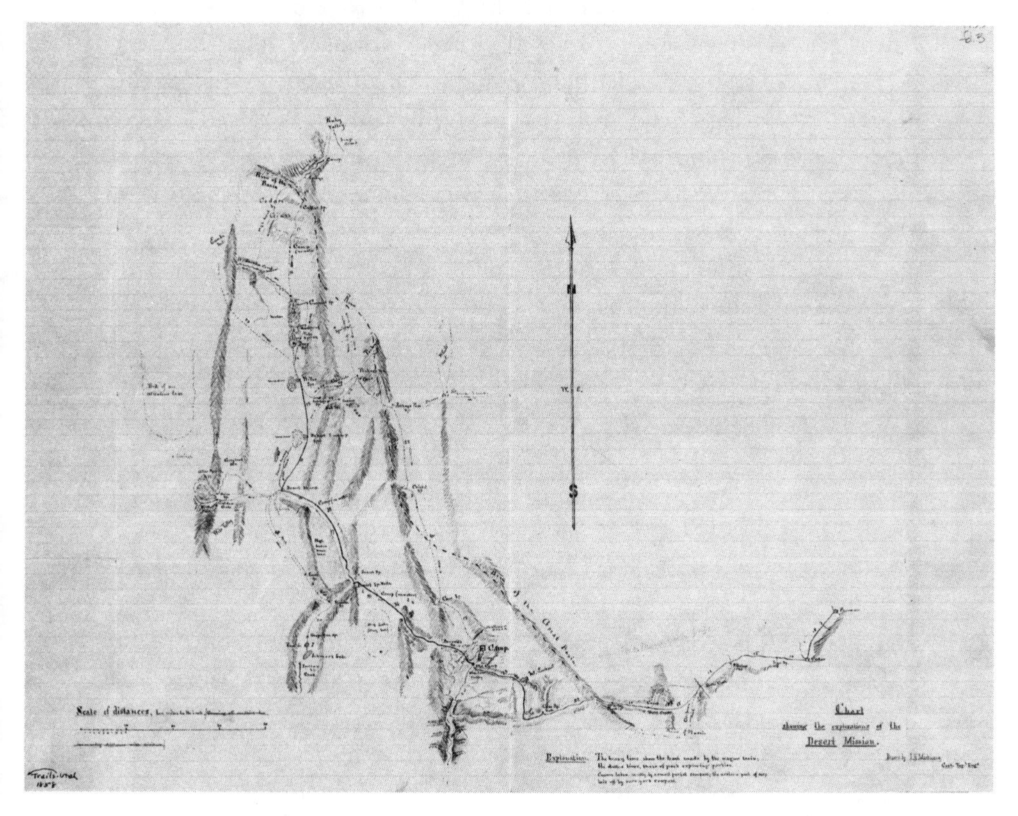

FIGURE 5. James H. Martineau's Chart Showing the Exploration of the Desert Mission (1858) was made during the expedition in which Mormons sought a new, safer sanctuary in present-day Nevada, far from federal authorities in Salt Lake City. At this time, Martineau was quickly learning surveying techniques and would become one of the most important Mormon mapmakers, drafting plat maps of Mormon communities in varied locales—including Utah and Arizona—as well as preparing railroad survey maps.

(1828–1921), reached the status of accomplished Mormon mapmaker.[24] Martineau was born in New York, but would shape Utah and the West. As revealed in his account of the White Mountain expedition, which searched for new lands to settle far from federal authorities, Martineau drafted several maps that helped church leader Brigham Young and other Mormon leaders better understand the countryside near what would later become the Nevada-Utah border. Of special interest is Martineau's map of the mission to the southwestern deserts (see figure 5). Building on his knowledge of the land that he had helped survey in the vicinity of Parowan and other Utah towns near Cedar City in the early 1850s, Martineau now added considerable information to a master map of the area that was virtually unknown to either Mormon or federal expeditions. Unlike Phelps's earlier map made from Mt. Nebo—and for that matter John Steele's maps of Las Vegas—Martineau's is more accurate. Its purpose of guiding the Saints to sanctuary is evident as it provides a blueprint for how to access this remote area—information that Brigham Young hoped to keep from antagonists.

Always scientific but also poetic and highly reverent, Martineau conveys no doubt about the role of religion in sustaining this venture. His accompanying report is replete with poems and hymns praising God, whose presence he feels in the locales that he maps. In his personal journal, Martineau included the words of a popular hymn titled "For the Strength of the Hills We Bless Thee" in recognition of the Mormons finding a haven from persecution in the mountains far to the southwest of the Great Salt Lake. The country here consists of fault block mountains of light-colored Paleozoic limestone that rise about five thousand feet above the sparsely vegetated desert valleys. These mountains are honeycombed with caves and pine-covered at their summits. Some of the tallest peaks in the eastern Great Basin even reach above the timberline, which is lashed by strong seasonal winds. It is striking country indeed, and it mightily impressed Martineau. With that in mind, the first stanza of the hymn is especially revealing:

For the dark resounding caverns
Where thy still small voice is heard;
For the strong pines of the forest
That by the breath are stirred;
For the storms on whose free pinions

Thy spirit walks abroad.
For the strength of the hills we bless Thee,
Our God, our father's God.[25]

In addition to conveying these strongly religious sentiments, Martineau was also scientifically inclined and his journal is packed with astute observations about natural history. Science, however, was only part of the equation: an even bigger part was spirituality. His journal entries often include prayers and shed additional light on the ethereal process of mapmaking. For example, while triangulating to determine positions using various mountains, Martineau noted a change in the weather, "leaving me in bright sunshine with the clouds below me like a vast illimitable ocean; the mountain peaks rising through them resembled islands . . . and at length a hole appeared in the cloud through which I could see the earth." As Martineau put it, "I seemed to be on another planet, and had the strangest feelings, until the cloud cleared away." To his joy, Martineau added that "below me lay the [Great Salt] lake—in fact—hundreds of square miles were spread out like a map."[26] In this endeavor, Martineau knew he was helping Brigham Young realize a dream of bringing the transcontinental railroad to Utah—a feat that would ultimately facilitate the arrival of thousands more Latter-day Saints to their "gathering" in the West. This may come as a surprise to many people who assumed that the Mormons dreaded the coming of the railroad; some may have expressed that fear, but only until they heard Young preach sermons in favor of it. For his part, Martineau regarded his stint with the transcontinental railroad as one of his more important accomplishments. Ever aware of the disdain that many non-Mormons had for his people, Martineau was both pleased and surprised when the non-Mormons praised his surveying work for the Union Pacific Railroad.[27]

At this point I would like to return to John Steele, who had more mapping in store in his service to the church. By the 1860s and 1870s, the Mormon West was a fact of life even though the federal government had lingering concerns about their polygamy, dubious loyalty, and potential for sedition. Undaunted, the Mormons spread their gospel throughout the West and were in fact spreading it worldwide at the same time. In filling in the blanks on maps of the West, their expeditions into southern Nevada and northern Arizona were instructive. These explorations

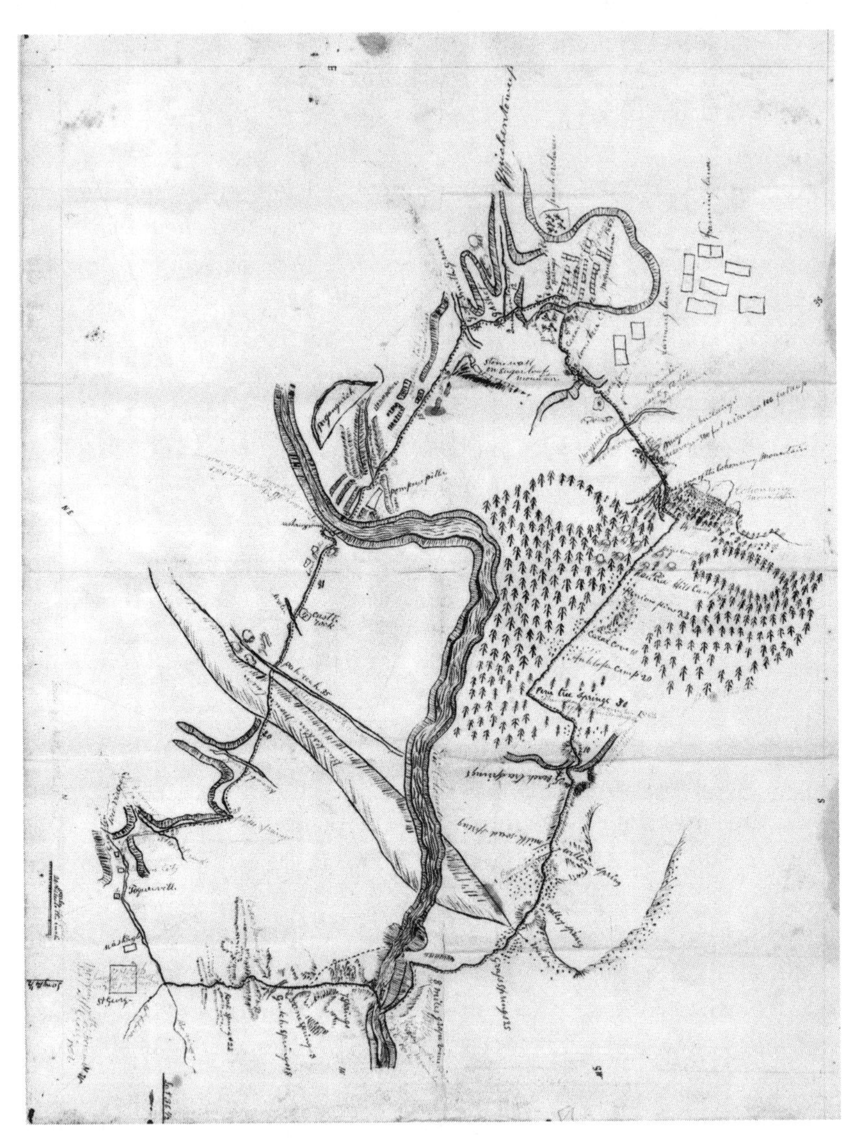

FIGURE 6. In his 1863 map designed to help Mormon missionaries from Utah reach the Moqui (Hopi) villages of Arizona, John Steele prepared what may very well be the first vegetation map of this part of the Southwest. In addition to the stylized pine tree symbols depicting timber, Steele also shows numerous springs and Indian villages as well as the prominent Colorado River drainage.

were well underway by the early 1860s and lasted through the 1870s, when Mormons finally began to settle the Little Colorado River country in hopes of actually involving Moqui Indians as settlers. That did not happen exactly as planned, but exploration of the Little Colorado River country and its Hopi settlements were prerequisites to both Indian missionary activity and later Mormon Anglo-American settlement of the same area. It is here that John Steele again emerges as regional mapmaker of considerable talent. Steele's map of the convoluted, labyrinthine topography encountered in the Colorado Plateau is noteworthy (see figure 6).

It was in this particularly difficult terrain that Mormon missionaries sought converts among the Moqui and other tribes. The challenge was to lay out a route by which other missionaries could reach the villages using southern Utah as a jumping off point. Steele's map of a route to these Arizona villages not only diagrams the major trails and river crossings, but also delineates the vegetation. Given the fact that water and wood were two of the resources that could sustain the missions, Steele's map is more than a statement of faith. It is also a shrewd inventory of the environment. In addition to delineating the hydrology, it uses stylized vegetation symbols to indicate the location of pines and other trees. This, it should be noted, may very well be the first vegetation map of Arizona, and Steele executed it with his peculiar combination of scientific observation and imaginative creativity. This map gives one the impression of soaring high above the landscape, and from that point being able to discern patterns that are not visible to those on the ground. It is simultaneously a navigational device, for it shows routes of travel, as well as a thematic map that depicts the distribution of whatever phenomena the mapmaker deems important—in this case coniferous trees. On Steele's map, and in the landscape itself, these forested lands stand like islands surrounded by a sea of sagebrush-covered open space.

Clearly, the maps drafted by Steele and many other Mormons differed from maps of the same period drawn for the U.S. government surveys. The main question here, though, is why? First, it should be noted that many Mormon mapmakers lacked formal training of the kind that U.S. military and scientific expeditions possessed. This, in part, may explain the vernacular, and sometimes quite innovative ways in which Mormons mapped what they encountered. Although several of the Mormon mapmakers were competent local surveyors who helped lay out villages (often in direct competition with federal surveyors), they did not hesitate to tackle larger areas on behalf of the church using a blend of pragmatism and faith. This lack of formal training may have been problematical at times, but it proved a boon to cartographic historians who now encounter a diverse set of highly original maps. The variations in style and technique are remarkable. They range from the competent and very disciplined cartography of Thomas Bullock and James Martineau, who could lay out communities with considerable precision, to the maps by W. W. Phelps and John Steele, which had an undeniable flair perfectly in keeping with their flamboyant personalities and outspoken religious zeal.

The maps discussed above hint at the power of creative expression to capture the character of place generally on behalf of religion. However, one last question remains: how can we be sure that these maps actually reflect religious belief rather than, say, artistic imagination? This is a crucial question that gets to the heart of the matter. In answering it, I should first note that very few mapmakers ever tell us what they are feeling when they make maps. Even Joseph Smith himself, who famously provided guidance on how to lay out the ideal "city of Zion" using a map, never claimed that God had commanded him to do so. Nevertheless, as historians contextualize Smith's map and his other writings, they concede that it was indeed inspired in the broadest sense of the word. We need only recall Smith's claim that the plat of Kirtland (Ohio) was the result of a revelation, and so it appears reasonable that the more abstract version of that plan—the city of Zion—was also revealed, broadly speaking. In a sense, *everything* Smith did at this time was spiritual rather than purely practical, and thus can be viewed in the context of his passionate religious inspiration dictating the way Saints would live and find eternal glory on earth and in the hereafter. Similarly, rank-and-file Latter-day Saints have long believed that the plats of their cities in the American West, from Salt Lake City to the smallest Mormon farm village, were a result of divine inspiration. In retrospect, the maps that charted how those places would materialize can be considered a tangible aspect of that spiritual inspiration. By putting the inspiration on paper, these maps became as much a part of Mormon material culture as the places they represented. If, as I have suggested, the intense spirituality of these Mormons mapmakers was a factor that enabled them to emotionalize the landscape and capture it innovatively, then that can be added to the long list of things that made Mormonism—and the new homeland that it created in the American West—so distinctive.

AMERICAN ZION

Mormon Culture and the Creation of a National Park

Betsy Gaines Quammen

WELL-LOVED ZION NATIONAL PARK (ZNP), established in 1919, is an anomaly in the history of land conservation in southern Utah. Set within a region that has become notorious for burnings in effigy, vandalism, death threats, and armed confrontations over public land issues, Utah's first national park came into being through a largely noncontroversial collaboration between locals and federal players.[1] The reasons behind this united support in developing the park are multifold. Many local Mormons dreamed of economic prospects brought about through tourism and became swept up in the patriotic notion of the national park. Others felt pride in the recognition of beauty within the geological wonderlands that marked their homeland. Federal officials also went above and beyond protocols to invest in the priorities of the local Mormon community in addition to engaging locals in park planning and employment opportunities. Although some locals were affected by federal resource restrictions that came with the new national status (first as national monument, then as national park), even those who lost the ability to use the lands within the park helped transform the region from remote Mormon settlement to a place of leisure, ease, and shared space. In the imagining of and investing in the creation of ZNP, a Mormon sensibility became reflected in relationships with the landscape, as local actors imprinted their own values on the park. ZNP reflected aspects of Mormon homeland, not like the place Joseph Smith founded in Jackson County, Missouri, but as an idea brought with the Latter-day Saints (LDS) into the Great Basin. This park, named Zion, became a space that encompassed Mormon cultural priorities and stood alongside America's romance with "wild" landscapes.

There are aspects to the story of ZNP that played to Mormon cultural sensibility. Most obviously, the word Zion, in addition to other names of regional geographical features, embodied Mormon notions of the sacred. Additionally, in the making of the park and other tourist opportunities, development of the landscape helped to enshrine Mormon priorities such as new roads atop old Mormon wagon trails. With the changes to the land and the influx of tourists, local Mormons reevaluated their region, borrowing from broader American values in regards to wild land, even while incorporating their own conceptions of the sacred. Regional Mormons also participated in rituals on the landscape, such as the Easter pageant, in their efforts to sacralize the park. They also came to embrace outdoor activity as a spiritual standard—as practiced by the Beehive Girls. ZNP represents a place of cultural cross-pollination, where the early twentieth-century American ideal of "wild nature" blended with a Mormon appreciation of developed space.

Romantic, spiritual stirrings among visitors to America's first national parks in the early twentieth century have been detailed in the works of Roderick Nash and Alfred Runte.[2] But little literature exists on the Mormon impressions of national parks in the early part of the twentieth century. The history of ZNP is one of mutual interest and investment between the federal government and local Saints. Together these parties built a place in the canyonlands designated solely for human enjoyment, not resource exploitation—the very idea of which ran counter to traditional LDS culture.

Before the establishment of ZNP, the region was already a unique space, a land of Anasazi, Southern Paiute, and Mormon cultural layers across labyrinths of polished rock. A Mormon homesteader named a remarkable feature Zion Canyon (the namesake of ZNP) before the land was "discovered" by the federal government. Following this discovery, and subsequent national enthusiasm over the region's beauty, there came a period of major restructuring of land. As boundaries were drawn for a federal conservation effort, locals lost land use privileges and this place shifted from pasture to playground. The transition led to new impressions of the canyon as visitors, both Mormon and non-Mormon, approached the park in pursuit of fun, not subsistence.

In this rearranging and reconceiving, marketing brochures for the tourism industry captured the imaginations of would-be visitors with flowery prose and breathless musings on sacred handiwork in nature. Some materials invited sightseers to see God in the rocks. Within these pamphlets, adventure was peddled to urban Americans with an interest in engagement with the wild. Roads were built to bring in crowds. And with these visitors, American enthusiasm for nature began to work on the imaginations of local Mormon people. Many Saints joined non-Mormons in the park to camp, fish, and play; however, lots of locals remained apathetic or unconvinced of the park's merit. To entice these skeptics, a religious rite was created within the canyon walls—a production so extraordinary and ambitious that it continues to make ZNP very distinctive in the history of American national parks. This was the annual Easter pageant, a short-lived but significant enterprise set within majestic rocks that attracted thousands of people to witness the staged resurrection of Jesus.

ZNP's story is a unique facet of Mormon environmental history. It is a story of interwoven narratives that knit together the priorities and imaginings of both Mormon and non-Mormon visitors. ZNP is a great American institution built within Mormon culture.

Zion Canyon is a geological feature that represents an amalgamation of human historical use and religious perspectives. Cultures became intertwined in the mid-1850s, when Brigham Young sent families south from the Salt Lake Valley to convert the Southern Paiute to Mormonism and to launch their Dixie Mission, an initiative to grow cotton, among other crops, along the Virgin River. This LDS settlement effort encompassed northern Arizona, southern Utah, and parts of Nevada. Here they pursued livelihoods very much at odds with the indigenous people already occupying the region.

Before the arrival of the Mormons, migratory bands of Paiute moved seasonally from place to place, hunting and gathering to make their living on what they viewed as their own sacred homelands.[3] To the newly arrived Mormons, this was their hallowed land, earned through struggle, providence, and prophecy.[4] These homesteaders settled a territory along the Virgin River, whose dusty waters carved the sinuous red walls of red rock country, including Zion Canyon. In this land of devastating floods, scabby pastures, insect infestations, and drought, the Mormons set about manifesting Zion, making the desert bloom through irrigation, cultivation, and development. As the Dixie Mission members sought to establish Zion, they did so on top of Paiute motherland.

Mormons had lived in southern Utah for a little less than two decades by the time the federal government sent an exploration team to map the Colorado River watershed. John Wesley Powell (1834–1902), an officer with the U.S. Geological Service, came to survey the Virgin River, leading his team on a quest through rivers, rocks, and canyons that ran adjacent Mormon domesticity. To Powell and his men, the lands of Dixie were nothing short of a wonderland, a sentiment evident in the accounts of expedition members Clarence E. Dutton (1841–1912) and Frederick Dellenbaugh (1853–1935). Both men's stirring words and Dellenbaugh's exquisite paintings helped inspire the designation of ZNP almost fifty years after the men first laid eyes on Utah's geography.[5] Powell called one of the most notable canyons Mukuntuweap, a term he said was the Paiute name of the place. It is unclear where this word actually came from and there remains no sharp translation.[6] The local Mormons called the canyon Zion or Little Zion Canyon.

The Mormons that Powell encountered, like many other American pioneers who headed west to settle, were interested mainly in the landscape as a place to produce food to feed their families. In settling southern Utah, these families planted orchards, cotton, and even vineyards while raising livestock. The value of the land was in its ability to produce, not in its dramatic scenery. By contrast, some members of Powell's party, influenced by a Thoreauvian spirit, came to regard this region as special due to its "wilderness" attributes. An American romance with wild places, influenced in part by the writings of Henry David Thoreau and Ralph Waldo Emerson, came into vogue as the western frontier became increasingly settled and urban areas more densely populated.[7] With the disappearance of uncultivated and unsettled lands, broader American attitudes on nature changed, and a movement based on the yearning for places of untamed wild emerged. Nature was seen as more wholesome than urban areas and became infused with spiritual significance. This value was not universally shared by the Dixie Saints. To the Mormons, with their history of communal settlement in marginal landscapes, cultivated places offering community and safe haven were preferable to wilderness.

But something so extraordinary about this region was the way these two sentiments came to reside together. Isaac Behunin (1803–1881), an early Mormon settler who felt safety within the steep red walls of the upper Virgin River, named Zion Canyon in 1863. According to family tradition, "Isaac was sitting with his neighbors one evening talking about the grandeur of their surroundings. He said a man can worship God among these great cathedrals as well as he can in any man-made church in Zion, the biblical heavenly 'City of God.'"[8] Safe within the walls, Behunin could worship his own way, without threat or the harsh judgment of others. He had followed Joseph Smith from New York to the Midwest, and Brigham Young to Utah. Finally, after thirty years of searching for a place to settle and practice his religious beliefs in peace, Behunin found Zion Canyon, a place of beauty and a God-given retreat. Behunin's perception of Zion was celebrated by Powell's team member, geologist Clarence Dutton, in the oft-quoted passage, "Nothing exceeds the wondrous beauty of Little Zion Valley. . . . No wonder the fierce Mormon zealot, who named it, was reminded by Great Zion, on which his fervid thoughts were bent—'of houses not built with hands, eternal in the heavens.'"[9] To Behunin, the canyon was impressive, but more importantly, it was far from the angry mobs that had chased the Saints from Daviess County, Missouri, in 1838 and in 1844 killed his prophet in Nauvoo, Illinois.[10] Dutton's impressions, on the other hand, were romantically hued; the canyon was untrammeled nature just as God had made it and divine in its natural splendor. It was the wild beauty that was the work of God. In the walls of Zion Canyon, the beauty of the wild, a sense of the sacred, and the security of homeland overlapped in mutual admiration.

In addition to providing refuge, Behunin and his Mormon neighbors relied on this place and the ability of the land to produce. The Gentiles (non-Mormons) who visited the region, drawn by spectacular geography, also acknowledged loveliness in Mormon agrarian domesticity. The rocks were not the only sight to behold. One member of Powell's party, artist Fredrick Dellenbaugh, wrote in an article in *Scribner's Magazine* about the aesthetics of Mormon Zion within the villages and farms in the foreground of the red rock:

> The Mormons being apt masters in irrigation, the land contains a number of districts that, by contrast with the surroundings, rival the Garden of Eden. Here grapes, peaches, almonds, figs, pomegranates, etc., are yielded in abundance. . . . imagine meadows, farms and shady brooks to be a mere phantasmagoria—when lo! the magic turn of the road reveals a sweep of emerald with ditches of dashing water plume like poplars of Lombardy, fan spreading cottonwoods, vineyards, roses, peach and apple orchards, fig-trees and all the surroundings of comfortable country life.[11]

The cultivation practices of Dixie put "green fields and foliage" and "bright oases" in the desert region of the Great Basin.[12] Though land and soil proved insufficient for large cotton production and the last mill closed in 1910, Mormon settlers created sophisticated irrigation systems and made the arid land productive.[13] In the era of Mormon settlement, featuring communal work and ongoing encouragement from the church, the region produced enough crops to export to Mormon settlements beyond the boundaries of Dixie.[14] Over decades and differing Church policies, the region yielded great quantities of fruit, wine, castor oil, indigo, tobacco, sugar, olive oil, and almonds among other crops.[15] Mormon orchards and orderly towns were admired, not just by Dellenbaugh, but by many writers including Wallace Stegner, Ed Abbey, and explorer Richard Burton.

As the Mormon pioneers worked the dust of Dixie, the American conservation movement was emerging and with it came the national park system.[16] By the turn of the twentieth century, Zion Canyon and the region's other rocky marvels garnered the interest of the federal government beyond mapping its contours.

In addition to the Mormon imprint on the land, the Paiute Indians had also used the region, including the development of irrigation for squash and other foodstuffs during years when natural food sources were scarce. But their imprint was marginalized in the making of ZNP. Mark David Spence detailed the erasing of Native American culture from the American national park narrative and the reduction of their presence to that of "visitor"—a form of denialism, conscious or not, that helped create the sense that parks are part of an unpeopled land.[17] Known in its first iteration of federally protected status as Mukuntuweap National Monument, the place later would be called Zion National Park. The supposedly Paiute word "Mukuntuweap" was deleted from the landscape, just like much of the history of the tribe's historical use.

In 1916, the National Park Service Organic Act created an agency to manage the growing number of national parks. The first two officials who ran the fledgling operation, Director Stephen Mather (1867–1930) and Assistant Director Horace Albright (1890–1987), invested a great deal of time in the Dixie region and worked locally to spread excitement over new tourist destinations. Albright was the first park service employee to visit Zion Canyon, in 1917. At the time, the road from Cedar City to Zion Canyon was still in terrible shape. Although Albright was very impressed with the canyon and would indeed recommend it for national park status on his return to Washington, D.C., he knew the region was still very difficult to access and needed major roadwork in order to open up the isolated region to an eager American public. [18]

Stephen Mather had already created relationships with railroads interested in benefitting from tourism to national parks—the Union Pacific and Santa Fe together advanced over a million dollars in 1915 for the development and marketing of national parks.[19] The land of the Mormon homesteaders had been bony and hard-won, but if people were going to come see the Grand Canyon, ZNP, Bryce Canyon, and other scenic destinations, an accessible southern Utah-Arizona tourist circuit had to be carved into tough country. In other words, this region needed to shift from a place of hardscrabble existence to a place of comfort and convenience. And in doing so, the place had to be built—and it had to be reorganized.

As Zion Canyon became an American park, local farmers and ranchers, as well as errant livestock, were removed from within its boundaries. The last Mormon family, the Crawfords, left the canyon floor in 1931. The displacement of people and the federal curtailment of land use are detailed in Karl Jacoby's work on the trend of dislocation, restrictions, and hunting bans wrought by the government and the American conservation movement.[20] Restrictions may have disconcerted some locals, not only because an unutilized landscape seemed to them a squandered opportunity, but also because Zion Canyon could no longer be used to move livestock from the upper canyon to the lower section. As Mark Spence pointed out, "what tourists, government officials, and environmentalists fail to remember is that uninhabited landscapes had to be created."[21] The park model was of a place unpeopled, a principle that impacted both indigenous people and the Mormons.

As documented in the ZNP archive, locals lobbied Mather and Albright to make exceptions for their sheep and cattle. Yet in spite of the park being created and the subsequently imposed federal restrictions on land use, local sources from the time (interviews, journals, letters, and media) reflect the fact that Mormon residents around ZNP, even if inconvenienced with federal land use restrictions, coalesced around the federal plans. The promise of economic boons, the glow of nationalism, and the warmth of state pride swayed the local communities to rally behind park designation.[22] But just as importantly, as they embraced this new idea of an uncultivated tourist attraction, locals devised several ways to make ZNP reflect Mormon sensibilities. Roads, rockwork (think embellished landscape vs. wild) and ritual all helped resacralize ZNP within a Mormon worldview. Zion had provided refuge for decades before the tourists arrived, and would again be reimagined with a Mormon perspective amidst the onslaught of dudes and sagebrushers.[23] As hotels and chauffeured automobiles replaced plows and livestock, as park infrastructure improvements brought electricity and telephones, pampered visitors dined in catered camps and heard amusing stories around the campfire. This parched corner of Utah became a new vision for visitors and locals to behold, fostering a new relationship between residents and homeland.

In his essay, "The Paradox of the Park," David Quammen discussed the "paradox of the cultivated wild" in Yellowstone National Park. This premise stemmed from contradictory notions of wilderness as untouched hinterland vs. a developed park for entertaining tourists. Quammen's argument on Yellowstone focused mainly on managing wildlife within a realm developed for humans.[24] ZNP too is a study in the paradox of the cultivated wild, though there aren't the same wildlife issues to manage. The paradox in Zion lies in the development of backcountry. Like many national parks, tourists can access some of the most dangerous and thrilling adventures in the region by driving to them. ZNP makes a desert canyon quest convenient.

In looking at how Mormons viewed sacred landscape, the historian Jedediah Rogers argues that southern Utah residents viewed roads as "expressions of ideology" and that the building of routes became an extension of Mormon dominion.[25] Through settlement in southern Utah, Mormons moved through the land only in arduous forays. Early access to Dixie was only accomplished through innovation and struggle. Paths, trails, and roads had a special meaning to the pioneer Saints—they were symbols of ingenuity that their ancestors carved into stretches of impenetrable country. The early cattle trails in Zion Canyon were so steep that cows and horses sometimes toppled to their deaths during seasonal drives from upper rangeland to valley floor.[26] Road and trail improvements served to enshrine the efforts of earlier generations who pushed into wild depths through ingenuity, perseverance, and patience.

Part of the push to open up ZNP came with the park's Zion–Mt. Carmel tunnel. When the park was designated in 1919, a trip from Arizona's Grand Canyon to ZNP, Bryce, and Cedar Breaks was inconvenient to say the least. In 1921, State Road Commission engineer Howard Means (1875–1951) was tasked with finding viable routes that would connect ZNP with the north rim of the Grand Canyon and Bryce Canyon. After numerous trips to southern Utah, hours studying maps, and many interviews with "old-time settlers," Means encountered John Winder, a rancher who had improved the old Paiute trail that ran from the valley floor to the east rim of Zion Canyon. Although Winder had never been east of the country between Springdale and Mt. Carmel (south and east of the park, respectively), "he had been into the country north of there and knew something of the terrain."[27] Together Means and Winder located a route that would come into being by blowing holes through the upper reaches of Zion Canyon.[28]

Former cattle country was soon adorned with bridges, arches, and roads as the park opened and regional automobile clubs and the Union Pacific Railroad continued to sell nature and draw crowds. The following passage, written for the U.S. Railroad Administration, conveyed an enraptured sense of place that rivaled the wonders of the world: "Zion Canyon is an epic, written by Mother Nature in her most ecstatic humor, illustrated by Creation in its most majestic manifestations, published by God Almighty as an inspiration to all mankind. I had become almost familiar with the miracles of Zion . . . it gives to one . . . a more profound impression of the wonderworks of God."[29]A trip to ZNP was marketed as transformative, spiritual, and even epiphanic. This was adventure upon hallowed ground. Drive, hike, camp, sightsee, and—while you're at it—meet God! In promotional materials, there were also invitations asking the would-be tourists to imagine their own interpretations of the sacred, adding to layers of cultural imaginings. As one brochure asserted: "It is a place to drink in beauty, to form new conceptions of the divine."[30] It echoed a sentiment expressed by Dutton and Behunin—that evidence of the celestial was found in the scenery.

Though Isaac Behunin saw glory in the rocks, other Mormon residents needed outside admirers to remind them of the beauty of their homeland. Lola Belle DeMille Bryner (1901–1974) was born in Rockville and baptized in the Virgin River. In an interview, she recalled being a child in a beautiful valley that was "untouched by civilization," busying herself bird-watching and flower-gathering. She said: "Our rides to Springdale were especially beautiful because we were going towards Zion Canyon. I didn't realize as a child how beautiful the ledges were until later when so many people came to see. Thousands of people travel through there each year to see those beautiful mountains of Zion."[31] Recounting her childhood in an interview given almost fifty years after the park's opening, Bryner made her observations through the lens of tourists. She grew up near the canyon but it was due to strangers that she finally appreciated the beauty of her home.

In spite of the growing popularity of the region and the steadily growing influx of tourists, regional admiration for the southern Utah canyon country didn't happen overnight. At nearby Bryce Canyon, also in the process of becoming federally protected, former park employee Arthur Stevens described a scouting mission to Bryce in the early 1920s. He explained in a 1982 letter:

"So impressed were we with the outstanding beauty of the region, and of Bryce Canyon in particular, that we started a movement to publicize it. Mr. J. W. Humphrey, superintendent of the (then) Sevier National Forest, was mildly amused at the idea that anything under his jurisdiction could be of national interest. The local people were indifferent."[32] Mormons, in large part, have appreciated their lands for production values, not scenery. But there were those who came to understand why their remote and rocky homeland held such allure for a world beyond. Inspired by this insight, there were locals that revisited their own backyards and reevaluated the space through culturally relevant tropes.

Maurice Cope (1885–1979), a Mormon park ranger who worked in both Bryce and Zion Canyons from 1928 to 1957, described his very original interpretations of landscape while walking through the eroding sandstone pillars of Bryce Canyon, which at the time was managed by ZNP:

> As you descend the next slope—you will follow a Comanche trail, which, mutely speaks of Indian legends and lore, and of dark skinned uncommunicative people who have inhabited this land, for ages, the brave chieftain by the trail is solemnly pointing towards the majestic Escalante mountain

> or Mt. Liahona. . . . He could tell you a wonderful story of strife and conquest, prosperity and decay, during the past thousands of years, how his forefathers left Jerusalem 600 years B.C. . . . He could describe their plights as they wondered in the desert . . . guided and instructed to build ships and landed upon this promised land . . . story of the rise and fall of two mighty nations . . . in the rise and fall of this mysterious civilization, all this is gracefully represented in the towering "Mt. Liahona."[33]

This is an interpretation based on Cope's understanding of the Book of Mormon. To see The Book of Mormon in the rocks was not unprecedented. Although Cope was not literal in his descriptions of seeing actual stories in the Book of Mormon manifested in landscape, there have been LDS literalists who attempted to prove their own history within the "archeological" evidence in southern Utah's sandstone. Dr. James E. Talmage, a professor of geology at the University of Utah, followed in the footsteps of John Wesley Powell and his men to chart regions near what would become ZNP. An 1895 entry in his journal noted that "small specimens of this stone have been brought to me: and several scientific men to whom they have been shown have pronounced them artificial production. Some of our people whose zeal for *The Book of Mormon* has actually clouded their judgment, pronounced this, as every other occurrence having any resemblance to archaic work as Nephite origin" (italics in original).[34] Unlike Talmage's example of people looking for Mormon history in the rocks, Cope was musing about regional formations and his religious beliefs. But his imaginings conveyed a cultural worldview inspired by the Book of Mormon, which he applied to landscape.

J. L. Crawford, a park amateur historian who was part of the last family to leave ZNP boundaries, mentioned perhaps a more typical Mormon approach to relating religiously to canyon country. In the prayer given each evening in Crawford's home, Crawford's father asked God to "bless the water and elements," a request to sanctify earthly resources that aided in agriculture or production.[35] Crawford recalled, "When I was very young I interpreted his—'bless the water and the elements' as 'bless the watermelons.' The Lord must have heard it too because as long as we lived at Oak Creek he grew great melon crops."[36] This sweet story demonstrated a more typical Mormon focus on environment—a place that provided fundamentals for agricultural production. For the Mormon family, the canyon's water and the elements were themselves manifestations worthy of God's attention and blessings as they put food on the table.

Maurice Cope's Mt. Liahona was never an official designation. But even if it was only Cope who recognized the mountain as Liahona, others also sacralized the landscape by giving features names with religious value. Kolob Canyon, a national monument annexed into ZNP in 1956, was named for the star Kolob, which the Saints maintained was near where God lived. ZNP's Mount Moroni was named for the angel who visited Joseph Smith. Mt. Carmel, the town just east of the park, was named after the place where the prophet Elijah challenged the priests of Baal, as described in the Bible's book of Kings. The diverse nomenclature resonated with Mormons and helped non-Mormons envision the LDS association between spirituality and place. Park officials kept many local names for formations in addition to adding other religiously influenced names to features, including Southern Paiute.

Centuries after the Anasazi and Paiute had ascribed their own labels and imaginings to this landscape, the park became a reflection of both Mormon and non-Mormon spiritual sensibilities. Fredrick Vining Fisher, a Methodist minister from Ogden, Utah, was credited with giving famous ZNP regional rock formations their biblical identities during a trip he took through the park with local guides. He named the Great White Throne, Angels Landing, and the Three Patriarchs, three gorgeous geological features that remain central attractions at ZNP. In addition to Fisher's contributions, others named the Altar of Sacrifice, Cathedral Mountain, and the Guardian Angels. The park also brought a member of the Southern Paiute Shivwitz band to help name formations. Tony Tillahash, who had been taken from his tribe in Utah and raised at the Carlisle Indian Industrial School in Pennsylvania, returned to his homeland and named, among other features, the Temple of Sinawava.[37]

As a ranger, Maurice Cope explored this regional landscape of eroding stones and mused about their religious significance, but most Mormon residents did not indulge in outdoor recreation. Leisure was not foreign to the Mormon pioneers, who had engaged in outdoor picnics, concerts, and dances throughout the history of this settlement. But the canyoneers and rock climbers who eventually flocked to ZNP differed from most local Mormons, who did not view the area with the sensibility of an outdoor enthusiast. Local William Flanigan was one exception. In 1900, he hiked through The Narrows, a section of the North Fork of the Virgin River, now a popular backcountry route in ZNP. When Flanigan attempted to go through the tight passage, many local inhabitants believed it was likely an impassible canyon with quicksand, waterfalls, and whirlpools. Even if it were navigable, the area residents had no interest in subjecting themselves to potential peril. But Flanigan was undaunted. Asking a friend to wait thirty-six hours at a rendezvous point, Flanigan requested that if he failed to show up, the friend should alert people in Rockville that he was missing. As he recorded in his journal:

> For several years now I have been trying to get someone to go down the Virgin River Narrows from the Crystal Ranch to Zion [Canyon], as no man, that we know of, has been down through them, they seem to think it would be impassible. . . . I started out at about seven or eight o'clock on the morning of June 30, 1900. I had good heavy shoes, good overalls, and I had a good exciting trip, and landed clear down to Springdale in about nine or ten hours, made the whole trip on foot.[38]

He may have accelerated the pace due to the fact that he saw mountain lion tracks and was worried he might run into a grizzly bear. The last bear had been shot in 1884, a short distance from The Narrows. Flanigan concluded that he was "not sure I could go clear through and it was a shaky place to spend the night alone. But I made it through and was thrilled."[39] Today, The Narrows is one of ZNP's most popular trips.

As tourism increased in Utah, more Mormons began to embrace recreational opportunities within their homeland. The year after ZNP's designation, locals too showed growing interest in the new tourist destination. In 1920, the St. George troop of Beehive Girls visited ZNP. This organization, founded in 1913, served as a Mormon version of the American Girl Scouts or Camp Fire Girls. Beehive Girls "endeavored to have faith, seek knowledge, honor womanhood, understand beauty and taste the sweetness of service."[40]

When the Beehive Girls visited ZNP with a visiting troop of Boy Scouts, they participated in recreational opportunities such as swimming, "hiking and enjoying the scenery."[41] In experiencing the outdoors, the Beehive Girls were following instructions to become model LDS women. Recreation and the knowledge of national parks became a Beehive Girl value in 1919. The *Beehive Girl Handbook* offered a wide variety of cultural experiences for girls aged twelve or thirteen. The first section of the 1919 handbook was devoted to religious obligations. According to the instructions, the Beehive Girl needed to practice "faith promoting" activities first and foremost. But it also, surprisingly, included an emphasis on cultivating an interest in outdoor activities. Beehive Girls were encouraged to participate in recreational activities—a reflection of the shifts taking place in Mormon culture. The handbook's prescribed pursuits included "row or paddle twenty-five miles in any six days, . . . cover twenty-five miles on snowshoes, . . . tell briefly the geological history of the geysers in Yellowstone National Park," and "tell briefly the geological history of Little Zion Canyon."[42] These goals reflected the idea that recreation, the National Park Service, and the environment had entered Mormonism as a value. LDS culture was embracing the great outdoors by instructing young females to be active as part of their development in becoming a Mormon woman. The edition of the handbook quoted above was written even before ZNP was designated, indicating that Mormon attitudes in the state of Utah began to mirror the larger American interest in nature and national parks.[43]

Recreational pursuits in ZNP continued to attract local interest. King Hendricks (1900–1970), a librarian, professor, and administrator at the former Branch Agricultural College (now Southern Utah University in Cedar City), travelled to ZNP on many occasions. In a short memoir left among his papers, he wrote of one particular fishing trip along the Virgin River in 1925. Over the course of two days, Hendricks and his party caught 143 trout. He also made observations of vegetation and rock formations, noting cougar and deer tracks as well as the presence of water ouzels and a rock wren.[44]

Hendricks's camping story captured the silly fun of friends, camping out, singing, fishing, and eating pork and beans by the campfire. In fact, his story sounded good-naturedly ordinary. But there was one reference in his piece that made his trip uniquely Mormon. While fishing and camping with his buddies, Hendricks referred to singing "Welcome, Welcome Sabbath Morning," a traditional Mormon hymn containing lyrics indicative of the Mormon settler worldview and the belief of being "other."[45]

> We are earnest in our labors. To God's kingdom we belong.
> Trials make our faith grow stronger; Truth is nobler than a crown.
> We will brave the tempest longer. Tho the world upon us frown.[46]

Although Hendricks was out having a good old American time, his boisterous group of songsters were reminding themselves of three essential

Mormon truths: their birthright, their fortitude, and their legacy of historical injustice. In spite of the cultural osmosis taking place in Zion, Mormons remained a culture apart in a region being flooded with non-Mormon American sightseers.

Though there were those who had zero interest in exploring these new tourist destinations, Hendricks and his pals represented a growing interest in recreation and park tourism among the Mormons. Utah senator Reed Smoot, one of the architects of the National Park Service, wrote of a visit to Yellowstone National Park with Horace Albright, who had gone on to become the superintendent of Yellowstone after his earlier work in the establishment of ZNP: "Thursday July 14, 1927. Supt Albright took us to breakfast and showed us the principal Geysers and wonders of Old Faithful and others. . . . We all delighted with the beauties of Jenny Lake at the foot of the Grand Tetons. . . . We met a number of bears on the way and took time to feed them."[47] Smoot later noted, "One large bear was swimming the river and we called to him, he turned and swam towards us and as soon as he was out of the water came to the Auto and I fed him some candy and he was happy."[48] This experience with the bear was the type of interaction encouraged by Superintendent Albright, who saw wildlife as entertainment rather than as ecologically significant in their own right. In an interview, writer David Quammen noted, "Albright embraced the idea that in order for the national parks, and Yellowstone in particular, to have support from the American people and from politicians, there needed to be wildlife as spectacle. And for wildlife to be available as spectacle, Horace Albright said, wildlife needs to be abundant and tame. So he sanctioned the idea that grizzly bears would be fed handouts from cars."[49] The national park was a human playground first and foremost, and wildlife was part of the romp.

Smoot's journals celebrated another new American construct, so essential to park tourism—the modern highway. During his family's vacation to Yellowstone made possible because of his Packard automobile, Smoot drove the northern stretches of Highway 89 from Logan to Idaho Falls, Idaho, then to Ashton, Idaho, and on to West Yellowstone, Montana. Smoot wrote of enjoying the improved route, an experience that his ancestors could never have imagined, but for which they no doubt would have yearned. At the end of his first full day in the park, he concluded his daily journal entry with, "All feeling fine and the Auto running perfectly."[50] His route took him over roads that made possible the forays of America's tourists into remaining pockets of developed backcountry.

Although Albright had moved up to Yellowstone, park personnel in ZNP continued to honor the early park tradition of community partnership with the local Mormons. Livestock had of course been banned, but both Albright and Stephen Mather created supportive friendships in the region that helped engender local support for the park. Sometimes a rule was bent in a way that suggested a willingness to invest in the locals' welfare rather than strictly adhere to rules of federal bureaucracy. During an interview, J. L. Crawford recalled locals asking in 1932 if they might harvest alfalfa still growing in the park. He said "the park did allow the people of Springdale to go in and harvest the hay. At least they made one general big cutting of all the hay and then they divided it up among the people of Springdale."[51] Though the park was no longer an agricultural space, it was still a Mormon space. And in allowing a harvest, park personnel favored a departure from official policy as a way to support the needs of locals who had once depended on the park space for growing feed for livestock.

Superintendent Eivind Scoyen (1896–1973) supported the town of Springdale by helping to build a new school and a Mormon chapel. He also played an important role in the transaction between the National Park Service and the Crawfords, the last family to live within the park boundaries. He recalled that testimony during an interview with J. L. Crawford, in which Scoyen paraphrased a conversation that he had with Crawford's father: "Everybody agreed to the deal. But I think your father came up to me, just after I thought everything was arranged, and said, 'Mr., we've been talking about it, and we can't sell you that property.' I said, 'What's wrong? Is the price alright?' 'Yes. The price is alright. And the conditions and everything are all right. But Brigham Young called us to settle here, and we can't go until we are released.'"[52] This conversation took place more than seventy years after the canyon was first settled. In response to the Crawford family's misgivings, Scoyen travelled to Salt Lake City to petition church president Heber J. Grant to officially release the Crawford family from "the call."[53] Although Scoyen could have dismissed the Crawford family's concern, he didn't. In fact, by appealing to the president of the LDS Church, he took the family's trepidation and sense of duty very seriously. In the honoring of Mormon culture, the National Park Service made it easier for the locals to welcome the park and esteem the actors involved in its manifestation and early years.

A last and profound example of cooperation between Mormons and non-Mormons in ZNP was the short-lived Zion Easter Pageant, an event that established a powerful ritual in sacralizing landscape. An abbreviated version of an Easter play first took place in the park in 1935, preceding its evolution into a production with several hundred cast members by 1938.[54] The last production was held in 1940. The pageant involved the faculty at Branch Agricultural College and Dixie College, many civic organizations, and residents of towns around the park. The production came to embody several meanings, including: the promotion of tourism, an engagement of reluctant regional residents in the idea of the park, an occasion for holiday entertainment, the celebration of a Mormon and non-Mormon Christian sacred day, and the sacralization of the landscape. The pageant's impact was diverse and reinforced the perspectives and objectives of the various groups invested in the park. By 1940, the event was drawing almost ten thousand visitors who came to watch a "passion play" staged at the foot of the Great White Throne.

The act of putting on the play consumed a great deal of time and resources. Costumes were elaborate, and photos show Herod's dancing girls, arms overhead, wearing scarves and embroidered tops that showed bare midriffs. Herod is pictured with a high, cylindrical crown, looking on at his harem, staff by his side. Jesus, white-robed and bearded, stands solemnly among King Herod's court. Mother Mary and Mary Magdalene, in embellished robes and gowns, cover their faces as they mourn. The landscape in these photos suggests an otherworldly place of ancient dramas—an abstracted version of the holy land.[55] The local newspaper exulted: "undoubtedly no setting in the world . . . compare[s] with this for the reenactment of the scene depicting the life and crucifixion of Him who has been called Savior."[56] In its final year, ZNP was the stage for six hundred people from neighboring communities collectively dramatizing the resurrection of Christ. Actors, members of the orchestra, singers in the chorus, and the audience used the park as a space to evoke the most significant event in Christianity.[57]

The pageant must have been profoundly relevant to the Saints who came to ZNP either to watch or to participate in the play. As a staging of the resurrection of Christ, the experience worked to reinforce Mormon beliefs presented in chapters 11–30 of 3 Nephi in the Book of Mormon, which depict the appearance of Jesus in the Americas after his resurrection in Jerusalem. To watch the play in a sacred homeland was in a sense a reaffirmation and validation of the Mormon worldview. The audience could see their own understanding of history and sacred text manifest within the dramatic scenery of the park.

Many locals participated in the various aspects of the production, which further invested them in the event. Springdale resident Phil Hepworth was a member of the Civilian Conservation Corps in 1940 when he became involved: "I saw the pageant. The last one though, done down there on the old Crawford farm. . . . I guess the reason I liked it so well was because I helped set up for it, get ready for it."[58] As regional residents became involved in the production, there emerged a sense of ownership and pride in both the event and the park, reaffirming the importance of this space. But in this view, we also see a distinction that becomes a thorny issue in the story of Utah public lands—tourists enjoyed the region as transients visiting a "wildland." The pageant was fun for them, but it was not the main reason to visit the gorgeous scenery of ZNP. At the time of the pageant, many regional Mormons, though certainly not all, derived an affinity with their landscape through the use of land. The pageant was a way to make ZNP relevant to them, in both the ritual and the actual production of the event. After all, land was meant to produce, and in this case the production came in the form of theater, entertainment, and religious rite.

Many Mormons living in the region held positions over the years on road construction crews or in park infrastructure. But many area residents still hadn't bothered to explore their own magnificent backyard, and the pageant seemed a way to remedy this indifference. According to a letter to Gilliam Advertising Agency, the Easter pageant was developed to lure regional Saints: "Naturally on Easter we have a religious angle to consider, especially for the local people and churches, but primarily this [pageant] is a thing for publicity, and to get not only advertising for southern Utah, but to get Utah people to see Zion National Park. It is after all rather humiliating to have as many California people visit Zion as Utah people do."[59] The letter is incomplete and unsigned, though its context suggests that it came from a local person interested in marketing southern Utah to other southern Utahns. It is interesting to see how the religious importance of the Easter pageant, the main hook in engaging local people to visit ZNP, was less important to the writer of the letter than the need for park publicity.[60]

The pageant ran only a few years and it drew thousands of people to ZNP. But eventually church authorities discontinued it. According to a letter written by members of the First Presidency and the Quorum of the Twelve Apostles, the church had serious concerns about the event. The letter expressed disapproval over local people leaving their own chapels on Easter Sunday, as well as a fear that the festivities were inappropriate for the Sabbath. Moreover, there was anxiety on the part of the church leaders that Jesus should not be impersonated, no matter how reverential the depiction. And finally, the church authorities were worried about risks involved in a Mormon collaboration with other Christians: "There is more and more tendency in the church, as we more and more mingle with non-members of the Church, to take on the activities of non-members, particularly where they have a religious character, and these accretions have a tendency to change the simplicity of our ordinances and of our faith. . . . The decision of the Brethren was that we could not give this Zion Pageant a Church approval, since there lurked in such celebrations certain grave dangers."[61]

ZNP was a place of Mormon and non-Mormon collaboration, though the idea of non-Mormon Christian theology potentially polluting local beliefs was unacceptable to the LDS church leaders. Although the pageant was conceived, produced, directed, and enacted by regional Saints, the ritual was meant to be shared with locals and broader national audiences. ZNP was at a cultural crossroads where worldviews and values collided. With the Easter pageant, the church feared that the park had become an insecure portal, wherein the act of watching something so sacred in company with non-Mormons could lead to the contamination of Mormon ideals. Church leaders wanted to maintain their autonomy and, as a result, they shut the pageant down.

The cancellation of the annual play came at a time when the region was experiencing growing tensions between local Mormons and federal land managers. The pageant was loved by park personnel as well as residents and might have provided an ongoing forum for continued communication and relationships in the region.[62] With its costumes and music, the play engaged professors, actors, students, and musicians from regional colleges and communities, giving them a chance to present their talents. Park staff and the college faculty both put great effort and enthusiasm into the holiday. Locals and visitors to the region were invited to appreciate scenery, share an aspect of their common Christian culture, and experience spirituality in landscape.

Mormons and non-Mormons celebrated the most sacred Christian holiday of the year together as portrayed and embodied within the jutting rocks.

In subsequent decades, further consideration of national parks and national monuments would be pursued in Dixie, but without broad local support. A rift began between some of the residents of southern Utah and federal land managers regarding proposed grazing, mining, and road building restrictions that came with further federal conservation efforts. In looking at the Maurice Cope collection at the Utah State Historical Society, I was struck by a photograph. An amateur photographer, Cope took hundreds of pictures of rock formations, fellow park rangers, and lodges in southern Utah. Among these shots is an undated photograph of a barbed wire garlanded "grazing limits" sign stuck on a fence line. This picture, so much a departure from his other work, eerily acts as harbinger of things to come—the story of post-ZNP Utah. Boundaries around protected designations and the subsequent restrictions became the bane to many locals. To many Saints, land was meant for production, and limiting its use was seen as heretical.

This is what makes ZNP paradoxical. The creation of the park went against Mormon sensibilities but helped to bring Dixie into the American imagination. It brought crowds of non-Mormons into the Mormon homeland. It asked a culture to shift and value landscape as scenery instead of just for resource use. The formation of the park involved circumstances that made the early years of tourist development in southern Utah exciting, participatory, inclusive, and nonthreatening. The making of the park occurred when many Dixie Mormons wanted to heal historic rifts with the government. The federal agents and visitors who visited Dixie were taken with Mormon culture and worked to create strong relationships and support. Roads were built that benefitted both locals and visitors. Projects were jointly undertaken, such as infrastructure developments and the Easter pageant. And finally, Mormon and non-Mormon visitors found ways of interpreting the lands that suited both their respective understandings of landscape and their worldviews. The park came about through a gentle and inclusive process. Given the climate today over public lands in Dixie and local ire over federal restrictions, sadly we are not likely to see another story like the creation of ZNP soon repeated.

ZNP is a rarity in Mormon country. It came into being before the polarity over public land issues pitted environmentalists fiercely against those pursuing or fighting to maintain resource extraction—mining, logging, and grazing "rights." Compared to the adjacent Grand Staircase Escalante National Monument, which provoked regional fury among federal agents and local residents, ZNP was established in a less contentious time. Pride, nationalism, and the spirit of collaboration buoyed the American preservation movement and led local Saints to get behind the idea of a national park. Among the many reasons for its success is the fact that the park size is small. At the time of the park's creation, the country was remote and its production value marginal. It also lacked the minerals and pasture (unlike in Grand Staircase Escalante National Monument) so its "withdrawal" from local use was largely uncontested. That said, nothing like the creation of ZNP would be possible today. Many Dixie locals fight every federal effort to protect public lands and regularly, sometimes violently, contest regulations for the protection of species, Native American sacred sites, and roadless areas (such as designated wilderness). Across the decades of contention over land use, regional priorities and temperaments have changed. In today's climate, even in a cultivated wild, there is too much wild.

PART III

Agrarianism and Urbanism

BEFORE THE BOOM

Mormons, Livestock, and Stewardship, 1847–1870

Jeff Nichols

IN AUTUMN 1846, the Latter-day Saints faced the worst crisis in their short history, as some twelve thousand men, women, and children huddled in crude shelters on the banks of the Missouri River. Their founding prophet, Joseph Smith Jr., had been murdered in Carthage, Illinois, in June 1844. The pressure on them to leave their city of Nauvoo had only grown since then. Smith's successor, Brigham Young, tried to negotiate with non-Mormons who continued to threaten lives and property. Young eventually agreed to abandon the city, and Mormons hurried to sell their homes and farms, sometimes taking pennies on the dollar or nothing at all as they fled west. As they waited around Winter Quarters for spring weather so they could travel to the Salt Lake Valley, they faced poverty, disease, malnutrition, and an uncertain future.[1]

The Mormons' means of escape lay in their animals.[2] They had gathered thousands of oxen, cows, and steers, as well as horses, sheep, chickens, mules, pigs, and goats.[3] The cattle were especially crucial, as they constituted money on the hoof, but also served many other roles: most importantly, oxen could haul wagons to their planned new haven in the Great Basin. Once there, cattle could do all the things they did for humans in that premachinery day: pull plows, haul loads, and give milk, meat, and leather.[4] But livestock also exemplified the tensions between communitarian ideals and market success that Mormons would face. Mormon leaders wanted autonomy and self-sufficiency in their western refuge. But they found themselves dependent on outside people and products, and they discovered that they could profit from outside demand for their own animal products. Livestock helped build Mormon communities and serve common needs, but they also represented products with a price tag, including an ecological one. As Utah moved toward a market economy, ever-larger herds wreaked damage on valleys and mountain slopes.

The Saints were not alone, of course, in their dependence upon cattle or their desire to profit from them. Livestock had constituted vital tools as well as substantial wealth since the Europeans came to the Americas.[5] Draft animals and improved plows allowed farmers to break heavy soils and haul their harvests to market. That effort and expense were justified. Wherever white settlers pushed into new territory, their livestock constituted part of their living cultural baggage (what Alfred Crosby named their "portmanteau biota") and a powerful material advantage.[6]

Mormons also brought ideological baggage, including the proper way to exploit the environment. Like the domestic animals themselves, most of those ideas came from the larger Anglo-American culture that spawned their church, but LDS leaders contributed their own specific strictures. Their prophets declared that animals were for human use, and thus by extension the range those animals grazed was too. LDS leaders insisted that the Saints had a moral responsibility to properly care for their animals, and again by extension a responsibility to care for the land.[7]

Despite these teachings, after a half century of Utah settlement it was clear that the range was in serious trouble. Albert Potter, a federal grazing advisor, surveyed Utah's rangelands in 1902 and described a landscape devastated by too many livestock. The resulting national forests helped to mitigate the worst excesses of overgrazing, mostly by forcing herd reductions.[8] Potter and others blamed cattle and especially herds of sheep that had grown very large since livestock numbers began "booming" around 1870. Those observers were almost certainly right about mountain slopes like those above Ogden and Logan, which had not been intensively grazed before that time period. Thomas Alexander has suggested that after the 1870s Mormons forgot earlier stewardship ideals and over-exploited the rangeland, especially when they began to move up into mountain pastures.[9]

Of course, not all graziers after 1870 were Mormons. But the evidence discussed in this essay suggests some other conclusions about the first quarter century in Mormon Utah. First, Mormon environmental stewardship was always selective and anthropocentric: animals and resources useful and profitable to people were protected, and those deemed useless or harmful were ignored or actively attacked. Second, when animals were mostly for local subsistence use and their numbers were relatively small, the range could sustain them. But when early settlers crowded too many animals on confined range in the 1850s, they caused the same kind of damage that Potter ascribed to the big herds after 1870. Historians writing about the years 1847 to 1870 have tended to emphasize farming rather than livestock raising. But the dynamic roles that livestock played in the Utah economy also demonstrate the tensions between sustainability and the market as Mormons confronted the realities of their new environment and reacted to the pressures and allure of the larger American economy.

Brigham Young understood the importance of livestock in 1846 and encouraged his followers to buy or trade for as many animals as they could afford.[10] Protecting and feeding the large herds were among the Mormons' main concerns. As their herds chewed up grass, Omaha and Sioux Indians around Winter Quarters (established on Omaha land) sometimes stole or killed Mormon animals.[11] Young wrote Thomas L. Kane that his choice of a place for the pioneer company to winter was "to secure the rush and pea vine for our flocks."[12]

In April 1847 Young began leading an advance company of 147 Saints to the Salt Lake Valley. The pioneer company, which moved fast and light, reportedly brought 19 cows and 66 oxen with them, as well as 93 horses and 52 mules.[13] Parley P. Pratt had calculated that each family of five going west would need three yokes of cattle or oxen, two cows, and one or more beeves.[14] Although few contemporary sources go into detail, the Mormons' animals were probably shorthorn cattle of mixed breeds acquired in Ohio, Missouri, and Illinois.[15] The later companies brought much bigger herds. These were not the first cattle in Utah. Feral animals likely wandered into the territory from Spanish settlements to the south, and Mexicans drove stock along the so-called Old Spanish Trail as early as the 1820s. Jim Bridger grazed cattle on the slopes of the Uintah Mountains, and Miles Goodyear had a ranch on the Weber River, where the city of Ogden is today.[16] Overland travelers to Oregon and California trailed animals through Utah, and some of them undoubtedly strayed or were stolen by Indians. But Mormons brought the first large herds to Utah.

Thousands of cattle had caused profound effects on the lands they passed through. A single steer might eat as much as twenty pounds of grass in a day,[17] and since immigrants on the overland trails stayed within a narrow strip of land, cattle decimated the grasses and drove away deer, elk, pronghorn, and bison. Those depredations were a constant source of conflict between whites and Indians, who sometimes charged tolls for passage or pilfered immigrants' cattle.[18]

The Saints had reason to believe that their proposed home in the Great Basin would be a good place for livestock. The Spanish friars Domínguez and Escalante, who traveled through Utah in 1776, never got farther north than Utah Lake. But those representatives of a pastoral people wrote enthusiastically of the grazing potential they saw in what became known as Utah Valley.[19] Mormon leaders studied John C. Frémont's reports of his 1840s expeditions, one of which predicted that "in the rearing of flocks and herds [the Great Basin] would claim a high place. Its grazing capabilities are great; and even in the indigenous grass now there, an element of individual and national wealth may be found."[20]

Although it is difficult to determine the ecological state of the land upon the Mormons' arrival in the Salt Lake Valley, the pioneers were enthusiastic, despite their later claims about finding a "desert." The lake level averages 4,200 feet, while the site where the Mormons would plat their city is a few feet higher, with "benches" to the east where ancient Lake Bonneville's shoreline stabilized for long periods.[21] The valley and benches lie within the "transition zone," characterized by bunch grasses and occasional shrubs like big sagebrush, rabbit brush, and some pinyon and juniper.[22] Some pioneers were clearly thinking of grazing potential as they recorded their first impressions. Thomas Bullock, who arrived on July 22, 1847, wrote "wheat Grass grows 6 or 7 feet high. Many different kinds of grasses appear, some being 10 or 12 feet high. . . . we camped on the banks of a beautiful little stream [now Emigration Creek] which was surrounded by very tall grass."[23] William Clayton, with Brigham Young and Heber C. Kimball, wrote on July 24, "there appears to be a unanimous agreement in regards to the richness of the soil and the good prospect of sustaining and fattening stock with little trouble."[24]

It seems evident from these early descriptions that relatively lush bunch grasses and forbs grew in the valley and on the benches. Those plants would indeed make good cattle feed in the short run, but many scholars conclude that the native perennial grasses, especially the *Agropyron* species (wheatgrass), had not evolved to be intensively grazed by large herds for season after season and were rapidly displaced by alien annual grasses and woody shrubs. Mormon domestic animals took the place of the relatively few native grazers: bison in unknown but probably not very large numbers as well as elk and pronghorn.[25]

For Brigham Young, farming and herding were not just the proper routes to prosperity, but also markers of civilization and a way to distinguish between whites and Indians. In this he echoed early European settlers of New England, who contrasted their own permanent villages and intensive cultivation with Indians who did not use the land "properly" but rather passed over it, harvesting a natural bounty of wild plants and animals.[26] Thomas Bullock reported that Young was not dismayed when hunters had bad luck supplying game, because that meant the Saints would have to concentrate on establishing fields and flocks. In his Indian diplomacy, Young would often use gifts of cattle to persuade Utes and other nations to adopt a sedentary lifestyle (and to stop them from preying on Mormon herds).[27]

The Saints worked out their grazing practices as they elaborated their theodemocratic governance. For the time being, they had a near-monopoly on land and power once they had broken Indians' ability to contest their claims.[28] The Salt Lake Stake Presidency and High Council initially acted as a city government, along with the Council of Fifty—a group of primarily high church officials—and in 1849 settlers stablished a permanent city government. In 1850, the U.S. Congress created Utah Territory as part of a larger compromise dealing with the entrance of states and territories into the Union. Mormons created a Utah territorial legislature firmly under the direction of Governor Brigham Young and the church hierarchy.[29]

The isolation of the Salt Lake Valley gave the Saints freedom to experiment with their communal ideas. The Mormons had practiced varieties of communal economics with varying success since Joseph Smith founded the church in 1830. Smith received a number of revelations, most importantly the one recorded in Doctrine and Covenants Section 42, which laid out the doctrine of "consecration and stewardship."[30] Under this doctrine a Mormon "consecrated" or deeded his private property to the church and received a "stewardship" in return: the use of a farm, animals, or other enterprise (often the very things he had consecrated). Early attempts to establish consecration in Ohio and Missouri failed through a combination of jealousies and outside pressures, and Smith instituted the "lesser law" of tithing, by which a Saint was expected to give at least ten percent of his "increase" to the church. For most of the Utah period, the tithing system predominated.[31] The principle of common property, however, continued to shape Mormon belief and practice, and Brigham Young attempted to reinstitute communal economics until his death in 1877.[32]

Brigham Young often repeated Joseph Smith's teaching that all creation belonged ultimately to God and that man was only granted temporary use rights, and then only if land and other resources were put to beneficial use or "improved."[33] In Jeanne Kay and Craig Brown's fine phrase, Mormons considered themselves "tenants obligated to a divine landlord."[34] Church leaders reiterated in countless talks that they were involved in no less a project than building the kingdom of God, and that required a self-sufficient farming economy based on patient, careful, communal labor under church supervision.[35]

The layout of Salt Lake City (loosely based on Joseph Smith's plat of Zion gridiron) called for animals to be housed close to home—but not too close. Leaders directed people to build barns and stables at the edge of the city, far enough from homes to avoid unpleasant smells and potential disease but close enough for security and easy care.[36] Young warned against the establishment of a market in real estate, although one quickly emerged nonetheless. As Charles S. Peterson and others have noted, although their animals were vital, the Mormons were primarily farming, not pastoral people. Both their experience in the East and their desire for community cohesion worked against the establishment of sprawling ranches. A typical Mormon family might have a couple of milk cows and beeves for their needs but rarely a large herd.[37] The second group of Mormon immigrants arrived in fall 1847 with over 1,000 people and thousands more animals.[38] In late 1847 a party was sent to California, where they purchased about 200 cows, the first of many such cattle-buying trips to both east and west.[39] And Mormons raised their own animals in Utah. The 1850 census counted 5,266 working oxen, 4,861 milk cows, and 2,489 other cattle in the territory—a little more than one such animal for each person.[40]

Like their Puritan ideological forebears, Mormon leaders had no quarrel with individual material success if it was righteously achieved and if it helped advance the kingdom of God. Certain essential natural resources, especially timber, water, and (sometimes) grass, should remain communal property under management by trusted church officials. Again like the Puritans, the Utah Mormons from the start experienced tension between self-sufficient, communal economics and a capitalistic striving for individual wealth. From the start, the management of livestock mixed individual and communal practice. Typically, after morning milking, designated "herd boys" or girls would collect animals from the various farms, drive them to a nearby common pasture along the bench, at creek mouths, or between the city and the Great Salt Lake, and return them to their owners in the evening. However, some towns employed professional herdsmen.[41] The Saints sometimes worried about the moral impact of herding on boys and girls. Daniel H. Wells insisted that children should be in school rather than herding livestock, which even taught some of them to be rustlers and outlaws.[42]

Controlling animals was a major concern in the new communities. The animals Mormons brought with them were both a necessity and a threat to other food sources, since hungry beasts sometimes wreaked havoc. Brigham Young was worried enough about the problem that during his trek back to Winter Quarters to meet the bulk of the Saints, he stopped to write a long letter to Salt Lake urging settlers to carefully fence their fields.[43] Just as crops were ripening in autumn 1847, cattle and horses broke through the fence and ditch that surrounded the city and its fields and ate much of the harvest.[44]

The Council of Fifty created a fencing or herd committee to oversee the work of separating animals from planted fields.[45] This was a constant headache in these pre-barbed wire days, as the area lacked timber and wooden fences are labor-intensive to build and maintain. Another piece of early legislation mandated that all Mormon-owned cattle be branded "LDS." That was evidently taking common property too far, and in late 1849 the territorial legislature passed a branding act that established a system for private individuals to record their marks. Authorities also established an estray pound that kept busy returning lost animals and assessing fines on careless owners. Both brands and pounds are signs that private property concerns continued alongside communal economics.[46]

Mormon settlers learned to grapple with the sometimes harsh climatic reality of their new home. Fortunately the first winter of 1847–48 was mild and stock ate relatively well, but the winter of 1848–49 was very harsh and killed hundreds of animals. The Council of Fifty ordered a roundup of horses and cattle and the creation of a community herd to ensure adequate care. John D. Lee, part of the roundup committee, noted that some people complained about encroachments on their property rights. The council was ready to call off the drive, but Brigham Young argued that "duty Says if they will not take care of there catle [*sic*], we must do it for them. . . . Then don't be bluffed off by insults or abuse."[47] A committee investigated the Utah Valley as an alternative winter pasture, but reported that the snow was about as bad there. Reports from northern valleys reported even more snow, and the Saints decided their herds should ride out the winter in the Salt Lake Valley.[48]

Mormons also explored other places to graze. Rush and Tooele Valleys to the southwest of Salt Lake City offered rich pastures.[49] In early 1851, they introduced livestock to southern Utah with several hundred animals in Parowan.[50] The largest island in the Great Salt Lake offered another alternative to fencing fields. The island, which Frémont had named Antelope Island,[51] is about fifteen miles long and four miles wide, constituting between twenty-four and thirty thousand acres, depending on the lake level. One white man, described as an old mountain man known as Daddy or Father Stump, may have already been living on the island with a few cattle, but he was evidently willing (or persuaded) to share.[52]

In September 1850 the legislature set aside Antelope and Stansbury Islands for communal use. They created the Perpetual Emigrating Company to finance the emigration of poor Saints to Utah. The company was authorized "to receive, either by donation on deposit, or otherwise, money, gold dust, grain, horses, mules, cows, oxen, sheep, young stock of all kinds, as well as any and every kind of valuables, or property, whatsoever; . . . the entire proceeds of the business of this Company, shall inure to the Perpetual Emigrating Fund [PEF] for the Poor."[53] The company put Fielding Garr in charge of herds on the island.[54] Most of the animals were part of the "church herd," common property replenished by tithing contributions, unclaimed strays, and natural increase. Garr also cared for Brigham Young's and Heber C. Kimball's substantial private herds, and when U.S. Army Captain Howard Stansbury conducted his survey of the Great Salt Lake in 1851, he left his government animals with Garr.[55] Public and private herds sometimes got tangled; Briant Stringham complained in 1855 that "President's Cattle & Church C. are in a mixed up fix without mark or Brand," a forecast of the jumbled estate Brigham Young would leave behind.[56] The LDS Church eventually claimed other ranges and established ranches, including land west of Utah Lake and parts of Cache Valley.[57] The PEF sent "traveling agents" to man stations along the emigrant route, either with oxen from the Salt Lake Valley or animals they purchased locally.[58]

Cattle—valuable and mobile—were both part of the communal experiments and well-suited as a kind of currency for the Mormons, especially when specie was in short supply. As LDS wards and stakes were created, bishops and stake presidents established tithing houses with stables, barns, and pastures to house the animals they received in tithing or to help pay general church debts. In 1852, Brigham Young called on the Saints to contribute livestock: "let every man look up his spare stock of this description, and with *them* help to liquidate these debts. Stock will now pay debts. . . . I will take every kind of spare stock I have, except my cows and sheep, and wipe off these debts until they are cancelled; and now every man who will do the same, let him rise upon his feet."[59] Young reported a year later that the debts had been paid.[60] He sometimes complained, however, that Mormons underpaid their tithing or that bishops used the system to build their own herds, saying, "When a good, handsome cow has been turned in on tithing, she has been smuggled, and an old three-titted cow—one that would kick the tobacco out of the mouth of a man who went to milk her—would be turned into the General Tithing Office, instead of the good cow. If one hundred dollars in cash are paid into the hands of a Bishop, in many instances he will smuggle it, and turn into the General Tithing Office old, ringboned, spavined horses, instead of the money."[61]

The church herd on Antelope Island constituted a sort of central bank upon which authorities could draw.[62] Brigham Young, always striving for self-sufficiency, created the Deseret Currency Association in 1857 to give the Saints their own circulating medium. The paper notes were backed by livestock and helped finance the "move south" in early 1858. When Gentile merchant houses were reestablished later that year, the notes were called in, mostly in the form of tithing—which furthered the process of the "church herd" becoming the property of the PEF.[63]

The church herd also proved to be a revenue-producing enterprise, and it got some unexpected deposits when gold seekers began swarming along the trails to California. Many stopped in Salt Lake, the only large settlement for hundreds of miles, to recruit their animals or trade broken-down ones for healthy animals, produce, or supplies. Many of their jaded beasts found their way into the church herd, along with livestock from the dissolved Brigham Young Express and Carrying Company (YX Company).[64] When the herd seemed big enough, a party would trek the animals to California and sell them for a more convenient currency.[65]

Brigham Young acknowledged the importance of the cattle trade in his 1853 governor's message, which highlighted the tension between self-sufficiency and market economics. Young admitted that for the time being, the livestock trade would of necessity play a larger economic role than he had originally envisioned, and that "so long as the California markets remain dependent upon foreign supplies, we may naturally expect large accessions will be made to our flocks and herds."[66]

Antelope Island appeared an ideal place to keep the church's cattle deposits. When the water was high enough, the lake acted as a natural fence. The grass, especially on the east side, was reportedly thick and lush. Heber C. Kimball noted that the winds kept snow off the grassy hills, and the lake waters may have somewhat moderated winter temperatures.[67] Both Kimball and Brigham Young introduced blooded horse stock to the island, which became as well known for its horses as for its cattle. But confining thousands of animals feeding on bunch grasses in a fixed place led to overgrazing, especially when natural forces conspired against the herds. By rough estimate, a cow needs about twelve acres in mixed bunchgrass and sagebrush country, so Antelope could sustain just over two thousand cattle. An average horse eats somewhat more than a cow.[68] Well over three thousand animals grazed Antelope Island by 1855 with disastrous consequences.

Swarms of insects had also plagued the Mormons since their second season in the West. Mormons usually called them grasshoppers or crickets, but scientifically they are classified as Rocky Mountain locusts, *Melanoplus spretus*.[69] The year 1855 saw the worst infestations yet as clouds of locusts chomped through fields and pastures. Brigham Young and others realized that the Antelope herds were in trouble and directed that they be moved to Cache Valley, northeast of the lake, for the winter. But that winter proved brutal, and of some 2,600 head of cattle in the church herd, only 300 survived the cold and snow. Some private herds suffered equal devastation.[70] The Saints had learned that their new home was not always going to yield a bounty. Herds were reestablished on Antelope Island, although a herder noted in August 1856 that "feed was scarce . . . all pasture on the Island was badly over-grazed."[71] When Fielding Garr died in 1855, Briant Stringham took over management of Antelope Island and the church herd until his death in 1871.[72]

Mormons also experimented with granting large ranges to church leaders. The Council of Fifty in January 1849 gave Brigham Young and Heber C. Kimball "the privilege of fencing in as much of the table lands and the spurs of the mountains east of the city, as they wish for pasturage."[73] In 1850 the legislature granted entire canyons to LDS leaders; for example, Brigham Young received City Creek canyon. The ostensible idea was that the grantees would act as careful stewards of water, timber, and grass. An alternative explanation is that the council was rewarding its most important members. These grants also helped create an effective Mormon monopoly of essential resources against outsiders. Throughout the 1850s the legislature granted herd grounds to individuals and groups across the territory.[74] Brigham Young had other suggestions that involved fencing off large chunks of public domain:

> I will propose a plan to stop the stealing of cattle in coming time, and it is this—let those who have cattle on hand join in a company, and fence in about fifty thousand acres of land, make a dividend of their cattle, and appropriate what they can spare, to fence in a large field, and this will give employment to immigrants who are coming in. When you have done this, then get up another company, and so keep on fencing until all the vacant land is substantially enclosed.[75]

A few non-Mormons did live in the Salt Lake Valley already, including merchants like Ben Holladay and members of the firm of Livingston and Kinkead, which established a lucrative trade driving Utah cattle to California. And of course many non-Mormon immigrants trailed livestock through the territory.[76] It seems clear that Brigham Young, so concerned about autonomy and self-sufficiency, meant to keep Utah grass for Mormon animals.

Most of those animals were oxen. Until the transcontinental railroad was completed in 1869, the greatest need for Mormon cattle was for motive power, and the Saints gathered oxen from east and west to serve their "gathering to Zion."[77] In the 1860s leaders organized official "church trains" that would leave Salt Lake in April, meet emigrating Saints on the Missouri River, and return with them in the fall.[78] Brigham Young constantly exhorted the Salt Lake Saints to contribute yokes of oxen for this vital gathering work.[79]

The federal government did not allow the Saints total autonomy in Utah. The Utah Expedition, which accompanied new governor Alfred Cumming, had been forced to spend the winter of 1857–58 near Fort Bridger. Part of the army's problem was Mormon raiders who burned supply trains and ran off cattle, some of which found themselves mingling with the herds on Antelope Island.[80] Once the conflict was settled and the army established at Camp Floyd, officers claimed Rush Valley for the federal herds.[81] Cattle played a small role in the grisliest of Utah stories: the murder by Mormon militia and some Paiute allies of about 120 immigrants at Mountain Meadows on September 11, 1857. As the Fancher party rolled through Utah in the weeks before, Mormons refused them permission to graze or charged them for the privilege, and the story of a poisoned ox carcass has been cited as a supposed motive for the murders. After the human slaughter, some of the Fanchers' animals ended up on Mormon or Indian farms or tables.[82]

Once Governor Cumming was established in office he reminded the territorial legislature that only the federal government could grant secure title to the public domain, so the legislature revoked the grants to individuals in 1860. Instead they substituted a wonderfully flexible law that gave existing (i.e., Mormon) towns the power to remove herds "found upon the range in the neighborhood or vicinity of any settlement in this Territory," which helped maintain the de facto Mormon grazing monopoly.[83] By then Mormons had settled all along the Wasatch Front, north into Cache Valley, and southwest along the trails to Southern California. Settlers often described abundant grasslands near these new towns, just as the 1847 pioneers had.[84] Jared Farmer has noted that expansion was not always the orderly, centrally planned "calling" described so often.[85] Settlers sometimes moved their herds onto new pastures in advance of leaders' wishes—for example to Utah Valley in 1849. Many new settlers established cooperative herds.[86] Some formally "called" Mormon colonies failed, despite (or perhaps because of) their livestock. Utes drove out the Elk Mountain Mission of 1855 (to what is now the Spanish Valley and the town of Moab); the Indians let most of the Mormons escape when they left their animals behind. The valley was successfully settled in the 1870s by individuals and small groups trailing herds of livestock.[87]

The Mormons' demographic increase meant more demand for cattle, and that inflicted damage on the surrounding lands. As early as 1849 Ezra Taft Benson reportedly hired herders to bring his cattle into Tooele Valley because the Salt Lake Valley was overgrazed.[88] By the mid-1860s, other people noted that valley grass was giving way to sagebrush and junipers. The *Deseret News* suggested in March 1865 that conditions were not what they had been in 1847: "When the settlements were few, scattered and small, and grass comparatively convenient and abundant, the random policy was not a question of so much moment. But denser settlements, consequent overstocked ranges and the late enactments prohibiting stock from being free commoners must now be considered in the question."[89] Apostle Orson Hyde warned in a LDS General Conference a few months later:

> I find the longer we live in these valleys that the range is becoming more and more destitute of grass; the grass is not only eaten up by the great amount of stock that feed upon it, but they tramp it out by the very roots; and where grass once grew luxuriantly, there is now nothing but the desert weed, and hardly a spear of grass is to be seen.
>
> Between here [Temple Square] and the mouth of Emigration Canyon, when our brethren, the Pioneers, first landed here in '47, there was an abundance of grass over all those benches; they were covered with it like a meadow. There is now nothing but the desert weed, the sage, the rabbit bush, and such like plants, that make very poor feed for stock. Being cut short of our range in the way we have been and accumulating stock as we are, we have nothing to feed them with in the winter and they perish.[90]

Observers noted similar damage to City Creek Canyon, and within a few years ranchers in Rush and Tooele Valleys were reporting the displacement of bunch grasses by shrubs.[91] As John D. Lee was brought to Mountain Meadows for his execution in 1877 for leading the 1857 massacre there, he reportedly could not recognize landmarks because of years of overgrazing.[92] The botanist Walter Cottam suggested in 1961 that throughout the Bonneville Basin, grasses had been largely replaced by big sagebrush within twenty years of settlement.[93] Brigham Young warned Mormons to "not keep so many cattle, or, in other words, more than you can well provide for and make profitable to yourselves and to the Kingdom of God."[94] George A. Smith worried in 1867 that Mormons had too many sheep on inadequate pasture and argued that fewer, better cared for animals would produce "twelve or fifteen pounds of wool from each one, instead of the bare backed animals, so common now that we might suppose they never had any wool within a mile of them. . . . If we suffer them to perish through cold and hunger we are responsible to God for the cruelty we inflict upon those animals."[95]

While Orson Hyde and others described the degradation of a once-lush landscape, Brigham Young stood that formula on its head, claiming at his last Pioneer Day celebration in 1877 that the Saints had encountered a "desert country, making it like the garden of Eden."[96] Wilford Woodruff had described the Salt Lake Valley upon his arrival as a "most fertile valley . . . clothed with a heavy garment of vegetation." Thirty years later he echoed Young, asking, "when we came to this country, what did we find here? A barren desert as barren as the Desert of Sahara; and the only signs of life were a few black crickets, some coyote wolves, and a few poor wandering Indians."[97] The scholars Thomas Alexander, James Allen, Richard Jackson, Dan Flores, and Jared Farmer have suggested that Mormons recast the story this way in order to strengthen their narrative that they had "made the desert blossom as the rose." While that seems persuasive, it may also be that by the 1870s they were looking at overgrazed benches and mountain slopes and contrasting them with the green of irrigated Mormon fields. Rather than redeem a desert, they had helped create one.[98]

Much worse damage lay ahead. The grazing "boom," as the 1940 Works Progress Administration's Utah Writers Project's grazing history labeled it, took off in the 1870s. Once railroads arrived, Mormons and their stuff could be moved much more easily and cheaply, and the demand for draft animals dropped rapidly. But the railroad also tied Utah to outside markets, allowing livestock to be easily imported and exported, and growing human populations demanded meat, leather, milk, and wool. The agricultural census (which is believed to undercount animals) counted 34,094 cattle in Utah in 1860, and even more sheep. Utah County claimed more livestock than Salt Lake County for the first time. The total represented nearly triple the number of animals in 1850 (although the human population had roughly quadrupled).[99] Sheep became a bigger story than cattle. Men such as John Seely in Sanpete County introduced superior breeds, including the Merino. Since sheep are somewhat easier to herd, need less water, and can eat a wider variety of plants than cattle, their numbers boomed and wool became a major Utah export.[100] In 1880 there were 132,655 cattle, and by 1900 numbers had skyrocketed to 264,750 while sheep numbers peaked at over 3 million.[101] New sorts of animals entered the territory too. The first Texas longhorns came in the 1870s, and after about 1880 some ranchers began to import superior British breeds like the Hereford.[102]

The railroad also brought an indirect end to the church herd. After Briant Stringham died in 1871, the Saints struggled to find a manager for the Antelope Island ranch, though it seemed less necessary. The Union Pacific Railroad claimed the island as part of its land grants and planned to lease it for grazing. The church tried to round up its horses but had little success with the virtually feral animals. After the church withdrew from the island, nearby settlers put a reported ten thousand sheep there. The competition caused some animals to starve to death. A private grazing company was created in 1884, which shot the remaining horses as pests. Private cattle and sheep herds grazed the island until the 1960s.[103]

Mormons did not completely abandon communal economics after the railroad arrived. Brigham Young pushed a revival of consecration and stewardship called the Order of Enoch, or the United Order. Towns across Utah launched a variety of communal enterprises, including cooperative livestock herds. The most thoroughly communal of the experiments was Orderville, on the Virgin River. Residents consecrated their property to a central corporation and received stewardships in return. But most such experiments were short-lived, and even Orderville abandoned communal economics after 1885.[104] Brigham Young's death in 1877 removed the most ardent advocate of communal economics. The leaders who followed him were understandably preoccupied with resisting federal antipolygamy campaigns, while many Mormons eagerly entered the market economy. Most continued to own their own animals, and after about 1880 the practice of communal herding faded. As the Mormons anticipated the passage of the Edmunds-Tucker Act in 1887, which would confiscate most of the church's property, President John Taylor ordered some $66,000 worth of cattle in tithing offices to be transferred to wards, and about three thousand horses and cattle owned by the church were sold to private individuals.[105]

Ranchers expanded onto lands that had been marginal to the Mormon core or that seemed so hot, arid, or infertile that they were likely unsuitable for farming. While cattle and sheep could live there, these Great Basin and Colorado Plateau lands were much drier and more fragile and took many more acres to support similar numbers than the lusher Wasatch Front.[106] For example, from about 1880 on, a variety of players contended for grazing lands north of the San Juan River. The Mormon settlers of Bluff, who trailed perhaps 1,800 cattle and horses over the horrendously difficult Hole in the Rock Trail in 1880, recapitulated the Salt Lake Valley story of mixed communal herding.[107] When farming seemed a failure, the Bluff settlers turned more fully to ranching. Francis A. Hammond, stake president of the San Juan Mission, helped create the "Bluff Pool" in 1886, a communally owned herd that paid dividends to stockholders and assessed them for herding labor based on the number of animals they contributed. The Bluff Pool had to contend for range with outfits like the Carlisle brothers, who were backed by British money. Both sides used tactics that had become familiar across the West, including claiming lands along watercourses (which guaranteed de facto control of the adjoining arid lands), fencing public lands, and having employees enter bogus land claims.[108] But the Bluff Pool could also draw on sociocultural advantages over their rivals. Like in the Salt Lake Valley settlements in the 1850s, Francis Hammond could exert his ecclesiastical as well as temporal influence over Pool members (and probably over the San Juan county court, which decided some water rights cases in the Pool's favor). The veterans of Hole in the Rock had a powerful Exodus-like story to draw strength from as they contended with outsiders encroaching on "their" grass and water.[109] The Carlisles eventually sold out, and by 1897 the Bluff Pool was disbanded in favor of private interests. The Bluff ranchers hired a tough young cowboy named Al Scorup to round up their cattle out of the scrub. Scorup and the Redd family eventually developed some of the biggest ranches in the country—privately owned animals grazing mostly on the public domain.[110]

All these factors—the railroad, population growth, new breeds of cattle and sheep, and a move from communal to market economics—coincided with the final displacement of Indian peoples and their confinement to reservations, which was mostly complete by 1870. For the first two decades or so Utah ranchers had avoided the mountains, both because of the cultural reasons discussed above and out of fear that Indians would steal or kill their animals (or the ranchers). Pushing herds up mountains helped displace the deer and elk Indians hunted, so livestock again assisted whites in strengthening their control of land. Indian removal allowed white graziers to run their animals on mountain slopes in the summer, then down to the deserts for winter.[111]

Ranchers also contested against predators, and for all their talk of God's creation, Mormon farmers and ranchers were as violent as any against those animals. In 1849, a "war of extermination" against "wasters and destroyers" took the form of a three-month-long contest against birds and animals—including wolves, foxes, bears, and mountain lions—that might prey on livestock.[112] As ranchers brought herds onto new ranges they waged renewed wars, especially on bears, wolves, and coyotes. Bears were greatly reduced, and Utah wolves were gone by the 1920s, although coyotes have persevered in the face of relentless shooting, trapping, and poisoning.[113]

Many more cattle and sheep combined with careless logging to produce ecological catastrophe for both herds and grass. The savage winters of 1879–80 and 1885–86 took a huge toll on herds and helped push ranchers like the Carlisles out of business.[114] By the 1890s people across Utah were documenting flash floods, arroyo cutting, and landslides on mountain slopes that had always seemed stable. Ranchers claimed that they could estimate the number of animals from miles away by the size of the dust clouds they kicked up.[115] Dan Flores describes the botanical changes after 1870 as "the unfolding of a new succession pattern," in which sagebrush, rabbitbrush, shadscale, and juniper replaced grasses. After about 1900, the exotic invasive species cheatgrass and Russian thistle (i.e., tumbleweed) also encroached on many pastures.[116]

Across the West, worried ranchers clamored for the federal government to do something. The result was the Forest Reserve Act in 1891, which later became national forests. Forest lands were withdrawn from homestead entry and grazing was managed by a permit system. Not everyone thought the new system had remedied the problem. In 1947, Walter Cottam asked "Is Utah Sahara-bound?" and answered his own question in the affirmative. While Cottam's worries were clearly overblown, grazing damage was and is long-lasting.[117]

Many people blamed sheep for this damage, especially herds owned by "outsiders," and clearly the enormous herds present between 1880 and 1910 deserve their share of the blame.[118] But there are problems with the outsider/sheep narrative. For one, Utah Mormon cattle ranchers introduced sheep on many ranges (which also helps account for the relative lack of cattle-sheep herder antagonism).[119] Mormon pioneers are often praised for their stewardship of natural resources, including in this volume. Stewardship worked well for nurturing human communities: Mormons founded hundreds of towns and raised generations of families. Livestock stewardship worked sustainably when human populations were relatively low, and when small herds of grazing animals could be moved around on valley pastures or islands. Mormon stewardship worked less well to protect grass on fragile mountain hillsides, and it was disastrous for animals that preyed on livestock. Prolonged drought, insect infestations, or a heavy winter rendered even the most careful stewardship moot. The sad ecological histories of Antelope Island and the east benches of the Salt Lake Valley, Rush Valley, and Tooele Valley before the "boom" demonstrate that when grazing animals were concentrated in a restricted area they caused extensive damage, whether their owners talked of building a kingdom or of turning a profit.

"THE PEOPLE CANNOT CONQUER THE RIVER"

Mormons and Water in the Arid Southwest, 1865–1938

Brian Frehner

SUBMERGED UNDERWATER FOR NEARLY seventy years, the Mormon town of St. Thomas, Nevada, reappeared in 2002. Approximately five hundred residents lived in the town at its peak—one of many founded along the Colorado River and its tributaries as part of Brigham Young's effort to construct a Mormon corridor from Salt Lake City to the Pacific Ocean. In 1937, water began pooling behind the recently completed Hoover Dam, inundating the town and forcing residents to flee and relocate further upriver. The town sat buried beneath Lake Mead only until the lake's shoreline began rapidly receding due to the over-allocated Colorado River's diminishing flow.

Both the subsidence and reappearance of St. Thomas serve as useful anecdotes to illustrate varying interpretations of humans' efforts to control water in the American West. On one hand, St. Thomas's disappearance underwater evokes Donald Worster's argument that "society in the region has been shaped by its advanced technological mastery of water"—resulting in "hydraulic societies"—and that residents of the town suffered from federal planners who imposed "a sharply alienating, intensely managerial relationship with nature."[1] In short, federal technocrats constructed dams that Worster calls "Leviathans," and these bureaucrats managed nature at the expense of the livelihoods of people who inhabited nearby towns. Other historians reject the notion that scientists and engineers employed by the federal government possessed such overreaching power. Donald J. Pisani contends that neither "science" nor "efficiency" did much to shape federal reclamation because new branches of engineering were just beginning to coalesce, and their practitioners relied more upon intuition and practical observation than laboratory science typically associated with twentieth-century universities and corporations.[2] Pisani enumerates a number of other factors that constrained the U.S. Bureau of Reclamation's ability to engineer an "empire" out of the American West's rivers—including maladministration, squandered opportunities, legal obstacles, actions of Congress, and even weaknesses in the 1902 reclamation law itself.[3] What is most notable about federal reclamation, Pisani argues, was *not* its ability to transform nature but the agency's numerous failures to accomplish its primary objective: to reclaim arid lands.[4] Like federal bureaucrats, Mormons who settled tributaries of the Colorado River in southern Utah and southern Nevada failed frequently in their efforts to irrigate this harsh, arid region.

The following story contends that understanding Mormons' efforts to reclaim the arid West may require historians to emphasize a middle ground between Worster's and Pisani's perspectives by considering how reclamation often produced a mixed legacy of success and failure. Historians have long held up Mormons as model irrigators, or "the Lord's beavers," for their proficiency at organizing themselves into colonies and overseeing the infrastructure to irrigate arid lands.[5] What follows is a story that complicates that narrative by relating the struggles, hardships, and failures Mormons experienced in their attempts to subdue the Virgin and Muddy Rivers that flowed through southern Utah and southern Nevada. While Mormons were remarkably successful irrigators, particularly in and around Salt Lake City, they encountered greater difficulties and often failure in the more arid region to the south where they settled in large numbers. Mormons who attempted to establish lives in this region from roughly 1865 to 1938 encountered a harsher environment where unpredictable rain patterns resulted in starvation, washed-away crops, and destabilized communities—sometimes causing them to collapse. Writing in 1941, shortly after water forced the abandonment of St. Thomas, the historian Juanita Brooks perfectly captured the perpetual struggle between Mormons who settled this region and their environment. Reflecting upon her youth in Bunkerville, Nevada, where the Virgin River flowed some fifty miles northeast of St. Thomas, Brooks recounted: "The people cannot conquer the river; it cannot shake them from its bank. It is like an endless war wherein first one side and then the other is victorious. The relationship is exactly as it was when the first settlers arrived."[6]

Founded in 1865, St. Thomas was the primary locale of a mission situated at the confluence of the Muddy, Virgin, and Colorado Rivers. The town's economy was based on agriculture and on its role as a railroad terminus where copper ore excavated from the Grand Gulch Mine was loaded and hauled away for smelting. A thriving community grew up where people lived, farmed, worshipped, and raised their children. In 1929, agents from the U.S. Bureau of Reclamation visited residents to inform them that their town would soon be submerged by a lake that would result from water pooling behind the Boulder Canyon Dam that the federal government would construct downriver.[7]

Like federal planners who built large-scale dams throughout the American West, residents of St. Thomas and of other Mormon communities along the Muddy and Virgin Rivers knew all too well the need to build infrastructure to meet their water needs. Following the pattern of the earliest Great Basin Mormon settlements, people who settled along Colorado River tributaries formed water districts and canal cooperatives in which they pooled labor and resources to build and maintain infrastructure that diverted water to their crops. By 1869, Muddy River settlers alone had constructed fourteen canals that ran more than sixty-five miles at a cost of $80,000 in order to irrigate just over seven thousand acres.[8] This was only one example of the transformations Mormon communities wrought throughout the Great Basin. Construction of such an elaborate water infrastructure suggests that their ability to pool labor and grow abundant crops enabled them to prevail in an effort one historian describes as converting a prehuman environment, or "first nature," into a landscape designed toward human ends, or "second nature."[9] This study demonstrates, however, that the features of a particularly arid landscape could undermine a given culture's ability to sustain second nature transformations over time.

Both the Muddy and Virgin Rivers proved volatile, at times overrunning their banks and washing away dams and bridges. Mormons who colonized tributaries along the Colorado River demonstrated remarkable ingenuity and skill at harnessing water from rivers to irrigate their lands. However, Mormons throughout southern Nevada and southern Utah sometimes struggled to establish enduring communities. This chapter will show that the aridity and the volatile rivers of this region presented them with unique technological challenges. Though they at times succeeded in building and sustaining their hydraulic societies, the unpredictability of nature frequently undermined their efforts.

"Technology" and "nature" are important albeit diffuse categories with meanings that vary over time and place. The dialectic between technology and nature possesses what some scholars have dubbed an "illusory boundary" because the relationship depends upon particular historical contexts.[10] For example, hydraulic civilizations emerged from 4,000 to 1,500 BCE in Mesopotamia, Egypt, and the Indus Valley. The technology these civilizations developed to practice irrigated agriculture represented a significant advance in humans' ability to control nature but also demonstrated that elites imposed rule on large numbers of citizens who performed the necessary labor.[11] Indeed, an examination of water control in the American West does not support the "comforting notion of progressive liberation of humans from their tools of desert conquest."[12] Rather, incipient attempts to irrigate the West from 1847 to the 1890s, when Mormons settled throughout the Great Basin, were "characterized by a general dependence on local skills and means" in which "small communities, living rather isolated from the rest of the world, diverted rivers to the extent of their limited ability."[13] Just as inequality emerged in response to ancient societies' efforts to irrigate, disparities in wealth emerged among Mormons located in the Salt Lake City core and peripheral locations where irrigation proved difficult. Power differentials existed within agricultural settlements in Greeley, Colorado (which was modelled on Mormon settlements), and in California's San Joaquin Valley. Thus, the "illusory boundary" between nature and technology in southern Utah and southern Nevada resulted in degrees of inequality among Mormons themselves as well as in failures to control nature.

The founder and leader of the Mormon religion, Joseph Smith, and his successor, Brigham Young, taught a theology infused with environmental values, but environment often prevented settlers from implementing their leaders' religious ideas. Scholars have argued that various Christian denominations possessed values that translated into stewardship over natural resources and, conversely, that religious values resulted in a paternalistic and hierarchical domination of the natural world. This paper steers a middle course between these extremes. Mormon religious values of cooperation and communalism greatly informed settlements throughout the Colorado Plateau and portions of the Mohave Desert, but the natural environment disrupted their lives so frequently and with such devastation that settlers often proved unable to realize the religious ideals that animated Brigham Young's geopolitical vision. Flash floods particularly devastated settlements. When clouds unleashed water, the physical environment transformed dramatically and with little notice. Engorged rivers washed away fields and undermined food supplies, carried off homes and workplaces, and potentially drowned human beings caught by surprise. Many early settlers suffered from severe food shortages when water inundated and destroyed their crops. Even the most committed, inspired, and zealous settlers often could not withstand or survive such conditions. Populations in many settlements either declined dramatically or entire communities simply abandoned their homes and moved to environments that offered better chances of survival.

The geographical and political space where the Virgin River flows constitutes a distinct environment, or bioregion, that shaped human settlement along its banks. A bioregion is a geographical space defined by natural rather than political boundaries and possesses unique characteristics derived from climate, hydrology, and ecology.[14] Bioregions vary based upon the qualities of their landforms, watersheds, plants, and animals as well as by the types of human cultures that subsist, thrive, and/or perish within the limits of the physical world they inhabit. The bioregional makeup of the environment through which the Virgin River flows shaped the lives and history of the people who settled there. The river originates in southern Utah, flows approximately fifty miles west where it crosses through the northwest corner of Arizona, and then runs another fifty-five miles through Nevada where it empties into the Colorado River. Mormons settled along its banks and along inflowing tributaries such as the Muddy and Santa Clara Rivers, all of which found their way to the Colorado River.

The space where the political boundaries of Utah, Arizona, and Nevada converge is also a space where two slightly different bioregions blend into one another. The first is the Colorado Plateau Province, a 130,000-square-mile region defined in large part by its cold winter temperatures at high altitudes and hot summer days. Precipitation averages twenty inches annually but typically less than ten inches in southern parts of the region. Rainfalls in summer most often in the form of thunderstorms that drop vast amounts of water in a relatively short time period. Much of the soil is "aridisol," or very dry and possessing a low concentration of organic matter with salt accumulated at the surface.[15] The second bioregion around the Virgin River basin is the Intermontane Sagebrush Province, marked by moderately cold winters, extremely hot summers, and annual precipitation ranging from five to twenty inches. This bioregion is even more arid than the Colorado Plateau. Sagebrush dominates but other plants grow as well, all of which tolerate alkali to varying degrees, a trait essential to their survival in soils that drain poorly and retain large deposits of salt.[16]

Vegetation and animals required millennia to adapt to this arid, alkaline terrain and the Mormons who rapidly attempted to settle, irrigate, and farm throughout the region encountered an environment that often frustrated their efforts to transform it. Mormons settled bioregions that marked southern Utah and Nevada as part of their desire to develop resources of the Great Basin and build a kingdom of God on earth. To realize this goal, church leaders dispatched settlers in 1852 to determine whether cotton would grow in a climate milder than that of Salt Lake City three hundred miles to the north. They established the town of Harmony at the juncture of the Santa Clara and Virgin Rivers, near modern-day St. George. Early reports to Mormon leaders indicated that crops grown in semitropical conditions would thrive in this environment, such as grapes, figs, flax, hemp, rice, sugar cane, and cotton. This positive report prompted church leaders to dispatch another fifty families in 1857 to establish the town of Washington near Harmony. Most of these settlers were new converts to Mormonism who had resided in the southern United States, and they were instructed to supply the Utah Territory with cotton. The environment soon thwarted their plans. The sandy bottom and banks of the Virgin River complicated irrigation by repeatedly washing away their dams. Even when they succeeded at getting water onto crops, minerals saturating the soil leached to the surface in the form of a hard crust that prevented crops from growing.[17] Mormon apostle George A. Smith visited the settlement and "learned that they were failing" in agricultural pursuits because "the waters of the Rio Virgin were poisoning the cotton" and because "the sun heated the sand, for the soil is nothing but the red sand of Sahara."[18] Cotton began to grow when settlers saturated crops with more water in order to reduce the soil's temperature. Eventually, cotton grew sufficiently that church leaders dispatched more people and charged them with building new settlements throughout the region.

To expand the cotton mission, church leaders dispatched approximately three thousand people to southern Utah and Nevada throughout the 1860s. One group consisted of 183 families that Brigham Young instructed to settle the irrigable valleys between St. George and Call's Landing, a distance of 125 miles.[19] They settled primarily into four communities along a 16-mile stretch of the Muddy River that flows into the Colorado River about 60 miles east of present-day Las Vegas. The first among these was St. Thomas, founded in 1865, and the others that soon followed were West Point (renamed Moapa), St. Joseph (renamed Logandale), and Overton. [20] One year after St. Thomas's founding, Nevada senator William M. Stewart introduced a bill to shift Nevada's state line one meridian of longitude eastward to ensure that the wealth generated from mines benefitted his state. Unaware that Congress had redrawn the political map, settlers in St. Thomas and in other communities along the Muddy River valley neglected to pay Nevada state taxes and accrued a debt they could not afford to pay, causing them to abandon their settlements in 1871.[21] Before departing, settlers along the Muddy River petitioned the Nevada governor and legislature to reduce or waive their outstanding debt because, they argued, of the hardships in cultivating an arid environment: "We have had to contend with great difficulties in trying to subdue these alkaline deserts having expended at least fifty thousand dollars in labor on water costs alone."[22] Back taxes pushed the Mormons out, but a dry, salty desert tempted them to leave as well.

The Intermontane Sagebrush Province presented harsh conditions for the settlers. Traversing the landscape to the Muddy River presented the first difficult encounter with the environment. The journey required them to drive heavy wagons over rocky trails, ravines, and streams, and their loaded wagons often became mired in the Virgin River's quicksand bed (which the trail crossed repeatedly). The steepness of one hogback required them to triple the number of animals on a team in order to make the climb and periodically block the wagons' wheels to rest their horses. Lack of water along the route meant rationing supplies to avoid dehydrating themselves or their animals.[23]

Environmental conditions did not improve even after arriving at the Muddy River. Although water existed in abundance, the appearance and taste of the Muddy River resembled its name. The river had a murky, milky color and drinking its water sometimes sickened settlers or caused cankers in their mouths. The surrounding desert environment of creosote bushes, cactus, and mesquite offered no shade from the scorching sun, which caused summer temperatures to rise as high as 120 degrees. Salt coated the naturally occurring grasses as well as the dirt, making for distasteful livestock forage and soil not conducive to agriculture. Winters proved no more hospitable because temperatures dropped to below freezing. Crude shelters partially shielded people from winds that whipped up desert sand, but dirt still found its way into homes—one woman swept nearly a bushel of it from each room of her dwelling.[24]

Controlling the Virgin River presented Mormons who settled all along its banks with problems and frustrations they struggled to overcome from the first days of settlement. Historian Andrew Karl Larson, a resident of St. George, argued that of all the problems settlers faced, "none was more persistent or frustrating than the struggle to control the Virgin River for irrigation."[25] Trying to irrigate crops from the river was a "constant battle" for communities near the waterway, from Shunesburg forty miles east of St. George to Bunkerville forty-five miles to the west.[26]

Settling along the Virgin River proved extremely challenging, but the experiences of one family suggest that hard work, familial relationships, and religious faith commingled to offer practical advantages and guidance while enduring difficult circumstances. Albert Frehner married Matilda Reber in Santa Clara, Utah, on April 20, 1888. Like many of the earliest Mormon settlers in southern Utah and Nevada, they lived a very sparse material existence in an arid environment that offered little sustenance. Immediately after marrying, they moved thirty miles west to Littlefield, Arizona, where Albert built a two-room adobe house with a wooden floor, planted crops, and grew a vineyard that he watered with the nearby Virgin River. Two years later, at the age of twenty-one, Matilda gave birth to the first of twelve children. Albert was absent much of the year working as a teamster, hauling freight from Bonelli's Ferry (where the Virgin River entered the Colorado) to the Eldorado Mine 135 miles southwest of their home in Littlefield. Each of the children worked from an early age to help provide for the family's needs by tending the crops, hauling water, or raising siblings. "Everyone in Littlefield worked in the orchards," recalled one of the Frehner children. As work progressed, "one could always hear the people (all relatives) talking, fruit dropping in buckets, fathers and mothers yelling orders to their kids. I'm sure Mama got the most of her brood."[27] Ten years into their marriage, Albert received a letter from church officials calling him to his native Switzerland, where he served on a mission for two years, leaving Matilda with four children and pregnant with twins. Describing their parents, two daughters recounted that "it never entered their minds or hearts to refuse this call" and that they "gladly accepted the challenge that lay head." Praising their mother, the women remembered that she "faced up to her responsibility," which involved two years of "tiring, exhausting work" running the post office, managing the farm, and boarding the local schoolteacher.[28] Matilda could take solace in the presence of her sister, Henrietta, who lived next door, separated only by a garden plot and some shade trees, to aid her in caring for the growing number of children. One of Matilda's daughters recounted that "mama and Aunt Henret did nearly everything together. Under the trees, they washed, combed hair, made soap, and both knitted and sewed and darned every moment of the time."[29] Cooperation and communalism were not necessarily easy or perfect among Mormon religious communities or within their families. In this particular case, however, the values and practices of family, work, and faith reinforced one another. Survival in very dire circumstances exacerbated by a harsh environment dictated that communalism prevail and that dissension and disorder had no place.

Matilda's children and grandchildren all proclaimed the matriarch's strong Mormon faith, and their memories of her highlight its practical advantages for survival. Her son, Merle, felt that her inability to read well resulted in her reliance on faith instead of church doctrine. Due to a lack of formal education, "her knowledge and reading ability of church books was somewhat limited," prompting him to help her read lessons published by the Mormon Relief Society and to write her replies.[30] A granddaughter who lived with Matilda toward the end of her life recalled, "Grandma's life was never easy." She concluded that her grandmother must have had "a very strong testimony of the truth" of the Mormon religion based on the fact that she agreed to let her husband depart on a mission while she remained at home, pregnant and with four children, with "only the land

and her strength to support them."[31] Such memories certainly reflect a strong dose of filiopietism prevalent among many accounts of early-day settlers by their descendants, but these statements also represent a hard-headed realism at the severity of circumstances Matilda faced throughout her life. She inhabited the land with very little to sustain her except a belief system shaped by her unique experiences. The necessity of surviving in such harsh conditions cannot be separated from her religion, a point that was true for many Mormons who settled this region.

Perhaps more than any other feature of this environment, water and the need to control it shaped people's lives. Juanita Brooks, another historian who was born and raised in southern Utah and Nevada, also emphasized the central role of water in shaping life in a Mormon community along the Virgin River. Brooks grew up in Bunkerville, Nevada, eighty miles east of Las Vegas. The arid, desert environment in which she lived dominated the experience of living in Bunkerville. The arrival of water in the irrigation canal that ran through town was an event worthy of celebration. "The water's in!" residents rejoiced, a sentiment that "all who live in the arid lands of the West will appreciate."[32] Brooks recalled that "the water, which meant life to us, seemed almost like a living thing" as it snaked its way through town.[33]

Water's arrival in irrigation ditches proved particularly memorable after prolonged periods of scorching heat. For nine months of the year the weather was pleasant, but for three months residents suffered from intense heat, "I mean the kind that thickens the whites of eggs left in the coop."[34] So pervasive was the absence of water and so significant when it arrived that children learned to organize their time around the availability of water. Brooks described how "the whole routine of my chores was determined by whether or not the water was in the ditch."[35] During episodes of drought, children escorted cattle and horses two miles so they could drink from the Virgin River.

Prior to digging an irrigation canal, people who settled along the Virgin River had to transport water by wagon to meet their needs. Residents in the town of Mesquite, Nevada, hauled water for two and a half miles by wagon, a cumbersome task that required at least two people per trip. The process required mounting a large barrel atop the wagon. One person scooped water from the river and dumped the bucket into the barrel, while the second person sat in the wagon resting a foot on the brake to prevent the horses from riding away. Once the barrel was filled, the family washtub was placed over it to prevent spillage and settlers then traversed the bumpy wagon trail back to Mesquite. Despite their best efforts, much of the water still spilled onto the ground.[36]

Shortages of food offered the most powerful testament to early settlers that transforming their environment would not come without costs. A scarcity of food plagued most Mormon settlements along the Virgin River throughout the 1860s.[37] Drought was the primary cause of food shortages. The definition of "drought," however, depends upon one's concept of "normal" rainfall.[38] Drought occurred so frequently on the Southern Plains that farmers there accepted it as a fact of life and were unsurprised when dry weather plagued the region in the 1930s. Settlers of the Intermontane Sagebrush Province and subsequent generations of Mormons quickly altered their definitions of "normal" and adapted their expectations of economy and diet to their immediate environmental circumstances. Prices escalated, food became scarcer due to drought, and few could afford to pay in either cash or labor for the food they required.[39] Mormon settlers adjusted their eating habits in order to receive the calories their bodies required. They grew alfalfa as forage for livestock, but began consuming it themselves to supplement their diet. They ate naturally occurring plants such as pigweed and redroot, or resorted to a concoction of "greens" derived from cooking the tops of carrots, beets, and other vegetables that normally would have been discarded.[40] Altering their diets may not have been their first choice, but it was their only choice during dry, lean periods and served as an effective strategy of gaining necessary calories to increase their chances of survival. Thus, their adaptations normalized their perceptions of environment and the choices necessary to sustain their bodies. In this respect, Mormons who settled in even the most acutely arid portions of the Great Basin behaved in a way consistent with many western settlers. Their attempts to make sense of their environments were interwoven with perceptions of their bodies and adjustments to new physical conditions.[41]

Church leaders who mandated the cultivation of cotton so that settlers could trade it for food failed to consider how environmental constraints might undermine such a plan. A drought from 1863 to 1864 devastated agricultural pursuits. Settlers could not adapt quickly enough to normalize such a severe lack of water. The inability to water crops for even a few days meant that plants withered in the southern Utah heat.

Drought also weakened farm animals. Farmers relied upon grasses that grew naturally to feed their livestock, and the lack of rain resulted in little or no forage.[42] Livestock that survived performed poorly because they needed more water and food than was available.[43] Church leaders quickly concluded that their decision to emphasize cotton over food crops had been a mistake they could not repeat.[44]

Mormon leaders who lived further north also displayed ignorance of southern Utah's environmental limits by imposing significant tithes on settlers during a time of drought and famine. The church called upon residents of Washington County to provide fifty-five outfitted wagons and teams in 1863 and twenty-eight in 1864 as a tithe to the Perpetual Emigration Fund.[45] The church planned to use these teams to transport starving people from Florence, Nebraska, to Salt Lake City. St. George residents also faced starvation, however, and Erastus Snow determined that there existed slightly more than one-fourth of a pound of food per day for each person until the next harvest occurred. Despite their own food shortage, St. George residents complied and dispatched the requested teams, wagons, teamsters, and supplies—but not without Snow taking the opportunity to remind the bishop who ordered the tithe about conditions in the south.

Harsh weather conditions particularly impacted grasslands and undermined St. George residents' ability to feed their livestock adequately. Snow expressed irritation when he explained to Bishop Edward Hunter that "it will be more difficult for us to send 28 teams this year" than sending fifty-five teams the previous year.[46] Unpredictable weather conditions presented new challenges for feeding animals. Dryness and cold temperatures during the past winter meant "little or no grass for our stock. No rain or snow to wet the ground since last September. Hay and grain for our teams have been entirely out of the question."[47] What seemed to irritate Snow the most was that he had previously informed Hunter of the difficult circumstances St. George residents faced: "As I have told you before I left Great Salt Lake City, our people have been constantly selling off their teams for grain."[48] Hunter's apparent indifference stemmed from his lack of familiarity with southern Utah. Snow considered him "not personally acquainted with" the area and took the opportunity to point out "a few facts which may be of service to you hereafter."[49] First, the Dixie Mission had lost population as wealthier people were "mostly released, or neglected to come."[50] Second, many residents who arrived departed for nearby Iron and Beaver Counties, which had larger populations. Third, people who resided in Salt Lake City owned much of the existing livestock that grazed nearby and used these animals to pay tithes to wards located there rather than in St. George. The conditions Snow enumerated cast the Dixie Mission in the role of a colony that existed at the periphery of Mormon settlement. Salt Lake City functioned as the core where Mormon religious leaders resided and dictated economic practices to the south. Cotton production was the primary function of settlements from St. George to the Muddy River valley, but settlers located in these peripheral locations suffered disruptions to their lives when environmental conditions undermined their ability to function in Mormon leaders' geopolitical vision.

A marked decline in population or total abandonment of Mormon colonies occurred particularly throughout the later nineteenth century for reasons related to the environment. According to one study, of the approximately five hundred communities established by Mormons in the United States throughout the later nineteenth century, 69—or 13.9 percent—eventually ended in failure. Adding settlements from Mexico and Canada increased the failure rate to 16.4 percent, or 88 out of 537 communities. External factors—the Mexican Revolution, conflicts with Indians, and tension with the federal government over polygamy—caused some of these failures. However, factors arising from within a settlement also caused many to fail. Most often settlers simply chose to leave, often due to the environment. Floods, insufficient water, and a poor geographic location typically prompted them to leave.[51] Environmental factors resulted in the failure of forty-three communities while the external factors listed above caused twenty-six to fail.[52]

Mormon settlements throughout the Colorado Plateau failed at particularly high rates, and they failed most commonly due to flooding. It is ironic that, in an extremely arid environment, torrential downpours of rain presented the biggest obstacle to settlement rather than merely a lack of water. Flooding did not present a major problem for Mormon settlements along the Wasatch Front, but plagued communities along tributaries of the Colorado River upon their founding and up to the present.[53] Gushing rain transformed rivers and streams into flooding torrents that threatened the lives and livelihood of people in the region. Historians have interpreted these floods as phenomena that both highlighted the "remarkable cohesiveness" of Mormon communities in times of crisis as well as catastrophes that "washed away the pioneers' determination," property, and sometimes their lives.[54] Indeed, the severity and frequency of floods presented settlers with an unpredictable force that threatened their communities and lives on an ongoing basis. Floods took Mormons' lives and destroyed their property, undermining their attempts to build stable and lasting communities.

The great flood of 1862 proved catastrophic to Santa Clara and Washington as well as to settlements farther up and down the Virgin River. Andrew Karl Larson wrote extensively about southern Utah's people and places and captured the sentiments of many in characterizing the Virgin River as "peaceful to the point of deceit" throughout much of the year but "a hussy to control" when spring runoff began.[55] Larson's description characterized the Virgin's tributaries as well. For example, the Santa Clara wreaked havoc for three days in January 1862. Approximately twenty families of ninety people lived in Santa Clara at the time. They were Swiss immigrants who had cultivated about twenty acres of peaches, cotton, and grapes as well as indigo, figs, and olives.[56] Daniel Bonelli was one of the leaders of this community, and he explained to Brigham Young the toll that "the disastrous and devastating flood" took on the community.[57] Bonelli recounted that "incessant rains and snowstorms" fell for two weeks, transforming "the little mountain stream Clara to the size of a monstrous river." He believed the precipitation "unprecedented," but considered it "a small thing" compared to the "dreadful inundation" that fell on January 17 and 18, which made the river overflow, "destroying orchards and field" and washing away Jacob Hamblin's mill. The massive influx of water transformed their environment. Uprooted trees and logs floated by "like arrows upon the terrible current [and] presented a spectacle of dreadful magnificence." Floods abated and currents slowed just long enough to give people who had climbed trees for safety the opportunity to escape. Just when they thought the worst had ended, the water-soaked ground shifted, buildings sank, and residents quickly began transporting food, furniture, and other possessions to "safe ground." At three o'clock in the morning the following day, their fort, schoolhouse, and nearby homes "disappeared and in their place roar now the wild torrents of the river." Banks of the river widened and crumbled to accommodate the increased flow. The road between Santa Clara and St. George became impassable because "dark masses of clouds hanging on the mountain tops" prevented "the monstrous river" from subsiding. Bonelli reported that the Swiss settlers would "soon feel heavily" a lack of food due to a "scarcity of flour and other provisions."[58]

Episodic flooding occurred along the Virgin River between the 1860s and the 1880s. A heavy downpour flooded the town of Orderville in 1881, leaving six to eighteen inches of muddy water in homes. The water carried timber, gravel, and rocks from nearby hills, piling debris onto the streets, gardens, fields and irrigation ditches. Four years later, floods destroyed dams and cut deep gorges into the valley. Willard Carroll, who lived in Orderville at the time, recalled that the river rose fourteen feet overnight and that floods devastated the town's infrastructure and food supply. When he went to sleep at 10:30 p.m., the waterline was ten feet below a nearby bridge. The next morning he awoke to a "roaring sound" of water that had risen four feet above the bridge. The swiftly moving current swept away all but one bridge as well as farmers' crops. Carroll's neighbor lost fourteen acres of wheat and only a small portion of a six-acre cornfield remained. His brother and father "lost heavily" in potatoes and carrots, and Thomas Chamberlain, a local church leader, lost crops of oats, barley, potatoes, and carrots.[59] The Virgin River and its tributaries presented settlers with a devil's bargain: abundant water for their crops yet an unpredictable and at times threatening menace when rainfall accelerated.

The river proved equally destructive to settlements downriver in Bunkerville, Nevada. Joseph Earl recorded on December 5, 1889, that "it is raining very hard." Two days later, "the largest flood came down the (Virgin) river that has ever come since the white people lived in this country. It was two feet higher than the flood of 1861. It washed away most all of our dam and part of our ditch." Two days passed and another flood of equal size occurred, causing him to lament, "We feel quite blue over the damage done to our ditch."[60] The river had risen two feet in height, which dug the channel deeper and wider.[61] The disaster that came in two waves devastated Mormon communities along the Virgin River, including those in the town of Washington, which served as the focal point of Brigham Young's cotton mission. People grew disheartened at what increasingly felt like a hopeless struggle to subdue the Virgin River.

The same flood threatened lives and destroyed the irrigation canals of Mormons who lived in the nearby settlement of Mesquite. In 1882, a downpour of water caused a flood that threatened the lives of the town's inhabitants, rushing through the windows of George B. Whitney's house and collapsing its roof. Whitney carried his wife and baby to higher ground and returned for his other children. His son Luke, who was eight years old, recalled vividly how he sat with his two sisters on a hillside, covered with a quilt: "The rain was pouring through those quilts; the lightning flashing; the thunder rolling and rocks rolling down the hillside." His father returned to their home in an attempt to salvage indispensable items buried by the mud. The following morning revealed that the flood had washed away fifty-eight segments of the irrigation canal. The community responded quickly and began "steadily working on the canal" to repair the damage. However, twelve of the men who labored became ill and nine had to cease working. A malaise fell over the town due to the scorching heat and "filthy water to drink," causing many to consider leaving permanently.[62] The flood took such a devastating toll on the material and psychological conditions of people's lives that within six months only ten families remained in Mesquite. The following year the religious community's bishop and counselors departed as well. Still, however, another family arrived and for six weeks built a dam made of logs and rocks. When another catastrophic storm destroyed their irrigation canal, they too departed Mesquite and moved further upstream. By 1891, the town was entirely abandoned.[63]

The same flood proved equally devastating to other communities along the Virgin River from Nevada to Utah. Many living in Bunkerville five miles away felt no recourse but to leave after the flood destroyed the irrigation ditch they worked so hard to construct. Hattie Leavitt Black also recounted that the flood came in two waves. Initially, increased water flow was small enough that levees kept the river at bay, but the second wave "lasted four or five hours breaking our ditch and carrying it almost entirely away."[64] People living in Washington, Utah, forty-five miles eastward experienced the same disruptive effects of the flooding river. Perhaps no greater symbol of Mormons' failure to settle the region can be found than canal company president and ecclesiastical leader Bishop Marcus Funk asking for and being granted permission to leave. Other settlers moved from Washington as well, and the town's population dwindled from approximately 600 to 312 by 1892.[65]

Flooding also occurred south of the Virgin River along rivers and streams that ran throughout portions of the Colorado Plateau into northern Arizona, washing away irrigation works and causing many people to abandon their mission. Roughly one hundred miles east of modern-day Flagstaff, residents in the town of Woodruff struggled to subdue rivers and streams for forty years beginning in 1877. People who attempted to live there built thirteen dams over this period, only to see eleven of them washed away by torrential downpours that caused waterways to overflow their banks.[66] Settlers believed they triumphed over the raging waters in 1919 when they constructed the Woodruff Dam just above the confluence of Silver Creek and the Little Colorado River, but the effort was only "a nominal victory" that merely "postponed defeat," according to one study of the effort.[67] The thirteenth and final structure withstood the river's flow but required that the exhausted residents confront the technical challenges of excavating irrigation canals along rocky canyon walls. Forcing water to obey their will had become such a challenge that by the 1950s the people of Woodruff had almost entirely abandoned irrigated farming.[68]

Flooding along the Virgin River so disrupted the lives of residents living in Littlefield, Arizona, that it eroded the farmland of the Frehner family and prompted them to relocate sixty miles southwest to St. Thomas, Nevada. In 1910, the Virgin River came alive once again as "swirling masses of red water" pushed their way through the town and onto their land: "Our biggest enemy in Littlefield were the floods that came down the Virgin. Everyone took with it a part of our farm. On Christmas Eve, the flood hit that was to drive us from our home."[69] The following year, the family's father loaded their belongings into two covered wagons and began a three-day trek to St. Thomas, where he had purchased sixty acres of land.[70] The loss of land that provided sustenance was surely severe, but Matilda mostly missed her friends, extended family, and the work that shaped the contours of her former life. One of her children recalled, "Mama missed Littlefield and the relatives very much."[71] The memory of the challenges incurred in forging a life amidst great struggle and powerful environmental hurdles have lingered in Matilda's mind, as she spoke of them even as her grandchildren entered the world. One grandchild recounted that "Grandma often spoke of the big flood in 1910 in the Virgin River and how some of the homes and a lot of the land washed away."[72] Memories of cultivating the land and the kinship enjoyed with family and friends seemed to haunt Matilda even as her children married, moved away, and had families of their own: "It took a long time for Grandma to get adjusted to the move from Littlefield even though she could see it was for the better. She missed the nice gardens and the fruit and the wonderful cold spring of water she and Henrietta used to sit by and chat during the day and mend or knit."[73] She missed the friends she made among the several Mormon families. Matilda and her family moved on to St. Thomas, where they would make more of the same kind of memories involving work, family, and friends.

While most Colorado River tributaries overran their banks and flooded nearby towns, the Muddy River was small by comparison, less prone to flooding, and eventually enabled an agricultural community to grow and flourish. One visitor noted the dramatic difference in water flow between the Muddy River and the juncture of the Virgin and Colorado less than thirty miles south. Daniel Bonelli operated a ferry where the Virgin emptied into the Colorado that he recounted was 610 yards wide and 63 feet deep and a "rapid current at full stage of water," thus requiring a ferry to cross. The visitor was "surprised at its magnitude," which contrasted greatly with the more sedate tributary to the north: "This stream called the Muddy is remarkable for its uniformity, remaining the same all year round."[74] So impressed was he with the slow-moving "stream" that he concluded the "valley would be a great feeder and would develop into something worth seeing." Mormons had vacated St. Thomas thirteen years prior, but he recommended the site as a place where "the bees of Deseret must of necessity swarm" again because "the sweet potatoe [*sic*] seems at home" here and the "grapes and fruit will do excellently."[75] Indeed, at the dawn of the twentieth century a more stable community arose in St. Thomas. The population increased from 93 to 194 people from 1910 to 1930.[76] The town's economy stabilized and expanded as farmers irrigated their fields to grow primarily cantaloupes but also alfalfa, watermelons, sugarcane, corn, sweet potatoes, peanuts, peaches, pears, grapes, and pomegranates, among other fruits and vegetables.[77] Relying upon teamsters to transport their produce, they capitalized on their proximity to the booming Grand Gulch copper mine that served as a ready market for purchasing food.[78] Given the central role of teamsters to the economy, the need for roads became imperative.[79] Completion in 1915 of the Arrowhead Highway enabled St. Thomas merchants to access markets in Las Vegas. [80] Even more important to accessing distant markets and growth of the town's economy was the arrival in 1912 of a railway spur from the Los Angeles and Salt Lake Railroad.[81] The growth of St. Thomas and accompanying infrastructure suggested that despite its isolation and rugged arid terrain, humans with their roads and railways had successfully constructed a second nature apparatus to overcome constraints imposed by prehuman first nature conditions.

Prosperity abounded for many St. Thomas residents as late as the 1920s, but flooding of the Muddy River undermined the technology and labor that helped community members transform an arid landscape into a burgeoning economy. Flooding has been endemic to the Intermontane Sagebrush Province that characterizes southern Nevada. In Clark County, which makes up the southernmost tip of the state, flooding occurred at least 169 times from 1905 to 1975.[82] Repeatedly, torrential rains resulted in heavy flood damage to bridges, railroad tracks, farmland, ranches, houses, barns, and irrigation canals. Large volumes of water fell rapidly onto sandy, alkaline soil resulting in severe erosion and disrupting communities throughout the county, including residents along the Muddy River living in towns such as Moapa, Overton, Logandale, and St. Thomas. Rain fell in such abundance and so quickly that the Virgin and Muddy Rivers overflowed and washed away people's homes and crops. Floods inflicted "considerable damage" in 1916 on property in St. Thomas and eroded a sixty-foot section of the bridge approaching town.[83] Again, in 1921, "torrential rains" fell for six days causing the Virgin River to divert course and remove a section of bridge, making it impassable.[84] Catastrophic floods in 1920 and 1932 destroyed grapevines, fruit trees, and gardens.[85] In a region plagued with flooding, St. Thomas fared considerably well and suffered less damage from naturally occurring floods than neighboring communities. The ultimate flood, however, that would inundate the town and compel all residents to find new homes occurred in 1936 and resulted from the efforts of the federal government to dam the Colorado River.

Volumes have been written about the Hoover Dam and the topic of water in the American West. The dam is significant as the first "great" construction project that instilled confidence in engineers to dam famous rivers throughout the world.[86] The reasons it was built included the need to provide farmers and settlers with a steady supply of water for agriculture and consumption, hydroelectric power, and flood control. To many critics, however, the dam serves as a metaphor for the time in which we live. Its construction inaugurated an age of great expectation, hope, and possibility for many, but the overuse, overallocation, and overregulation that have led to its current state of diminished flow—revealing a town like St. Thomas—may also serve as a reminder that not all people benefitted equally from its existence.[87] Donald Worster argued that political pressure to build the dam was not so much a mounting demand from an indigenous population but originated outside the Colorado River Basin as part of a strategy by urban entrepreneurs to stimulate a migration boom, by agricultural capitalists in California's Imperial Valley, and by engineers within the federal bureaucracy keen to demonstrate their technological expertise.[88] According to Worster, "The chief political lesson of the Hoover Dam is that a new concentration of economic, social and political power is the outcome of the domination of nature."[89] This inequality became most apparent in 1930 when the federal government began making plans for relocating St. Thomas's remaining 234 residents, which involved appraising their land and property that would soon be flooded in order to determine what constituted appropriate compensation. Federal officials perfunctorily involved the community in the process by including a local resident, Levis Syphus, on the board of appraisers, but the final report generated consternation and disapproval. In response to the appraised values, St. Thomas bishop Robert O. Gibson reflected that people felt hurt by the great variation, inequality, and seeming arbitrary prices attached to their property.[90]

Just as Matilda Frehner and her family fled Littlefield, Arizona, due to floods, they once again had to abandon their home and relocate due to uncontrollable water. This time, Matilda and Albert headed north to Alamo, Nevada, to live with a son who had grown and started a family of his own above the headwaters of the Meadow Valley Wash, which also constantly flooded but ran southward and emptied into the Muddy River near their previous home. They were not alone in their exodus from St. Thomas. Hugh Lord stubbornly refused to believe Lake Mead's water would ever reach St. Thomas and decided to go fishing one evening only to witness rising water swirl around his parked car.[91] Lord's story possesses a touch of humor, but the departure of the vast majority of people like the Frehners from St. Thomas reads like the ending of a sad movie. Mother Matilda and brother Vivian were the last to leave. Vivian expressed his feelings by kicking one of the walls the last time out the door, much to his mother's dismay, who "came running" when she heard the sound of his foot hitting the wall. He attempted to explain himself: "'We start tearing it down in the morning!' and she said, 'That doesn't make any difference to me. I want that home the way it is.'"[92] Her children understated their mother's grief when they recounted the necessity of moving once again:

"It must have been difficult, at her age, to leave her beautiful home and all of her friends and go to start over once again."[93] Difficult indeed, for when Vivian and his parents got into the car to depart, he recalled that "I never saw her cry so hard in my life. She was shook up really bad."[94] On their way out of town they stopped at the Mormon Relief Society to dispense remaining provisions and to say goodbye to longtime friends. Vivian minced no words in capturing the moment: "It was just like a funeral. They did nothing but divide up things and cry."[95] After saying their goodbyes, "finally it came time for Mom and me to get into the DeSoto and travel to Alamo, and she took one big look and said, 'Well that's the last home I'll ever have.'" Sensing great public interest in the story of the town's sinking, Bureau of Reclamation engineers took care to document an official date of subsidence: June 17, 1938.[96]

People who lived in communities along the Virgin River and nearby tributaries would eventually contain the raging waters by building larger and more sophisticated canals, dams, and catchment basins. By approximately 1900, they had constructed much of the infrastructure that enabled their population and food supplies to stabilize. However, floods would never entirely disappear and, in fact, continue to wreak havoc throughout this bioregion up to the present day. So much rain fell during October 2004, for example, that water in the Virgin River increased from zero to eleven thousand cubic feet per second in the course of two days. The Santa Clara River's flow increased from zero to nineteen thousand cubic feet per second over three days in January 2005.[97] As recently as 2015, flash floods took the lives of twenty people along the Utah-Arizona border, across which the Virgin flows. Containing the river proved possible for a time, but controlling the rivers of this bioregion indefinitely remains an illusion.

Environments are not static spaces wholly controlled, despite the prominence and proliferation of towering dams throughout the American West. The hydraulic societies that proliferated throughout the region present an illusion of anthropomorphic control over nature because they leave a heavy footprint, at times inundating and erasing from the map towns like St. Thomas, Nevada. What makes the story of St. Thomas so compelling for understanding water in the West, however, is the recent reappearance of foundations and artifacts that had been submerged for more than eighty years under Lake Mead. Remains of the town remind us that the loss of place is a common theme in American life and particularly in the American

West. Construction of the Glen Canyon Dam and the ensuing inundation presents the most notorious reminder that hydraulic societies can eliminate landscapes and communities and create environmental disputes about who profits and loses when dams alter a river's flow.[98] Such tales of loss leave us with a conundrum, to which we typically respond in one of two ways: either we exalt transformations of environment and the dams that created them or we castigate such technologies and lament the loss of nature incurred.[99] Such a narrow range of options does not leave much intellectual flexibility to assess relationships that existed between the environment of southern Utah and Nevada and the thousands of people who constituted the Mormon culture and religion.

"THERE ARE MILLIONS OF ACRES IN OUR STATE"

Mormon Agrarianism and the Environmental Limits of Expansion

Brian Q. Cannon

AMERICANS HAVE A LONGSTANDING AFFINITY for agriculture as a wellspring of character and democracy. Thomas Jefferson famously articulated American perceptions of the link between a virtuous citizenry and farm life when he wrote, "Those who labor in the earth are the chosen people of God, if ever he had a chosen people, whose breasts he has made his peculiar deposit for substantial and genuine virtue." Jefferson believed that small-scale farming fostered industriousness and resourcefulness. Due to their economic self-sufficiency, he reasoned, yeoman farmers were more likely to resist political bribery or coercion.[1]

Mindful of the Biblical commission to Adam and Eve to till the soil, many Latter-day Saints born in the nineteenth century espoused Jefferson's affinity for farming. "If you want virtue, go into the farming country," advised Apostle Heber C. Kimball in 1857. Apostle John A. Widtsoe articulated the Jeffersonian bias when he proclaimed, "Strength, vitality and power, and rich form of living come to those who deal with the earth as the Lord gave it to man." Many Mormons regarded farm life as the ideal laboratory for nurturing religious faith, morality, and industriousness. Church president Heber J. Grant declared, "There seems to be strength, physical, moral and religious, which comes to those engaged in cultivating the soil, which, on an average, is far superior to that of any other occupation."[2]

Mormons' conviction that agriculture was ennobling and divinely ordained intertwined with their belief that God had tempered the Great Basin climate to enable the Latter-day Saints to farm there. Thus abundant agricultural harvests in the Mormon heartland reified the Latter-day Saints' special relationship to God. Broader cultural convictions undergirding American agricultural expansion—including the notion of divinely ordained manifest destiny and the belief that rain would providentially follow the plow—reinforced the Saints' perceptions of divine approbation. Combining these strands, Mormon settler and autobiographer William Farrington Cahoon asserted that Providence "made the wilderness we came to inhabit to blossom as the rose," proving that God was the Latter-day Saints' "friend." Similarly, Nancy Tracy, who trekked west in 1850, credited "God's blessing" with the transformation of the Salt Lake Valley from a "barren desert" into "fine orchards and beautiful dwellings." Some asserted that God had extended the growing season in Utah's mountain valleys. Edward Phillips, who established a farm east of the Great Salt Lake in 1850, recalled, "At the time of my first settling here, we could not raise a peach tree, but the elements are so softened that now we can raise any kind of fruit." Similarly John Taylor reminisced in 1865, "Remember the time when we could not raise peaches to eat, and it was a doubt whether an apple tree would grow or not?" The climate had been too unpredictable, the killing frosts too frequent. But God "blessed the elements for our sakes, and also the earth" to the point where "there is not a better peach growing country in the world than this."[3]

Church leaders sharply distinguished between Mormon agricultural success in the Great Basin and agricultural settlement by non-Mormons elsewhere. Brigham Young taught, "There is not another people on earth that could have come here and lived. We prayed over the land, and dedicated it and the water, air, and everything pertaining to them unto the Lord, and the smiles of heaven rested on the land and it became productive." God had tempered the climate to sanctify the Great Basin, Apostle John Taylor taught in 1865, but "let the Saints leave this place and it would return again to the wilderness condition. The wicked could not live here."[4]

It is unclear how widely shared these beliefs were, but by the twentieth century the fond hope expressed by some members of the pioneering generation that the agricultural resources of Utah would sustain the growing population and never diminish seemed to be flagging among the Mormon rank and file. Water was available to irrigate only a tiny fraction of the state's land, and California and Colorado had surpassed the once-dominant Beehive State in irrigation investment, technology, and productivity. Apostle John Henry Smith lamented in 1909 that some towns and farming districts were regressing: "the canals are filled with weeds; the orchards are old; dead trees are seen; the fruits are wormy and unfit for use; the farms are without legitimate and proper cultivation." Ecologically fragile mountain and desert pastures had been gnawed into the ground, a consequence of miscalculation and overgrazing. On alpine rangeland near the headwaters of southern Utah's Escalante River, for instance, where the grass had once grown "so high you could hardly see the sheep for it," vegetation was sparse. Beginning in 1897 torrential summer downpours, a common phenomenon in the Southwest, darkened the river's waters with mud and scoured its banks. Each flood washed away more farmland along the stream until its bed, which had once spanned only a dozen feet, stretched out to nearly a tenth of a mile. The populations of most rural districts were still rising, partly because of new births. But population pressure, water shortages, erosion, declining soil fertility, and economic depression induced many of the rising generation to abandon the farm communities of their childhood for life in Utah's cities and towns or for opportunities in other states. The quest for greater economic security in the face of environmental limitations in rural Utah took some of the settlers to cities and rural districts where Mormon congregations were fledgling or nonexistent—a fact that alarmed some church leaders, who feared the migrants would lose touch with Mormonism.[5]

This chapter first examines Mormon leaders' attempts to check out-migration and perpetuate frontier abundance by encouraging new agricultural settlement in Utah early in the twentieth century, along with the environmental and economic consequences of that expansion. Thereby it illuminates facets of Mormonism's environmental history, highlighting the interplay between nature and cultural forces, such as the teachings of church leaders and the application of those teachings by Mormon settler-farmers. Clinging to traditional views of Zion building, church leaders from the 1890s to the 1920s encouraged the Latter-day Saints to "build up Zion" by settling new lands in Utah, farming more intensively, and patiently enduring privation. Partly in response to that counsel, Mormons moved onto fragile lands, with dire ecological and economic consequences. Next, this chapter explores another, subsequent cultural intervention on the land: the federal government's efforts to adjust rural land use. During the 1930s land-use planners and scientists employed by New Deal agencies documented the negative environmental impact of recent agricultural expansion and proposed permanently removing some farms from cultivation. In Utah, Mormons directed and participated in these reformist programs that sometimes conflicted with the ideals of their religion and with church leaders' counsel, and Latter-day Saint leaders pushed back modestly by pointing their followers to settlement opportunities on newly reclaimed lands. In the long run, market forces and environmental constraints determined the persistence of Mormon farmers on the land to a greater degree than either New Deal policies or sermons. Utah's farm population fell by over one-third between 1940 and 1959. By midcentury, while many Mormons and their leaders from farm backgrounds continued to favor farm life as an ideal, they recognized that few in the rising generation would farm. Over time the challenges of a largely metropolitan Zion effectively eclipsed agricultural settlement and farming in Mormon religious discourse. In the process, Mormon cultural identity underwent a fundamental transformation in response to environmental and economic forces.[6]

As early as 1898 Apostle George Q. Cannon vigorously decried Mormon wanderlust. Cannon regarded Utah as a sacred physical space, a promised land where distinctive Mormon identity and religious beliefs could be nurtured. Mormon identity was becoming attenuated now that some of the more distinctive markers of the group—including polygamy, theocracy, and consecration—were waning. Cities with their polyglot populations provided numerous social and cultural alternatives to Mormonism. Cannon counseled the rising generation to settle on a farm "some place in the State and build the State up, build up Zion." The seventy-one-year-old apostle voiced the pioneer generation's conviction that God would multiply the region's resources, advising, "There is plenty of room on every hand."[7]

Notwithstanding Cannon's counsel, Mormons continued to leave Utah for larger tracts of unclaimed or inexpensive land with better water rights in "Canada, Wyoming, Colorado and Mexico." By 1910 over sixty-one thousand people born in Utah—about one-fifth of all natives of the state—had moved elsewhere. "I have heard Utah people say, that we have no more farming lands, and that our boys must go somewhere else to secure farms. Many having farms have a spirit of unrest," observed Apostle Reed Smoot in 1910.[8]

Apostle Heber J. Grant took up the gauntlet in 1903, advising the Saints to give Utah a second chance and "create improvements whereby the land will produce more." By husbanding scarce resources and farming intensively, the Mormons could make the land yield more abundantly. "We think we are cultivating the land, why, we are simply scratching it over," he exhorted. Six years later Apostle John Henry Smith urged faithful Mormons to "secure lands—within the confines of our own states and neighborhoods." The following autumn, apostle and senator Reed Smoot pointed out that Congress (with his encouragement) had recently passed the Enlarged Homestead Act, thereby doubling the number of acres a homesteader could claim from 160 to 320. The techniques of dry farming promised to convert unirrigated, semi-arid rangelands into fields of waving grain. "There are millions of acres in our state capable of yielding an abundant crop of grain by the process of dry farming," Smoot advised. "There is plenty of land for every one of your sons to establish a home within the borders of the state of Utah." Apostle John Henry Smith in 1910 directed the Saints' attention to lands in eastern Utah recently opened for settlement on the former Uintah Indian Reservation.[9]

While church leaders' religiously motivated desires to promote rural settlement were particular to Mormon country, outmigration from the American countryside was a nationwide phenomenon that alarmed pundits in every state. Although America's rural population grew by 9 percent between 1900 and 1910, the urban population spiraled upward by a whopping 39 percent. Migrants from abroad and from the countryside were moving to the cities in droves. In Utah the rural-urban disparity in population growth was even greater. The rural population increased by 29,095, or 17 percent, over the decade, while the urban population rose by 66,967, or 63.5 percent. Members of the nation's Country Life Commission, convened by Theodore Roosevelt in 1908, decried the gravitation of rural Americans to cities, while proponents of the more amorphous back-to-the-land movement in the 1910s and 1920s advised city folk to take up farming.[10]

Church leaders were influenced by these national discussions. After attending the 1910 National Farmland Conference where the urban drift was discussed, Apostle John Henry Smith warned, "People who crowd into cities and live in rented homes, who are subject to every little change in the character of their employment, and who find themselves, in a great measure, the slaves of their fellow-men, cannot be fully patriotic and devoted to their country."[11]

In the 1910s it appeared that apostles' and pundits' warnings about urban migration and their paeans to farming were bearing fruit. Americans increasingly plowed up new lands, and Mormons joined in the trend. Over that decade Utah's farm acreage increased by nearly 50 percent, or 1.65 million acres, and the number of acres of key crops harvested (corn, wheat, oats, barley, rye, hay, sugar beets, and potatoes) rose by nearly 37 percent. High prices for farm products and the generous provisions of the Enlarged Homestead Act of 1909 spurred some of the expansion. Dryland farms accounted for about three-fourths of the new cropland (1.25 million acres), much of it in remote desert country where roughly five thousand homesteads were established, but over four hundred thousand acres were also brought under cultivation on irrigation projects. Acreage devoted to Utah's foremost irrigated cash crop, sugar beets, rose by 240 percent over the decade.[12]

Many Latter-day Saints who claimed new land in this era did so with the pro-settlement rhetoric of church leaders ringing in their ears. Over 80 percent of the settlers who flocked to former Ute reservation lands in Duchesne County were Mormons. Latter-day Saints were heavily represented, too, among the new settlers in many dryland farming districts in western Utah and eastern Nevada, including Tintic Valley, Dog Valley, the Escalante Desert, and Rush Valley. To a greater degree than has been generally appreciated, this twentieth-century wave of homesteading and settlement in Utah had ecclesiastical as well as economic underpinnings.[13]

Beyond their public, pro-settlement rhetoric in formal church venues, general church leaders facilitated and promoted some of the new settlements in Utah's hinterland, especially on irrigated land, between 1890 and World War I. Thereby they sanctioned and legitimized specific ventures. In the 1890s the church-backed Deseret and Salt Lake Agricultural and Manufacturing Canal Company filed on government land northwest of the Sevier River under the Desert Land Act and partnered with the Deseret Irrigation Company to construct a network of dams and canals for conveying water to the land. Church leaders named the new settlement

Abraham and recruited settlers. In 1902 when the privately owned Bear River Water Company—a corporation formed to convey water to more than forty thousand acres of new farmland in Box Elder County—fell on hard times, the church-supported Utah Sugar Company (led by church president Joseph F. Smith) purchased the company, extended its canals, and sold land to settlers. Thereby the church facilitated "the most stable and lasting agricultural development" in Utah in the Progressive Era.[14]

In response to lobbying from cash-strapped developers of the Hurricane Canal in southern Utah in 1902, the church purchased $5,000 in company stock, a vital infusion of capital that enabled the company to complete the project and open two thousand acres to settlement and farming. The church's investment infused an essentially economic enterprise with ecclesiastical legitimacy. As local booster James Jepson told the church's president, "the psychological effect" of the support would "be equal to the money itself, for the people will not be afraid to invest in the project," knowing that church leaders wanted it to succeed. Two years later church president Joseph F. Smith spurred Mormon settlement of eastern Utah by instructing Wasatch stake president William Smart "that he would like to see Utah settled up by our people instead of their going off to other states." The prophet urged Smart to "go forth and use arguments by way of persuasion to get our people to settle" in the Uinta Basin. Many of the recruiting arguments Smart subsequently employed were religious. Later in the decade the church president accompanied Quince Kimball and Carel Mangum, developers of a dry farming settlement in Johns Valley, to the church-owned Zions Bank and personally approved a loan of $8,000 to purchase Sweetwater Spring as a culinary water supply for the new town of Winder, named for the prophet's recently deceased counselor John R. Winder.[15]

Local church officers capitalized upon general church leaders' rhetorical and financial support for Zion-building to dignify the plowing of new farmland as a sacred endeavor. For instance, inspired by an apostle's sermon, Israel Bennion, a church leader in Rush Valley, urged poor Scandinavian converts to establish farms there, assuring them that God would enlarge the water supply if they shared their resources with each other. Bennion and his new neighbors trusted that they were doing God's work. "My motives are about thus," Bennion confided in his journal. "I want this waste place of Zion redeemed [an allusion to Isaiah 1:27]; I want the poor Saints provided with homes." William Smart, Joseph Murdock, and James Jensen, the presidency of the Wasatch LDS Stake, sent a letter to bishops and stake presidents throughout Utah in 1905 urging them to publicize settlement opportunities in the Uinta Basin for "young men and middle aged, who have little or no holdings where they now reside." Referring to the counsel of general church leaders for support, Smart and his counselors wrote, "You will call to mind that it is deemed inadvisable for our people to seek new homes afar off when such tracts as these await the reclamation of the husbandman, and are located within the confines of Utah, the center place of the Lord's establishment of his people in the West."[16]

Emboldened by this rhetoric, some Latter-day Saints who established new farms envisioned themselves as divine agents and Zion builders, and their religious sense of a mission fortified them in the face of privation, water shortages, and crop failure. At a Uintah Stake Conference in 1917, Stake President Don Colton urged hard-pressed farmers "to keep up their courage and not become discouraged." A member in the congregation stood and accepted the president's challenge, "express[ing] his determination to help build up the country." Similarly, Bishop Wilford Hyde told his fellow settlers on the Metropolis Reclamation Project in 1912, "It is God's will that we are here." He advised them therefore to "to struggle along and not give up," to exercise faith, looking forward to the day when "beautiful fields and a prosperous community" would crown their efforts. Carlos Lambert, the agricultural extension agent who advised the settlers, attributed their extraordinary tenacity solely to "the social and religious organization of the church" that gave them "hope and held them together.[17]

Although Mormon settlers' religious faith that they were building Zion intensified their commitment to hardscrabble farms early in the twentieth century, it rarely resulted in a distinctive cultural landscape. Distinguishing elements of nineteenth-century Mormon farming villages included homes set on a grid with fields surrounding the village (a legacy of Joseph Smith's plat of the city of Zion), irrigation ditches running alongside streets, cooperatively constructed irrigation canals, relatively small farms, and (for a time in the 1870s) communal farming under the direction of local church leaders. By contrast, in the early twentieth century, as Charles Peterson has noted, most dryland farming communities in Mormon country closely resembled their counterparts elsewhere in the West with their "section-line roads, ripgut fences, one-room schools, and isolated homesteads cover[ing] the country."

The new farms also followed national trends toward larger size, higher capital investment, risk-taking, and commercial orientation, with less emphasis being placed upon self-sufficiency or the commonwealth. Although small farms remained relatively common in the Beehive State, average farm size nearly tripled between 1880 (69 acres) and 1925 (192 acres). In constant dollars, average implement values per farm were 80 percent higher in 1925 than in 1890.[18]

To finance their expanded operations Utah farmers increasingly mortgaged their land—a trend decried by church leaders away from village-oriented fiscal conservatism and a subsistence economy and toward national norms. In 1890 less than 6 percent of the farms in the state were mortgaged, but that percentage increased to 11 percent in 1900, 22 percent in 1910, and a whopping 44 percent in 1925. The prospects for making money ballooned with higher levels of investment, but so did the likelihood of foreclosure.[19]

In the 1920s church leaders' warnings about the perils of farm debt materialized: markets for farm products evaporated following World War I, commodity prices fell, and foreclosures multiplied. The economic incentives for persisting on recently settled lands dwindled, and farmers retreated from the land. In a sophisticated study Conrad Taeuber, an agricultural economist employed by the Bureau of Agricultural Economics, estimated that close to 19 million Americans moved from farms to cities and villages in the 1920s. Over the same decade about 12 million moved from cities and villages to farms. Net outmigration from farms to urban areas was especially pronounced between 1922 and 1926. In Utah, Taeuber estimated, the countryside suffered a net loss of roughly forty-five thousand farm residents over the age of ten during the 1920s. The only state in the West that experienced a higher net loss than Utah was Montana (at fifty-one thousand).[20]

Church leaders continued to discourage families from fleeing to cities in the 1920s, advising rural Mormons to "stick to the farm" despite the rural economic depression. Agrarian beliefs shaped their counsel. "We can rear better citizens and better Latter-day Saints upon the farms than in any other place," advised church president Heber J. Grant in 1923. Countering the observation that farm life deprived families of electric lights, radios, indoor plumbing, and leisure time, Presiding Bishop Charles Nibley taught that toil was noble and that farm labor instilled a stronger work ethic than a "soft" urban job. Rather than coveting urban lifestyles farmers should cherish their unparalleled opportunity to "commune" with God under the open skies, Nibley's counselor David A. Smith advised in 1927. "From the very beginning there has been a relationship among God, man and the soil, making man dependent upon the soil for his very existence." Peace and happiness would flow from farm work, he promised.[21]

The counsel steeled some farmers' resolve to persist under discouraging circumstances. For instance, in October 1928 settlers in the Pahvant Valley west of Delta who faced falling income and crushing bonded indebtedness met in a special Mormon priesthood meeting. Earlier in the year Apostle Melvin J. Ballard had spoken in their stake conference and prophesied that the faithful would weather "storms in season" and have the ability to "pay the creditors and redeem this land for a heritage to this people." Recalling that prophecy, the men approved a resolution affirming that "the majority of our people have confidence in the promise of President Heber J. Grant and Melvin J. Ballard as prophets, seers and revelators." Southward in Washington County, a depressed region with tiny farms and sharply limited water, one government employee noted that Mormon residents stuck to the land because "they had been called to this locality by the church and promised that it should be indeed a home for them and their children."[22]

While church leaders in the 1920s were advising Mormons to "stick to the farm," a land-use planning movement with scientific rather than spiritual underpinnings was gathering steam in the nation's land grant colleges, the offices of agricultural extension agents, and the headquarters of the U.S. Department of Agriculture. The agricultural scientists and economists who championed this movement used soil samples, data on precipitation, and farm production records to criticize the movement of farmers onto what they branded "submarginal" land unsuited for cultivation—along with the resultant wind and water erosion, plant regression, and poverty. Rather than inducing impoverished settlers to trust that God would temper the elements on hardscrabble farms, they argued it was time to face environmental limitations squarely, acknowledge that the frontier era was over, and remove marginal farmland from cultivation. As Secretary of Agriculture Henry Hyde stated in 1930, "In the United States there are millions of acres of lands which, because of location, soil exhaustion or natural infertility, cannot be made to produce a living equal to the American standard. These are known as submarginal lands. From the standpoint of national agricultural efficiency, as well as individual efficiency, they ought not to be farmed." Land economist L. C. Gray defined submarginal terrain as land that would not yield "normal standards of return to labor and capital" even "under proper conditions of utilization."[23]

The idea that lands were inherently submarginal, a cultural construct backed with a degree of empirical evidence, potentially clashed with the Mormon spiritual conviction that with irrigation water, dry farming techniques, and faith in divine providence the desert would "blossom as a rose." Not surprisingly Mormon scientist Thomas L. Martin softened the concept for Mormon consumption. In 1928, Martin, an agronomist at Brigham Young University, introduced readers of the church periodical the *Improvement Era* to the concept of submarginal soil. Unlike land-use planners such as Gray, who argued that submarginal soils were inherently unsuited for cultivation and should be permanently removed from farming, Martin argued that poor farming rendered the soil submarginal and urged his readers to conserve and rejuvenate soil by rotating crops, applying manure, and removing weeds from irrigation ditches. It was not so much the soil as the people who used it that made the land submarginal. "Marginal people tend to collect on soils which yield a margin, and sub-marginal people accumulate where soils have been allowed to decline to the point of low and unprofitable productivity," he wrote.[24]

The Great Depression of the 1930s and demands for government action to provide relief to destitute farmers galvanized support and financial appropriations for land use adjustment. In 1934 the federal government began purchasing submarginal farmland and resettling struggling farmers. A succession of agencies supervised the buyout.[25]

The program of identifying and retiring submarginal lands from cultivation implicitly critiqued the Mormon perception that pioneering in harsh lands and bringing them under cultivation to build Zion was a noble enterprise, but New Dealers and Mormon leaders rarely commented on the ideological tension. Church leaders' sympathy for the farmers' economic vulnerability and poverty may account for their silence. Perhaps some had also been persuaded by the land-use planners' empirical arguments regarding environmental limitations. For their part, New Dealers pragmatically downplayed the tensions and instead compared the New Deal approach to Mormon pioneer values.[26]

Most of the men who administered the New Deal's potentially iconoclastic land use adjustment programs in Utah were themselves Mormons. A few, including agronomist and state land planning consultant Aaron F. Bracken, were less than fully engaged in the church, but others, including regional land planning consultant J. Howard Maughan (later called as a stake president and mission president) and stake president William F. Palmer, were devout, regionally prominent church leaders. The regional director of the Agricultural Adjustment Administration's land policy section, the agency responsible for initiating submarginal land purchase projects in Utah in 1934–35, was Philip Cardon, a son-in-law of Apostle Anthony W. Ivins and a former student of Apostle John A. Widtsoe. These Mormon New Dealers never publicly blamed the pro-settlement and pro-farming rhetoric of church leaders for the movement of Mormons onto marginal lands. In 1942 Palmer admitted that many of the places colonized by Mormons in the pioneer era could be termed "submarginal" at the time of settlement in the sense that they "cut squarely across every condition that our government today considers indispensable to successful colonization." But he defended the nineteenth-century pioneers and attributed their triumphal persistence on unlikely lands to divine intervention.[27]

Although they refrained from directly criticizing Mormon colonization and settlement, the New Deal's land planning specialists did flay some of the results in their reports. In a 1934 preliminary report on land use problems in Utah's agricultural districts (based largely upon data collected by the U.S. Census Bureau and other government agencies), Bracken pinpointed districts with farmland that "should never have been cleared" due to their "physical and economic" handicaps. Bracken identified seventeen areas of the state where land use adjustment was needed.[28]

Two years later, after touring the problem areas identified by Bracken and interviewing county extension agents, Howard Maughan, Bracken's supervisor, added sixteen more areas to Bracken's original list of seventeen. Maughan estimated that the federal government would need to purchase and repurpose 5,280 farms statewide in order to redress land use problems in these areas. Most of the farms to be purchased had already been abandoned, but 1,200 were still inhabited and the residents would need to relocate.[29]

In their reports Maughan and Bracken blacklisted 350,000 acres of dry farm country spread across the western half of the state that were "not adapted to farming" for environmental reasons. These included nearly 70,000 acres north and west of the Great Salt Lake; 11,000 acres west of Utah Lake in Cedar Valley; about 140,000 acres in Rush, Tintic, Dog, and Juab Valleys; nearly 600 dryland farms in the Escalante Desert of western Iron and Beaver Counties; 22,000 acres in Hamlin Valley northwest of Cedar City; 40,000 acres in Johns Valley north of Bryce Canyon; and the Curlew Valley bisected by the Utah-Idaho border. Most of this land had been settled in the 1910s under the provisions of the 1909 Enlarged Homestead Act, and none of it had been colonized as part of a church-sponsored venture, but church leaders had still encouraged Mormons to claim dryland homesteads. The movement into these dryland farming areas with insufficient precipitation or poor soil had been "disastrous for the settlers," most of whom were impoverished, and "bad for the land" because the protective grass cover had been plowed, leaving the land vulnerable to erosion, Maughan summarized. Subsequently livestock had been allowed to graze on the abandoned farmland, exposing the soil to severe wind erosion and creating dust storms.[30]

Bracken and Maughan also recommended permanently removing some irrigated land from cultivation. Some of that land had been settled during Brigham Young's lifetime, including severely eroded terrain in Pine Valley and marginal farms near Gunlock and Kanab. Farms settled in the 1870s and 1880s between Cannonville and Escalante and in the Fremont River Valley between Fruita and Hanksville were too small to be economically viable. Moreover, in many cases the water could be applied more productively to more fertile soil elsewhere. In the irrigated districts of Carbon and Emery Counties, settled by Mormons beginning in the late 1870s, water was adequate but topsoil had been washed away and huge gullies gouged out as flash floods encroached upon farms. The limited land base could no longer support all the residents. Similarly, in San Juan County farms in the pioneer settlement of Bluff, founded in 1880, were "gradually washing away" as the river bank eroded. "Little now remains of what was formerly a prosperous community," Bracken reported. He recommended resettling the remaining families.[31]

The scars of imprudent irrigated settlement were even more common on recently developed land, some of which had been settled with encouragement from religious leaders. The network of tracts in the Delta area, where irrigation had been attempted since the 1890s, included seventy-five thousand acres "which never did, nor never can economically produce crops in sufficient quantity to provide a living for a family." In these districts, Bracken reported, "Incomes reach the vanishing point, tax delinquency is almost 100% and abandonment is high." Injudicious application of irrigation water and poor drainage had brought alkaline deposits to the surface, necessitating expensive drainage works. As the alkaline-encrusted farms were abandoned, livestock often foraged in the fields, decimating native grasses and creating a foothold for toxic invasives such as whorled milkweed or plants with thorns and spines like Russian thistle. South of Utah Lake in Genola, water that began flowing from the Bureau of Reclamation's Strawberry Reservoir in 1915 was being unwisely applied to "gravelly" and "inferior" soil. Bracken identified the Uinta Basin, home to 1,600 destitute families, as the "largest and most serious [irrigated] problem area in the state." The maladjustment was especially acute in Duchesne County, settled largely by Mormons under William Smart's leadership between 1905 and 1912. Water supplies were undependable and the soil in many locations was "shallow," rocky, and alkaline. Although the district was better suited to stock raising than farming, few settlers possessed grazing rights on the public domain.[32]

Some problem areas possessed fertile soil but lacked sufficient water, including the western portion of the Salt Lake Valley, northeastern Utah Valley, the Elberta area south of Utah Lake, and the state-sponsored Hatchtown and Piute projects along the Sevier River. The Newcastle and Enterprise areas in western Iron County, home to over five hundred farmers, also suffered from chronic irrigation water shortages.[33]

A sweeping government buyout of over a million acres of farmland spread across these problem areas—nearly twenty percent of the state's agricultural land—and the government-funded relocation of over 1,200 farm families from those lands (as envisioned by the state's land planning consultants) would have been fundamentally transformative and would have likely ignited strident criticism from civic and church leaders. As it was, the sweeping reforms contemplated by New Dealers fizzled; only seventy thousand acres of farmland were purchased by the New Deal land reform agencies in Utah: thirty thousand north of Bryce Canyon surrounding the town of Widtsoe, and more than forty thousand between Benmore and Tintic Junction in Tooele County. The lands acquired in Utah comprised barely 0.7 percent of the 11 million acres acquired nationwide between 1933 and 1946 under the land use adjustment program. Only two already-dwindling communities, Benmore and Widtsoe, were phased out.[34]

Unlike some facets of the New Deal, this tiny buyout elicited negligible criticism in the Beehive State. Once, when the New Deal designated some recently purchased land in Tooele County for experimental revegetation studies, officers in the local grazing district fretted that the government land purchases might "eliminate thousands of acres of valuable grazing lands from grazing purposes." The government buyout of the town of Widtsoe got its namesake, John A. Widtsoe's, attention. Addressing a gathering of U.S. Department of Agriculture employees he reportedly quipped, "If I were to hate anybody, I'd hate your organization. You moved the town away that was named for me." But Francis Mortensen, a government employee who heard the remark, believed that Widtsoe was "joking" even though he recognized by the comment that the apostle "felt very seriously and earnestly about the town that was named after him." Widtsoe, an agricultural scientist who had once championed dry farming, also acknowledged its limitations. He eventually concluded that dry farming at its "very best" was "only a minor practice to be followed by people who live near the cultivated areas."[35]

Those farmers who sold their blacklisted farms to the government and resettled more acutely felt the contradiction between the New Deal's land use adjustment program and their religiously inspired sense of mission on the land. After all, the Book of Mormon taught that righteous living and prosperity on the land were correlated: "And if it so be that they shall keep his commandments they shall be blessed upon the face of this land, and there shall be none to molest them, nor to take away the land of their inheritance; and they shall dwell safely forever." After he deeded his farm in southern Utah to the government, Reed Beebe struggled to reconcile his failure with the memory of a promise made by Apostle Melvin J. Ballard in a local church conference. Beebe's neighbor Dorothy Elder also puzzled over Ballard's prediction. As Beebe recalled, the apostle likened the valley to the Garden of Eden and promised, "It [the valley] will remain thus so long as you stay out of debt, and keep the commandments of God; but if you fail to keep his commandments it will be cursed, and will be taken from you." "How had we failed to keep God's commandments?" ruminated Beebe. "Wherein had we failed, but we had."[36]

While the New Deal's land use adjustment program was removing farms in Rush and Johns Valleys from cultivation, drought and economic reverses were forcing out farmers across the state. In the process more land blacklisted by New Dealers went out of production.[37] After increasing early in the Great Depression, the number of farms in the state fell by 17 percent between 1935 and 1940 while the number of acres harvested in 1940 was 16.7 percent less than in 1930. Outmigration was especially heavy from the overextended agricultural districts of Millard, Duchesne, San Juan, and Emery Counties and the overcrowded farming and sheep ranching communities of Sanpete County.[38]

Settlers who had taken up land in good faith found reason to question that faith or their own standing before God. As the population dwindled in the Mormon town of Metropolis near the Utah-Nevada border, one resident wondered if the settlers' failure was a sign that "the Lord is displeased with the efforts that have been put forth in His behalf." A prophecy made by Apostle Melvin J. Ballard in Duchesne in 1919 stipulated that if settlers there would live righteously they would "prosper and have water," which impressed William Smart. Smart subsequently experienced financial reverses and referred to them as an "atonement," implying that they amounted to penance for a moral lapse or shortcoming on his part.[39]

In 1940 Apostle J. Reuben Clark reinforced the struggling settlers' sense of guilt. The church leader nostalgically recalled the 1910s "when our lands were in the heyday of their productivity." Deep snow blanketed the mountains in the winter and assured abundant irrigation water, "farmers and stockmen were prosperous," and "our homes and fences and barns and corrals were in repair." By contrast, little snow had fallen of late and Clark expected that "another fifty years of this will almost return us to the desert." Everywhere the lack of industry was evident, from "idle" fields to weed-strewn streets and tumble-down barns and corrals. Clark attributed the decline in precipitation and harvests to the failure of Latter-day Saints to live their religion and called for repentance.[40]

Even as farmers across the region struggled to eke out a living in the face of drought and depression, Mormon leaders persisted in searching for agrarian remedies for the destitution and hunger they observed in the state's cities and towns. It was hard to shed the Jeffersonian vision of the redemptive power of farming. David O. McKay in 1936 urged Mormons who had moved to the city to "go back to these farms, to retill them, and at least produce sufficient for our own needs." In 1938 mission president and future general authority Preston Nibley departed from the traditional emphasis on remaining in the Mormon culture region and pointed to the Columbia Basin in eastern Washington as an emergent agricultural frontier due to the construction of the Grand Coulee Dam by the U.S. Bureau of Reclamation, encouraging Mormons to settle virgin farmland there. As part of its welfare plan, conceived partly as a modest counterweight to the New Deal, the church's Agricultural Advisory Committee prepared and circulated lists of farmland available for purchase in the twelve western states in 1939. "Our people will always be a more stable people if they can keep their feet firmly planted on the soil," wrote Orval Stott on behalf of the committee. In the decade following World War II hundreds of Mormon families pursued this course and established new farms on federal reclamation projects.[41]

The migration of hundreds of Mormon families to farms on new irrigation projects following the Great Depression was more than offset by the outmigration of thousands of other farm youth and their parents. Utah's farm population fell from 105,000 in 1940 to 81,000 in 1950 and 69,000 in 1959, or less than 8 percent of the state's total population. Meanwhile, the number of farms fell by 30 percent between 1940 and 1959. Environmental constraints played a role in the decline, but even larger factors were World War II, industrial expansion in urban areas, and the cost-price squeeze in agriculture that imperiled small farms.[42]

In the 1940s and 1950s, as the percentage of farmers within the church's membership plummeted, the reality of a largely metropolitan church eclipsed the agrarian image of Zion, and the frequency and intensity of spirited tributes to farming from the Mormon pulpit waned. Through the 1930s, church leaders perceived Mormonism as a predominantly rural religion. Presiding Bishop Sylvester Cannon in 1934 observed, "We are, in considerable part, an agricultural people." Church president Heber J. Grant rejoiced in 1937 that "the majority of our missionaries come from farms" and Apostle John A. Widtsoe stated that he was "thrilled" that "perhaps two-thirds of the Latter-day Saints are farmers." After World War II that was no longer the case. As Ezra Taft Benson noted in 1945, "[We] do not hear as much now about these things [agriculture and farming] as was once spoken in the Church." Puzzled by the silence, "many young men and some older men" who had been taught to regard farming as the noblest vocation approached Benson to inquire about "the Church's interest in agriculture and farming." Likewise, a "regular" parade of young men approached John A. Widtsoe, the tireless and aging advocate of family farming, "want[ing] to know where to go to find new land," and wondering what had become of the church's Agricultural Advisory Committee.[43]

Church leaders had not rejected the agrarian ideal with its veneration of farming as an ideal source of virtue. In the 1950s they attributed social ills, including irreligion and narcissism, to rural depopulation. Apostle George Q. Morris denounced the trend toward disbelief among college students. "If there is anyone who wants to go to college and hasn't the spiritual capacity to take on a college education without losing his faith, he had better stay on the farm and do his duty in the Church and serve his fellow men and serve God and grow in knowledge and intelligence and power there," he observed in 1952. A year later Benson attributed rising materialism, selfishness, and divorce to "industrialization, specialization, concentration of populations in great cities, the great reduction in the number of people living on farms, [and] the change in our agriculture." But Benson and his colleagues—with the possible exception of Widtsoe, who died in 1952 at the age of eighty—admitted that small farms were no longer economically viable and that the majority of Mormons lived in urban areas. As production costs rose due to mechanization, the best way to raise money to pay for machinery was to till more land. When some farmers moved and sold their land to their neighbors, the remaining farmers enjoyed "a better chance to produce efficiently and live well," as Apostle Ezra Taft Benson, who was serving as secretary of agriculture in the Eisenhower administration, observed in 1956. Rhetorically he asked who in their right mind would "insist that farmers do away with modern equipment and go back to horses and mules, so that adjustments to justify modern equipment would not be necessary."[44]

The Mormon concept of a providentially founded agrarian Zion based in Utah helped spur agricultural settlement on dryland farms and new irrigation projects early in the twentieth century. Much of the land that was broken up after 1900 lacked the requisite soil fertility or water. During the First World War demand for farm products ensured farmers of a market for all commodities they could grow. But the agricultural depression that ensued following the war and deepened in the 1930s battered Utah's farmers. New

Deal-era economists and planners catalogued the dire environmental consequences of decades of agricultural expansion and recommended that the government purchase and retire one-fifth of the Beehive State's farmland to bring the state into line with its environmental limitations. The proposed buyout challenged the Zion-building outlook of Mormon agrarian expansionism. Church leaders offered negligible opposition to the buyout, which barely got off the ground in Utah. Instead they tacitly acknowledged that Utah's line of settlement had stretched beyond its environmental limits and advertised opportunities for farm settlement elsewhere in western America. No longer focusing exclusively upon Utah in their definition of Zion building, they also began downplaying Zion's agrarian dimensions as the Mormon countryside hemorrhaged its population in the 1940s and 1950s. By midcentury, Zion came to be marked by practices suited to metropolitan living, although most urban stakes retained a welfare farm where church members periodically hoed sugar beets or picked fruit by assignment. With the exception of these farms, Mormonism's standardized brick meetinghouses surrounded by cookie-cutter subdivisions eclipsed the older agrarian image of Mormon country. Environmental limitations and market forces had fundamentally reoriented the Mormon culture region.

"THE PROPHET SAID TO PLANT A GARDEN"

Spencer W. Kimball and the Transformation of the Mormon Agrarian Tradition

Nathan N. Waite

In March 2009, First Lady Michelle Obama planted the White House Kitchen Garden, the first garden at the presidential residence since Eleanor Roosevelt's victory garden in 1943. Owing in large part to an economic recession, the number of households growing food rose from 36 million in 2008 to 40 million in 2009. By planting a garden, the first lady was both highlighting this trend and trying to amplify a conversation about healthy eating and child obesity.[1]

As one Mormon blogger pointed out that same week, "The Mormons have been promoting gardening for many decades." While the nation was experiencing one of its periodic gardening revivals, Latter-day Saints had steadily kept up the tradition since the 1970s, and the agrarian impulse had been part of Mormonism from its earliest years in the 1830s. Still, the blogger was willing to offer an inclusive hand of fellowship into the gardening club: "If you're Mormon," she wrote, "teach your children the song, 'The Prophet Said to Plant a Garden.' If you're not, you might substitute the words 'the prophet' with the words First Lady and sing the song anyway."[2]

The person most directly responsible for the inclusion of gardening as a feature of current Mormon belief and practice is Spencer W. Kimball. Almost immediately after Kimball's appointment as prophet and president of The Church of Jesus Christ of Latter-day Saints in late 1973, he called on his people to clean up their yards, redouble their gardening efforts, and pay more heed to home food preservation. He brought the subjects up again and again in the church's twice-yearly worldwide General Conference for the better part of a decade. His emphasis on gardening remains an essential element of Mormons' memories of Kimball, and although rarely is food production mentioned over the pulpit in Salt Lake City or in the church's official periodicals anymore, the gardening thread is still prevalent among Latter-day Saints along Utah's Wasatch Front and beyond.

The roots of Mormonism's agrarian tradition run deep, as other essays in this compilation attest, reaching back to the earliest gathering places and the Saints' imagined heaven on earth, Zion. In 1833, two years after Joseph Smith designated the western border of Missouri—and at the time, of the United States—as the place where the Saints would build a city of God to prepare for Christ's Second Coming, he and other church leaders made that ideal concrete with detailed maps and descriptions of the city's layout. At the center of the city of Zion would be a collection of religious and civic buildings, with the "House of the Lord," the holy temple, at the very center. The farms and livestock were to be outside the city limits, to allow residents to live nearer each other and thereby foster a community. The residential area was not to be densely urban, however: "no one lot in this City," the plat mandated, "is to contain more than one hous[e]." Though lots were narrow in width, each would be a half-acre in size, allowing houses to be set back from the road and providing a deep backyard. Preserving a middle-landscape aesthetic straddling urban and rural, blocks alternated between a north-south orientation and an east-west one, so that each front door looked out not at another line of houses but at the sides and the backyards of the neighbors across the wide street. And every yard in Zion was to produce its own food: the plat called for "a small yard in front to be planted in a grove according to the taste of the builder the rest of the lot for gardens &c."[3]

Though the Zion in Jackson County, Missouri, was not to be, future Mormon settlements followed the plat's pattern, adapted to accommodate local topography. In his study of the typical Mormon village in the Intermountain West, Richard Francaviglia noted that the plat of the city of Zion "is perhaps the most important single document in the history of settlement of the West." Although there is no single community in the

Mormon West that followed the Zion plat exactly, components of the plan appear throughout the region—blocks oriented to the compass directions, wide streets, and flourishing gardens. Tracing the line from imagined community to actual landscapes, Francaviglia writes, "The stipulation of the front yard for a grove, with garden space around the house, coupled with the order to build in brick and stone, goes far in explaining some important visual elements in the Mormon landscape: the solid masonry homes, the clumps and rows of trees, the intensively cared-for gardens."[4]

The gardens and yards of the communities modeled after Zion were both productive, their fruits and vegetables supplying the pioneer kitchen, and decorative, beautifying the landscape in the tradition of ornamental gardens. Visitors to Utah took note of both aspects. Boston native Elizabeth Wells Randall Cumming, whose husband Alfred Cumming was appointed as governor of Utah Territory in 1858, glowingly described one home and yard. Without knowing it, she was essentially describing the specifications of the plat of the city of Zion: "I wish I had a picture of it for you—for it is very pretty. It stands about 130 feet back from the street—flowers &c in front—peach & other small trees on each side of the house & extending to the street—a large garden behind & on each side."[5]

Visitors often wrote of the neat and inviting appearance of Mormon residences. Mark Twain, who customarily spared no vitriol when it came to the Latter-day Saints, wrote approvingly in his 1872 *Roughing It* that Salt Lake City had "a limpid stream rippling and dancing through every street in place of a filthy gutter; block after block of trim dwellings, built of 'frame' and sunburned brick—a great thriving orchard and garden behind every one of them, apparently—branches from the street stream winding and sparkling among the garden beds and fruit trees—and a grand general air of neatness, repair, thrift and comfort."[6] A generation later, John Muir was struck by the Salt Lake Valley's "leafiness": "Most of the houses are veiled with trees, as if set down in the midst of one grand orchard. . . . They are set well back from the street, leaving room for a flower garden, while almost every one has a thrifty orchard at the sides and around the back. The gardens are laid out with great simplicity, indicating love for flowers by people comparatively poor, rather than deliberate efforts of the rich for showy artistic effects."[7]

Thirty years after Muir, non-Mormons were still noting with admiration the beauty and charm of pioneer villages in the West. In 1942, at a time when church leaders saw their agrarian way of life threatened with extinction, preeminent western historian and popular novelist Wallace Stegner wrote of "the characteristic marks of Mormon settlement: the typical, intensively cultivated fields of alfalfa and sugar beets and Bermuda onions and celery, the orchards of cherry and apple and peach and apricot (and it is not local pride which says that there is no better fruit grown anywhere), the irrigation ditches, the solid houses, the wide-streeted, sleepy green towns."[8]

Church leaders took pride in this aspect of the Mormon landscape. In their eyes, the Latter-day Saints had come to the wilderness and tamed it, and every rose hedge and peach tree was a visual confirmation of their triumph and improvement of the land. General Conference was often a venue for extolling the virtues of beautiful agrarian communities. "Will I plant an orchard?" asked Brigham's brother Joseph Young in 1857. "Yes sir, I will. . . . I will never cease my exertions here, but I will do all that I can to beautify the place. I have done my best to do so, according to my means: I have planted my grape-cuttings, and I have eaten some fruit; I have planted my peach orchard, and have eaten the fruit thereof; and I rejoice to see improvements among this people."[9] Brigham Young preached a similar message again and again. In 1863, for example, he instructed,

> Beautify your gardens, your houses, your farms; beautify the city. This will make us happy, and produce plenty. The earth is a good earth, the elements are good if we will use them for our own benefit, in truth and righteousness. Then let us be content, and go to with our mights to make ourselves healthy, wealthy, and beautiful, and preserve ourselves in the best possible manner, and live just as long as we can, and do all the good we can.[10]

In a prayer on the land where a temple was to be built in Cache Valley, Utah, Orson Pratt petitioned God's help in beautification: "Wilt thou bless their labors in their endeavors to beautify their habitations, and in planting out shade trees to make this a delightful place."[11] Gardens, orchards, and shade trees were an integral part of being Mormon in this period when the Saints endeavored to make the wilderness bloom as the rose.

Young and the other Mormon leaders were not the first Americans to speak of agriculture in such reverential tones; rather, they were participating in a long tradition that praised farming as the noblest occupation. In many ways Thomas Jefferson was the foremost proponent of this idea, and in *Notes on the State of Virginia* he famously wrote, "Those who labour in the earth are the chosen people of God, if ever he had a chosen people, whose breasts he has made his peculiar deposit for substantial and genuine virtue. It is the focus in which he keeps alive that sacred fire, which otherwise might escape from the face of the earth."[12] Jefferson—and many after him in the nineteenth century—felt that agriculture yielded not only a healthy crop but a healthy and virtuous citizenry, while industry and manufacturing were ultimately debasing activities. In this view, the yeoman farmer was and should be the quintessential American. While nineteenth-century Mormon leaders thought more highly of industry than Jefferson had—and, given their isolation from the rest of civilization, they had little choice but to pursue mining and manufacturing themselves—still they saw working with the earth as the quintessential Mormon activity. And certainly in the nineteenth century, farming was the way of life for most Latter-day Saints.

Researcher Emily Anne Brooksby Wheeler wrote:

> In Mormon culture, the garden was an important part of the religious landscape. It provided the Mormons' Zion with self-reliance and reduced its dependence on the "wicked" outside world, and its cultivation proved the faithfulness of communities and individual members to their assignments. In the larger sacred landscape created by the Mormons, gardens were an individual expression of faith with very practical benefits.[13]

Farming and gardening, working with and at times wrestling against the land to eke out a subsistence in a semiarid region—these lifestyles and values were instilled from the church's beginning and seeped into Mormon culture until Mormonism and agrarianism were inseparable.

It was concerning to the church hierarchy in Utah, then, when national and statewide agricultural trends shifted. According to historian Donald H. Dyal, "By the dawn of the twentieth century, something new was afoot in Mormondom. Mining, railroads, statehood, and the accompanying market economy were transforming Utah from an overwhelmingly rural polity to an increasingly urban one." The proportion of Utahns engaged in agriculture dropped from 65 percent in 1890 to 50 percent in 1910. From 1900 to 1910, Utah's urban centers grew by 60 percent, well above the national average of 35 percent.[14]

In many ways, the twentieth-century story of Mormon communities in the United States follows the agricultural trends of the larger American public, though with some delay. The number of farms in the United States fell for the first time in its history between 1920 and 1925, from 6,448,343 to 6,371,640. The Great Depression saw the number rise again, as job losses in other sectors drove Americans back to the land, but after 1935, every census for the next forty years recorded a drop in the number of farms, down to 2,314,013 in 1974. Nearly two out of every three farms had disappeared. In Utah, which was among the more rural states, the picture was more complex, with a different pattern of peaks and valleys, but the trend was unmistakably similar. After an upward trend reversed in the late 1930s, the number dropped from a high of 30,695 farms in 1935 to 12,184 in 1974, a loss of 60 percent.[15] Concurrent with this drop in farm numbers was the rise in acreage per farm. As industrialized agriculture increased productivity of land and allowed for fewer hands to manage larger fields, enormous monocultural, commercial farms crowded out smaller family and subsistence farms. When Spencer W. Kimball became president of church, fewer Americans—and fewer Mormons—had a direct connection to the land than ever before.

In the first quarter of the twentieth century, church leaders speaking in General Conference added to their praise for agriculture a note of alarm about its decline. Even though John A. Widtsoe, church apostle and dedicated agriculturalist, confidently noted in 1944 that "probably more than half of our Church membership is agricultural,"[16] the trend was clear, and church leaders were worried. "Let us stay on the farm," implored church president Heber J. Grant in 1923. "There seems to be strength, physical, moral and religious, which comes to those engaged in cultivating the soil which, on the average, is superior to that of any other occupation I know anything about. . . . I want again to assure you that the best place in the world to rear Latter-day Saints is on the soil."[17]

In 1940, First Presidency second counselor J. Reuben Clark struck a downright gloomy tone as he connected changes in weather patterns in the Intermountain West to God cursing a decreasingly agricultural people:

> The old time snows have not been falling in the mountains for many years. Our farms are run-down, many acres of them are idle, our fences and barns and corrals are falling down, our homes are unpainted, houses are shabby, we look too much like idling ne'er-do-wells. Another fifty years of this will almost return us to the desert. . . . Why should the Lord bless the land we do not farm? Why should He give us water for the land we are not

> using? Why should He give us crops when we do not plant? Why should He bless us with His blessings when our reliance is put not in Him but on government gratuities for not working, for not farming, for not doing the things that have been man's allotted part since Adam was driven from the Garden?[18]

Here again church leaders were drawing on the larger agricultural tradition of Jeffersonian America and adding a distinctly Mormon twist. Clark invoked the longstanding value that Latter-day Saints placed in work and industry. In the tradition of their pioneer forebears, who cooperated with God in making the desert a garden, the Saints again needed to return to the right relationship with the land: working together to improve and beautify and make it productive, lest God revoke his promise to bless them and their land with prosperity. The Lord would not simply give them abundant water and fertile valleys; they had to show they would use such blessings well.

By the time Spencer W. Kimball stepped into the role of church president in the 1970s, farming as a way of life was largely foreign to church members' lived experience. In contrast, agriculture occupied a large space in Kimball's memory and was still at the core of his and his family's identity. His family had moved from Salt Lake City to Thatcher, Arizona, in 1898, when Spencer was three years old, so his father could preside as stake president over multiple church units there. They lived on a family farm in Spencer's formative years, as he later recalled:

> We raised almost all of our own food. [My father] always wanted a garden—he wanted a garden to eat from and a garden to smell. I used to pump the water by hand to water the garden, and also I learned to milk the cows, prune the fruit trees, mend the fences, and all the rest. I had two older brothers, who, I was convinced, took all the easy jobs and left me all the hard ones. But I don't complain; it made me strong.[19]

Throughout his life, Kimball would return to stories and principles learned from that period: thrift, simplicity, hard work, and appreciation for water in a harsh environment.

Kimball's wife, Camilla Eyring, came from hardscrabble church colonies in Mexico, and she too learned to love and value the rural way of life. Although Spencer went into banking after their marriage and they no longer lived in a rural setting, the agrarian ideal figured prominently in their domestic life. In 1940 the family completed a new house, one for which they had planned every detail and overseen construction. According to Camilla's biographer, "The six-room home in pueblo style was not elaborate nor expensive, but for them it was 'a mansion and built of love and dreams.' . . . They planned and planted a garden and an orchard, had a barn with cows, chickens, and a pig—in fact, they felt they had established an estate where they could end their days."[20] Things turned out differently for the Kimballs, however; three years later Spencer was called to the Quorum of the Twelve Apostles and the family moved to Salt Lake City. Under Camilla's leadership, they adapted to their urban situation and continued to prioritize gardening and home food production. A family friend once quipped regarding the couple's manner of dividing up duties in the garden: "She plants and tends it. He eats it—and gives advice."[21] In Camilla's yard, "vegetables grew among the roses, beans climbed to the roof of her porch, and fruit trees shaded her yard. Canning the produce extended the satisfaction she had in living from the labor of her own hands."[22] According to the Kimballs' son, "They had, in their season, tomatoes, spinach, squash, string beans, potatoes, carrots, apricots, berries, and apples, all from their quite ordinary back yard."[23]

The modern American environmentalist movement came of age as Kimball rose in the ranks of the Mormon hierarchy. Crises of air and water pollution arose after World War II, fueled by the skyrocketing number of automobiles and the growth of the industrial sector. In 1962, Rachel Carson published a seminal environmental book, *Silent Spring*, awakening millions specifically to the hazards of pesticide use in industrial agriculture and more generally to public health dangers imposed by industrial food production. A series of laws passed in the 1960s and early 1970s sought to curb environmental harm, culminating in the Clean Air Act of 1963, the Clean Water Act of 1972, and the establishment of the Environmental Protection Agency in 1970.[24]

The environmental movement had many points of potential harmony with LDS doctrine and culture. Campaigns to clean up litter coincided with prophets' pronouncements back to Brigham Young. Protecting biodiversity and curbing deforestation matched Mormon (and Christian) reverence for Creation. Reducing consumption and pollution and conserving the earth's resources hearkened back to Joseph Smith's revelations and Brigham Young's sermons. But in the 1970s and 1980s, conservative politicians—and with them, conservative Christian denominations, among them Mormons—grew increasingly antagonistic to environmental causes, and the term *environmentalism* has never recovered from its fall from grace among the religious Right.[25] Regulating and curbing business and personal choices to protect the earth, which in the opening pages of the Bible God gave man dominion over, seemed downright unchristian. One dominant strain of American environmentalism proved especially disturbing to Mormons, since it was diametrically opposed to Mormon doctrine: the campaign against overpopulation.

The specter of an exploding population outstripping the earth's capacity to sustain it was not a new idea; from Thomas Robert Malthus's warnings in his 1826 *Essay on the Principle of Population* to Fairfield Osborn's 1948 *Our Plundered Planet*, scientists and philosophers had warned that rapid increases in population growth could spell disaster for humankind. But it was Paul Ehrlich's *Population Bomb*, published in 1968, that brought the idea of overpopulation to the mainstream. In his book, as well as widely publicized speeches and articles, Ehrlich predicted famines, epidemics, and other societal catastrophes in the near future—all caused by multiplying humankind the world over. Under Ehrlich's influence, the "zero population growth" movement rose to prominence in the late 1960s and early 1970s.[26] The movement advocated that for the health of the planet and assurance of humankind's survival, the number of births should not exceed the death rate, stabilizing society's population. In pragmatic, personal terms, it meant a woman should practice birth control and families should be limited to an average of two children. As Ehrlich famously said, "The mother of the year should be a sterilized woman with two adopted children."[27]

Church members responded with general horror at the idea of population control. Large families were not only culturally important to Mormons and connected to their agrarian roots, but also theologically significant. Church members understood that the earth carried "enough and to spare" for humankind,[28] and that there were innumerable masses of God's children waiting in heaven for an opportunity to be born on earth. To limit births or preach against large families was to run counter to God's plan of salvation. Addresses and the church's semiannual General Conferences spoke out against the zero population trend, including one by Marion G. Romney in April 1974 calling it a "false doctrine that we must limit the population of this earth." He continued, "The earth was made by the Lord and he made plenty for all."[29] The church's official periodical published an article in May 1971 by a Purdue agronomy professor and stake president, Phillip Low, who offered his debunking of the dire predictions of the zero population movement.[30] Perhaps most influentially, the villains of the Mormon musical *Saturday's Warrior*, which was wildly popular in Utah, were zero-population advocates, and they sang a song to that effect.[31]

Spencer W. Kimball was a follower of current events, and his files were filled with clippings from newspapers and magazines highlighting the issues of his day. Food shortages and a burgeoning population were on his mind in the early 1970s, as evidenced by two articles he filed together. The first, which he marked up heavily, was a *National Geographic* report lauding advancements in "research, mechanization, insect and weed control, credit, genetics, electricity, fertilizers, better communications and marketing, new food products, advances in soil and water conservation." All of this, the article promised, would increase productivity and efficiency of land use and bring a superabundance of produce "from a dwindling number of farmers working fewer and fewer farms." At the top of the article, Kimball wrote out a passage from Latter-day Saint scripture: Doctrine and Covenants 104:17–18, which begins "For the earth is full, and there is enough and to spare."[32] The other article, clipped from a popular evangelical Christian magazine of the day, warned of "a curse on the land" caused by industrial agriculture's dependence on laboratory-produced hybrid corn. It commended organic food production to readers.[33] Kimball's outlook seemed to span both views: he celebrated homegrown food but looked to technology and mechanization to improve lives and fill bellies. God had ordained the earth for his children, and rather than limit the number of children born, humankind could make the earth more productive. One way this would happen, he believed, was for individual Saints to return to growing food themselves.

With the unexpected death of Harold B. Lee in December 1973, Kimball became the prophet and president of the church. His opening address at General Conference the next April set his agenda as a conservative one, stating his intention to stay on the same trajectory as his predecessors and reemphasize policies and programs:

> In the press conferences an ever-recurring question has been asked us: "Mr. President, what are you going to do now that you have the leadership of the Church in your hands?"
>
> My answer has been that for the past 30 years, as a member of the

> Council of the Twelve Apostles, I have had a little to do with the making of policies and the formation of the present extensive, full, and comprehensive program. I anticipate no major changes in the immediate future, but do hope to give increased emphasis to some of the programs already established. This is a day of consolidating our efforts, and firming up our programs, and reaffirming our policies.[34]

One program that received increased attention and prominence during Kimball's tenure was the welfare program. Established during the Great Depression, the church's welfare program was meant as an alternative to government aid programs, which church leaders saw as destructive and unstable.[35] According to the Church Welfare Plan (originally called the Security Plan), which was introduced at the April 1936 General Conference, charitable donations were to be increased by church members so that local bishops and Relief Society presidents could take care of the needs of their congregants. Aid was "not to be normally given as charity; it is to be distributed for work or service rendered." The central church organization would be responsible for helping members find work, and a First Presidency–appointed Church Relief Committee would coordinate the program.[36] In historian Donald H. Dyal's estimation, the program was an attempt to care for the church's poor while preserving traditional values and combatting what it perceived as the dangers of the New Deal: "Basically, the Church wished to replace the dole (that is, receipt of relief monies without working for them) with a work-reliant program where people could work for their sustenance. The Welfare Plan was conservative in the sense that it attempted to 'conserve' the Mormon values of thrift, independence, Christianity, labor, and agriculture."[37]

The church had by no means discontinued the Welfare Plan by the 1970s, and local church units still incorporated home food preservation into their programs.[38] But after World War II, a long period of economic prosperity in the United States had made church welfare issues less urgent. Then, beginning in 1973, Western markets experienced a two-year recession, exacerbated and partially caused by the oil crisis of 1973, which occurred as a convergence of increased oil consumption in the United States, declining domestic reserves, and an embargo by oil-rich countries after the United States supported Israel in a war against Syria and Egypt.[39] At the same time, a combination of unfavorable weather, the energy crisis, and economic policy changes caused a sharp increase in grain prices worldwide, sparking fears of food shortages.[40] In these conditions, church leaders again faced swelling numbers of members in need of welfare assistance, and the tumultuous world events made it easy to imagine apocalyptic scenarios. A renewed focus on the Welfare Plan, self-sufficiency, and emergency preparedness was natural for the church.

A "Welfare Session" had been a part of General Conference since the 1930s, but during Kimball's presidency it was published for the first time in the official *Conference Report* in April 1974 and in the church magazines beginning in October 1975. In the first session published in the *Conference Report,* Marion G. Romney of the First Presidency remarked on the need to reemphasize church welfare: "In recent years we haven't had that intensive program of training and I would suppose there are many bishops here today who were never trained in the fundamentals of the Welfare Program."[41] While other specialized meetings were discontinued as part of Kimball's move to "consolidate our efforts," the Welfare Session of General Conference continued into the 1980s. The last conference that included a Welfare Session, October 1982, was also the last conference Kimball attended before his death.

At the April 1974 General Conference, his first as prophet and church president, Kimball set one of the principal themes of his tenure when he mentioned gardening during the Welfare Session. "I should say that in our little yard Sister Kimball is our farmer," he noted, "and she nearly feeds us through the year from that little yard in the back. . . . We just almost live on beans and it is good food, very good food. The little gardens and the few trees are very valuable."[42] Though he did not specifically instruct the Saints to grow a garden at this time, it was clear he saw home food production as a key component of self-reliance and a worthwhile activity.

In September 1974, the First Presidency issued a statement urging members to clean up their yards.[43] Cleaning up and beautifying one's surroundings was a national issue in the 1960s, with Lady Bird Johnson championing the "Keep America Beautiful" campaign from the White House. Less about an environmental ethic than the belief that such improvements promoted a healthy and productive society, the cleanup campaign fit well with Mormon ideals reaching back to Brigham Young's pronouncements on beautifying Zion. According to President Kimball's son and biographer, Edward Kimball, the immediate impetus for Spencer W. Kimball's statement was his travels in his calling as church president. As he visited the Saints throughout the church, he would note the deterioration of structures and fences and say, "They may be poor, but that is no excuse for not having things neat and tidy."[44] In October 1974, at Kimball's second General Conference as church president, he announced the "cleanup campaign" to the entire church and explained:

> We are a throw-away people. Trash piles grow faster than population by far. Now we ask you to clean up your homes and your farms. "Man is the keeper of the land, and not its possessor."
>
> Broken fences should be mended or removed. Unused barns should be repaired, roofed, painted, or removed. Sheds and corrals should be repaired and painted, or removed. Weedy ditch banks should be cleared. Abandoned homes could probably be razed. We look forward to the day when, in all of our communities, urban and rural, there would be a universal, continued movement to clean and repair and paint barns and sheds, build sidewalks, clean ditch banks, and make our properties a thing of beauty to behold. . . . We urge each of you to dress and keep in a beautiful state the property that is in your hands.[45]

The broken fences, dilapidated barns, and overgrown irrigation ditches were the result of the steady transition in Utah away from agriculture. What Kimball was seeing and bristling at as he visited wards and stakes was the end of an era. Like past church leaders, he carried with him agrarian values and idealized life on the farm. But Kimball proved to be more pragmatic. Repairing and painting old barns may be preferred, but he was not opposed to removing them. Doing so would remove the vestiges of a bygone era and erase the Saints' agricultural heritage, but Kimball was looking forward, adapting to the times and seeking beauty in whatever residential circumstances the Saints found themselves. He recognized that the farming days were over for most families in the United States, but he still saw the potential to redeem the land (and souls) by renewing interest in gardening. Here was farming that could be adapted to the skills, time, and space of suburban or even urban Saints.

In the next General Conference, Kimball reiterated the need for the Saints to clean up their land, and he tied it explicitly to another widespread environmental campaign of the time: reducing pollution. "We recommend to all people that there be no undue pollution," he urged, "that the land be taken care of and kept clean to be productive and to be beautiful."[46] Here he echoed Brigham Young from a century earlier, who declared, "The soil, the air, the water are all pure and healthy. Do not suffer them to become polluted with wickedness. Strive to preserve the elements from being contaminated by . . . filthy, wicked conduct."[47] Like Young, Kimball connected moral and environmental decay. On another occasion he wrote, "I am not opposing the proper efforts of the ecologists. I am hoping that while we improve garbage disposal and clean flowing rivers and take from our breathing air the soot and dust and chemicals, that we cleanse our moral and spiritual environment and make it also clean and endurable."[48]

The 1975 General Conference talk that included a reference to pollution was also the first time Kimball enjoined the worldwide church to grow a garden. He did not do so as a new announcement but spoke of it as a natural extension of welfare principles long taught in the church: "We are pleased that many people are planting gardens and fruit trees and are buying canning jars and lids. City officials here and many other individuals are planting patches of soil almost equal to the days of the 'victory gardens' in World War II. We congratulate those families who are listening and doing."[49]

As church president, Spencer W. Kimball spoke in seventeen biannual General Conferences, and he mentioned cleaning up yards and growing gardens and fruit trees in all but two of them.[50] For the most part, he gave only short reminders and briefly commended those who had followed his counsel. He often read from letters he had received from those who had found success in gardening. In 1976, he reported on gardening statistics clipped from an unidentified magazine, noting that for the first time since World War II, "51 percent of the households in the United States plan a garden for this year." Kimball called this "garden fever" and encouraged his listeners to contract it.[51]

In this first of many exhortations to grow food, Kimball showed his willingness to depart from the agrarian ideal in order to focus on the principles of self-reliance and emergency preparedness. He quoted an article from *U.S. News and World Report* that encouraged people to plant a garden even if all they had was a little spot to do so: "It can be the play area that doesn't get played on anymore, a sunny plot behind the garage, a 10-foot strip that runs across the back of the lot, or the adjoining lot that was bought to grow grass and play catch on." The article, and therefore Kimball, framed the discussion in economic terms. Growing a garden would cut inflation and cut down on grocery bills. But it also pointed to larger environmental concerns, emphasizing that a backyard garden was the simplest way to "ease the world food crisis."[52] Instead of relying on industrial-scale food production in faraway places, the Saints could secure their own food—and presumably free up commercially grown food for other mouths. Perceived food shortages and the energy crisis of 1973–74 provided useful images for driving home the point that food storage and home preservation may be desperately needed soon: "Can you think how the highways could be made desolate? When fuel and power are limited, when there is none to use, when men will walk instead of ride?" Kimball asked.[53] Marion G. Romney painted a similar picture at a Welfare Session of General Conference, warning, "It may be that some time in the future we will survive or starve on what we can produce ourselves. I want you to take this matter seriously."[54]

Kimball's speeches coincided with a resurgence of interest in gardening throughout the United States. In 1971, the proportion of households planning a garden was 37 percent, and by 1981, the number had grown to nearly half.[55] In some ways, the motives of the Saints and the larger American public were related. Though perhaps not entirely reducible to economic forces, the trend in gardening in the United States follows economic indicators, and periodic financial downturns spark renewed interest in gardening as a cost-saving activity. The larger national trend, however, was also tied to environmental and health concerns. Many came at gardening through a different train of political thought, through causes like the back-to-the-land movement, the organic food movement, and later, the slow food movement. But they arrived at the same place as the Mormons, and they gardened for many of the same stated reasons—they believed working the land and growing one's own food was healthy for the body and for society.[56]

Another demographic trend throughout the nation facilitated a renewed interest in gardening in Mormon country and beyond: suburban growth. Dwight Eisenhower signed the Federal Highway Act of 1956, and the United States subsequently poured half a trillion dollars into constructing high-speed interstate roadways over the next five decades, decreasing commute times to urban centers and leading more affluent, mostly white, families to outlying areas. In the "Mormon Corridor" of the Intermountain West, this meant the steady transformation of Utah communities such as Bountiful and Farmington from farmland to housing developments. The Latter-day Saints' ideal of an agricultural Zion could live on in the new American ideal of an attractive house with a white picket fence on a quarter acre in the suburbs.[57]

Kimball encouraged home food production among a wider demographic spectrum, however, and he was aware of his international audience with varied economic circumstances and living conditions. He quoted from letters he had received from church members in Frankfurt and São Paulo. His files were filled with other responses from Buenos Aires and Tokyo, and he made notes of individuals in other parts of the world adopting and adapting his instruction to keep a garden.[58] Though Kimball's primary lived experience was in the American Southwest, he saw his words as relevant to church members everywhere, and through his correspondence and travels he was gratified to see members both in the United States and other countries heeding his call and reaping the benefits.

Members were told to adapt their gardening to their individual circumstances, but also to maximize the land available to them: "We encourage you to grow all the food that you feasibly can on your own property. Berry bushes, grapevines, fruit trees—plant them if your climate is right for their growth. . . . Even those residing in apartments or condominiums can generally grow a little food in pots and planters."[59] In a later conference Kimball reiterated: "Many of the numerous gardens are found in hanging baskets, in containers on stairways, on trellises, and in window boxes."[60] Recognizing that the average Saint had less and less space available for planting crops, Kimball was willing to meet the members where they were—in cities and suburbs—rather than simply wringing his hands at the changed landscape. The core of his message was that growing something was better than growing nothing.

Yet, despite Kimball's efforts to accept the demise of the Mormon family farm, the long tradition of church authorities privileging traditional agriculture and the ideal of the plat of the city of Zion unquestionably influenced Kimball's vision of a Mormon community. "I was with President Tanner in the city of Cardston [Alberta] a few weeks ago," he remarked in one Welfare Session meeting, "and noted a clean city; and I mentioned it time and time again as we drove through the streets, a clean city." He continued:

> I noticed no backyards filled with trash and other waste, and I could not help but mention to him again, "Look at the row there, this whole row of homes, and as far as you see there are gardens, corn and beans and squash." There were little fruit trees in between, and nearly every yard, as far as we could see, every backyard was cultivated; and I am sure the good people there were living considerably out of their yard, rather than out of the store. I was pleased indeed to see that there are many of our people who have not forgotten the lessons of yesterday, and are still listening to the words of the leaders.[61]

This Mormon town retained the look and feel of yesteryear. It resembled what Mark Twain and Elizabeth Cumming had seen in Utah a century earlier, and President Kimball rejoiced in it.

As the years went by, Kimball expanded his reasons for advocating for gardens to supersede the usual themes of economics and emergency preparedness. In April 1977, after speaking at length on the practical benefits of a garden, he remarked, "We deal with many things which are thought to be not so spiritual; but all things are spiritual with the Lord."[62] Building on nineteenth- and early twentieth-century panegyrics to working with the land, but also fitting comfortably within the land ethic of more recent environmental voices like Aldo Leopold and Wendell Berry, Kimball declared that one simple reason for gardening was "so that we do not lose contact with the soil."[63] In a later address he reiterated this belief: "Staying close to the soil is good for the soul."[64]

Along with enumerating the virtues of working the land, Kimball also counseled more generally about the value of work. In doing so, he drew again on traditional ways of life and positioned himself against the prevailing cultural attitudes he saw around him. As noted earlier, he remembered working on the farm as a boy and the value of work it instilled in him: "Little did I know as a boy that daily chores in the garden, feeding the cattle, carrying the water, chopping the wood, mending fences, and all the labor of a small farm was an important part of sending down roots."[65] Looking through the Mormon lens at the expulsion of Adam and Eve from the Garden of Eden, Kimball saw the Fall as fortunate, and the consequent divine decree that they labor and sweat for their food was something to be celebrated, not lamented. "We must bring dignity to labor in sharing the responsibilities of the home and the yard," he said.[66] The Lord, he continued, "rebukes our acquisitive society. He rebukes our comfort-loving, take-it-easy philosophy. . . . Ours is a comfort-loving society. We equate comfort with civilization. Thanks to our Heavenly Father and his Son that the program is austere."[67] And later in that same conference, he observed: "I see no disadvantages in work. I believe it was one of the clever and most important and necessary creations of our Father."[68] Work was the great leveler for Kimball. Speaking of the church's volunteer-operated welfare farms, he exulted in seeing "banker and merchant, wealthy and poor, go into the fields, into the orchards, into the gardens and work together to produce."[69] Here he invoked both the idea of "improvement" and the ideal of community, hearkening back to the communitarian principles of early Utah and the cooperation required to build the Mormon Zion, whether in Joseph Smith's Missouri or Brigham Young's Great Basin.

Two of the most frequently quoted passages from Kimball on gardening point to the deeper purposes he associated with the activity. First was its capacity to unite individuals and generations in a common cause and shared abundance:

> I hope that we understand that, while having a garden, for instance, is often useful in reducing food costs and making available delicious fresh fruits and vegetables, it does much more than this. Who can gauge the value of that special chat between daughter and Dad as they weed or water the garden? How do we evaluate the good that comes from the obvious lessons of planting, cultivating, and the eternal law of the harvest? And how do we measure the family togetherness and cooperating that must accompany successful canning? Yes, we are laying up resources in store, but perhaps the greater good is contained in the lessons of life we learn as we live providently and extend to our children their pioneer heritage.[70]

In the next conference, he reiterated the symbolic value of growing food. Here, not only did he downplay the cost-saving component of gardening, he praised the act even if it made no economic sense at all:

> Even if the tomato you eat is a $2.00 tomato, it will bring satisfaction anyway and remind us all of the law of the harvest, which is relentless in life. We do reap what we sow. Even if the plot of soil you cultivate, plant, and harvest is a small one, it brings human nature closer to nature as was the case in the beginning with our first parents.[71]

For Kimball, then, a principal reason staying close to the soil was good for the soul was that planting, caring for, and harvesting crops taught of spiritual truth and brought people nearer to God.

The call to gardening that came so frequently from the General Conference pulpit in the 1970s subsided in the 1980s, although it did not entirely go away. Because of declining health, Spencer W. Kimball stopped speaking in General Conference after October 1982. After that, though they did not reiterate or reissue the mandate to garden, other church leaders continued to present gardening as the ideal or use it as a metaphor for gospel living. When it did come up, gardening was placed in a symbolic context, as it had been in Kimball's later sermons. Rather than speaking of gardening in terms of food security or environmental health and beauty, church authorities now used it as a metaphor. In October 1985, General Relief Society President Barbara W. Winder gave a General Conference talk titled "Draw Near unto Me through Obedience," in which she used gardening as an example of the blessings church members could expect when they heeded the words of their prophet: "Several years ago, our newly married daughter and her husband . . . determined they would follow the prophet's advice and have a garden. Their first attempts at gardening were pathetic. The weeds grew much better than the vegetables." Yet as the family developed their skills, she continued, the results improved. Winder concluded, "Now their gardens are attractive, worthwhile 'survival' projects, as the family enjoys and shares the produce. They preserve the excess for later use. Besides the practical lessons they learned, they found peace and assurance in keeping the commandments."[72]

By emphasizing the emblematic nature of gardening and the importance of obedience over the practical needs a garden met, Winder signaled a shift in thinking about gardening. Perhaps with the improving economy of the mid-1980s, and as the food and gas shortages of the 1970s faded into the background, the need for self-sufficiency felt less urgent, the coming apocalypse a little more distant. Even Winder's setting off of "survival" in quotation marks distanced her from the very real need Kimball felt a decade earlier to prepare for a time when survival would be the reality.

Apostle James E. Faust invoked Kimball's appeal to gardening in the April 1986 General Conference when, like Winder, he mixed the practical and symbolic importance of working with the soil, ultimately emphasizing the latter. Speaking of "the ability and skills to make things with their hands," Faust noted that such ability "is not only an economic advantage, but it also provides much emotional resilience," and he quoted Kimball's observation that gardening offers not only vegetables but the opportunity for a "special chat between daughter and Dad."[73] This was the last time Kimball's words on the subject were quoted at General Conference. From then until the present, talks occasionally mentioned gardening as something Latter-day Saints were supposed to do or used gardens as a metaphor for gospel lessons.[74] But no longer would a prophet or other general authority explicitly instruct church members to plant gardens.

The use of gardening tropes was nothing new, of course, and parables drawing on planting and harvesting feature prominently in the Old and New Testaments. Yet when invoked over the pulpit at LDS General Conference, the tropes carried additional significance since they drew on the themes of beautification and improvement emphasized since the earliest days of the church. Even if agriculture had faded as a way of life, it still could carry value in the hearts and minds of Latter-day Saints through metaphor and story.

Yet gardening and home food preservation did not fade away and become just the stuff of stories. On the contrary, these activities remain significant aspects of Mormon belief and practice. Two primary reasons for this endurance are that the institutional memory of Kimball's words has been slow to subside and that his call to gardening filled a gap in Mormon teaching and theology. Because Kimball spoke of gardening for so many years, it gave ample material for church curriculum designers and periodical editors to draw from. Manuals for Sunday school, priesthood meetings, and Relief Society have all included gardening excerpts from Kimball's talks. Articles on gardening have appeared dozens of times in church magazines since the early 1970s. On the church's website, "gardening" receives a prominent topical treatment and is thoroughly covered on the church's welfare page.

Among the more conspicuous placements of Kimball's gardening exhortations is in the *Children's Songbook*, the collection of hymns and songs for use in the church's organization for children, the Primary. Penned by Mary Jane McAllister Davis in 1982, the year Kimball stopped speaking in General Conference, the song tries to teach children the benefits of both following the prophet and working the soil:

> The prophet said to plant a garden, so that's what we'll do.
> For God has given rich brown soil, the rain and sunshine too.
> And if we plant the seeds just right and tend them carefully,
> Before we know, good things will grow to feed our family.
>
> We'll plant the seeds to fill our needs, then plant a few to spare,
> And show we love our neighbors with the harvest that we share.
> Oh, won't you plant a garden, too, and share the many joys
> A garden brings in health and love to happy girls and boys![75]

Although the song receives less attention today than it did when first published, it enjoys semi-canonical status as an official children's hymn. Because the *Children's Songbook* is in constant use, unlike a magazine that is quickly discarded or a manual used only one year in four, there is ample opportunity for a new generation of church members to discover the call to gardening through the songbook.

Further, digitization of church materials has granted newfound prominence to older, more obscure articles and talks. LDS.org and the Gospel Library reading app have flattened the hierarchy of church content, so that an article or manual from the 1980s is as likely to come up in search results as newer information. Whereas in the early 2000s, the only ready access the average church member would have had to Kimball's words on home food production would have been through Sunday curricula, now transcripts and videos of all his talks are readily accessible, as are the many articles and essays on gardening that were published in church magazines.

The continuing appeal of gardening, however, is attributable to more than simple availability of church material. Spencer W. Kimball struck a chord with Mormons from across the ideological spectrum. The gardening message appeals to the ultraconservative "prepper," who is convinced the apocalypse and Second Coming are just around the corner and feels the need for food self-reliance.[76] At the same time, other Saints grasp onto Kimball's words as a rare example of prophets taking a stand on environmental issues.[77] In between, the average church member simply seeks to follow the prophet.[78] Though different, all three groups would agree on the importance of keeping a garden.

The 2010s have again seen a resurgence in gardening in the United States, the result of both the recent economic troubles and increased demand for "real" and "slow" food. Though the ongoing injunction to grow gardens ended with Spencer W. Kimball's death, the principles remain an article of faith for modern Mormons. Unable to shake loose completely from the agrarian ideal so fundamental to their history, yet living in worlds of urban concrete and suburban lawns, church members return to the soil in great numbers each spring, following Kimball's advice to make do with what they have, whether it be a large vacant lot or a window-box planter. To each, the words of Kimball invite: "The planting of gardens . . . is health-building, both from the raising of crops and the eating of them. . . . We hope this will be a permanent experience of our people, that they will raise much of what they use on their table."[79]

"FOR THE STRENGTH OF THE HILLS"

Casting a Concrete Zion

Rebecca K. Andersen

I GREW UP NEAR A GRAVEL PIT. I associate the sound of a summer morning with the metallic whir of rocks crushing and grinding. During searing afternoons we balanced on our bikes and squinted up at the gravel pit trucks as they rumbled by, gears whining and grating. Chrome-lined windshields glinted a hard blue in the unforgiving sunlight; trailer hitches jumped and clanked. Every evening cool canyon breezes whipped through the darkened trees and bushes. Leafy shadows danced crazily on the bedroom wall as fine sand blew through the open windows, encasing the piano keys in powdered grit.

Drive along Utah's Wasatch Front from Brigham City to Spanish Fork and you will see the gravel pits. They look like white craters against the brown foothills. These sand and gravel operations, many of which are owned by Mormons, are controversial and compete with real estate developers for prime bench land property along the Wasatch Front. In one sense, Mormons have always been in the construction business. Early Mormon settlers understood their colonizing mission in terms of the redemption and improvement of a people and a land. In building and designing the spatial dimensions of their kingdom of God on earth, Mormons removed resources from their mountain ramparts with an almost worshipful regard. In many respects, it was the presence of these resources in the first place that gave the mountains their sacred quality and value. By wisely utilizing these resources, Mormons hoped to redeem their arresting alpine backdrop. Whether today's Mormon sand and gravel operators view their activities in a similar light is hard to tell. But what is apparent is that the Wasatch Front's plentiful sand and gravel resources have played a crucial role in building the twentieth-century kingdom of God by providing a relatively low-cost aggregate base for Mormon suburban sprawl, churches, and temples.

Nowhere is the paradoxical relationship between what historian Jared Farmer would call Mormon mountain veneration on the one hand and resource extraction on the other more patently visible than along Beck Street, Salt Lake City's northern industrial strip.[1] Oil refineries line the street's west side; to the east gravel pits unmask the hillside's pink and red striations. The street itself winds along the base of a large alluvial cone, known as the Salt Lake salient or the City Creek spur—an odd tectonic structure that consists of sediments from recent geologic time and sedimentary bedrock.[2] The salient's most prominent feature is Ensign Peak, a small knoll that nineteenth-century British travel writer Richard Burton referred to as "the big toe of the Wasatch Range."[3] Ensign Peak acquired its significance early on when Brigham Young and others hiked to its top shortly after entering the Salt Lake Valley and waved a flag, or "ensign to the nations," establishing the new Mormon homeland as a gathering place for Israel.

In the twentieth century, accelerated sand and gravel extraction just north of Ensign Peak played an important role in how Mormons memorialized their past. While Ensign Peak never entirely lost its significance for Mormons, attention towards the peak seemed to have waned following World War II, when, in response to the postwar building boom, the need for roadways and infrastructure accelerated sand and gravel extraction north of the peak's base. Ironically, despite its proximity to gravel pits, Ensign Peak remained closed to automobile traffic (no asphalt roads). Because of its inaccessibility, fewer people trekked to the peak's top and it quickly became a neglected landmark. It was not until the late 1990s, when a few concerned Mormons sought to rehabilitate Ensign Peak, that interest was renewed. Their efforts coincided with other Salt Lake residents, who were intent on preserving the salient from corrosive gravel excavation. As a result of these combined efforts, a well-maintained trail now allows visitors to ascend Ensign Peak and see the legacy of Mormon city building. From the top, looking southward, they see the valley floor spread out before them, divided into neat residential blocks; the interstate highway, a single white, concrete ribbon, curves gently out across the Wasatch Front's base. Turning to the north, visitors see the massive gravel pits from which the modern city grew.

On the surface the story of Beck Street's quarries has much to say about the development of Utah's vibrant sand and gravel industry and built environment. Yet, excavated cross-sections reveal a much more nuanced and striated past. The history of Beck Street's gravel pits and their relationship to nearby Ensign Peak taps into a larger literature that reconsiders nature as an intellectual category separate and apart from human activity and manipulation.[4] More importantly, however, the pits ask historians to reflect on the changing social values behind resource utilization and exploitation—the ways in which this utilization shapes and, in the case of the Mormon Wasatch Front, may literally form the constructive base for memory making.[5] Furthermore, quarrying activity along Beck Street over the years has inadvertently produced an entirely new geologic space. Once the backhoes and rock crushers fall silent, planners and architects will have to decide what to do with the old industrial space, laying down yet another social and political layer to an already complex site.

The spot where Mormon leaders chose to raise their ensign, dedicating the mountain valleys as a gathering place for the nations, contained even more geological meaning than they might have supposed. Ensign Peak and today's gravel pits sit astride the Warm Springs Fault, a portion of the Wasatch Fault zone, which is part of the larger Intermountain Seismic Belt, and serves as a dividing line for the Rocky Mountain and Colorado Plateau geographic provinces—a kind of geological gathering place whose deep time eclipsed the Mormon arrival by 600 million years.[6]

The Wasatch Front's sand and gravel resources, however, are largely a product of the Pleistocene Lake Bonneville. In 1852, not long after the Mormons arrived in the Salt Lake Valley, Howard Stansbury seems to have first articulated the idea that an ancient lake inundated the region.[7] Beginning in the late 1860s, G. K. Gilbert conducted an exhaustive study of Lake Bonneville, finally publishing his results in 1890. He reasoned that receding glaciers inundated present-day northern Utah, northeastern Nevada, and southeastern Idaho with water. At some point, Gilbert believed, Lake Bonneville topped Red Rock Pass in southeastern Idaho and tumbled into the Snake River drainage system. Over a long period of time, lake levels receded further, reaching just above today's Great Salt Lake, Lake Bonneville's placid puddle.[8]

Bonneville's constant lapping left behind a series of sandy benches, bars, and terraces easily visible from the valley floor.[9] The most plentiful sand and gravel resources are found along Lake Bonneville benches and bars, carried there by glacial streams and rivers, washed and refined by the ancient lake's currents and waves.[10] Although Gilbert's theory has been revised and expanded, his basic premise remains largely correct and revolutionized the field of geomorphology and structural geology. Lake Bonneville quickly became the textbook example for pluvial formations and processes.[11]

When Mormon pioneers first entered the Salt Lake Valley they saw tall grasslands and canyon streams meandering towards the Great Salt Lake, a mirror of water stretching across the valley floor; when they looked down at their feet, they couldn't help but notice the gravel. Orson Pratt wrote, "Streams from the mountains and springs were very abundant, the water excellent, and generally with gravel bottoms." The soil seemed fertile, a "friable loam with fine gravel." For many, the location's only drawback was its lack of trees. Thomas Bullock, however, believed the mountains themselves would more than make up for this. He wrote, "There is an ocean of stone in the mountains with which to build stone houses and walls for fencing. If we can only find a bed of coal, we can do well, and be hidden up in the mountains unto the Lord."[12] Howard Egan journalized, "Nature has fortified this place on all sides, with only a few narrow passes, which could be made impregnable without much difficulty."[13] After their first week in the Salt Lake Valley, Mormons came to view mountains in whatever ways seemed most beneficial to them: as protective ramparts guarding against Gentile (or non-Mormon) invasion, as sacred spaces, or as repositories of rich resources.

Ensign Peak in particular stood out to the new settlers. Shortly after Brigham Young's arrival in the valley, on July 26, 1847, he and several others went on a short reconnaissance mission, hiking to the top of this small knoll. Still battling a recent illness, Young struggled to make the ascent as the mid-morning summer sun beat down on the men's necks and backs.[14] Apostle George A. Smith later explained the climb's significance. "President

Young had a vision of Joseph Smith, who showed him the mountain that we now call Ensign Peak." He continued, "There was an ensign fell upon that peak, and Joseph said, 'Build under the point where the colors fall and you will prosper and have peace.'"[15] Associated with the recently martyred prophet, the peak immediately acquired strong, symbolic referents, legitimizing and marking the new Mormon homeland as a gathering place for Israel. Church leaders later dedicated Ensign Peak as the most sacred of all Mormon places in the valley. On July 21, 1849, for instance, Addison Pratt received his endowments there, a religious ceremony that usually takes place only inside Mormon temples.[16] As late as the 1930s, Uinta Basin colonizer William Smart hiked to the top of Ensign Peak and spent time contemplating and praying, even constructing an altar for those purposes.[17]

Not unlike other frontier Christians, nineteenth-century Mormons believed God had given them the Salt Lake Valley and its resources so that they might redeem the earth through settlement and industry. Although Mormons believed the earth was the Lord's, if settlers were wise, obedient, and faithful stewards, the earth would yield forth its resources for their beneficial use. Just as Thomas Bullock anticipated, the canyons yielded high quality building stone, such as limestone that could be quarried and processed into plaster and mortar, and of course the ever-present sand and gravel, "these pebble stones that are so abundant here."[18]

In quarrying stone from nearby canyons, Mormons hoped to redeem and exalt the very mountains that guarded their villages and farms, developing and utilizing mountain resources to literally build the kingdom of God.[19] Brigham Young envisioned the perfect city as having graveled streets and paved walks, all "nicely swept, and everything neat, nice and sweet."[20] In an 1861 address Daniel H. Wells, echoing Young's counsel, told members living in Logan, Utah, to "go to with your might and build up the kingdom of God, by quarrying the rock, by bringing the timber from the canyons and making it into lumber, by making adobies, mixing the mortar, burning the lime, and drawing from the elements around us the material necessary to beautify and build up, and to exalt in every way those principles that essay to establish righteousness over the whole earth."[21] Consequently, limekilns soon dotted the mountainsides and small intermittent sand and gravel pits served local and individual construction needs.[22]

Perhaps living at the base of such a dynamic mountain range caused some to ponder the Wasatch Front's origins and speculate on the processes that created it.[23] In the October 1859 General Conference, Brigham Young expounded on some of his own geological theories: "The elements, of which this terra firma is composed, are every moment either composing or decomposing." Young stated, "They commence to organize or to compose, and continue to grow until they arrive at their zenith of perfection, and then they begin to decompose. When you find a rock that has arrived at its greatest perfection, you may know that the work of decaying has begun."[24] For this reason, Young believed it prudent to build with adobe and even suggested for a time that the Salt Lake Temple be made of mud. Adobe, tempered with gravel, was "better than a brick house," Young explained. "For when you burn the clay to make brick, you destroy the life of it, it may last many years, but if the life is permitted to remain in it, it will last until it has become rock, and then begin to decay."[25] Gravel and clay were only on their way to becoming rock, and thus would somehow outlast all other available building materials. As living, evolving elements, gravel and clay possessed a particular strength the others did not.[26] In a sense Young seemed to reconcile the dichotomy between mountain veneration and sand and gravel extraction: in constructing adobe structures, humans served as geological agents and aided in the cyclical creation of stone and rock, contributing to and enhancing the mountains' existing sacred qualities. Thus while Young counseled Mormons not to turn their plowshares into mining picks, quarrying never attracted the same official criticism, associated as it was with community building and self-reliance, even an act of creation.[27]

When Brigham Young voiced his theories about adobe and gravel extraction, most quarrying activity in the United States and along the Wasatch Front remained small, individually owned and operated. The gravel was more often than not used for railroad ballast or for local road improvement projects.[28] The most plentiful sand and gravel resources were generally found along the banks of rivers, streams, and harbors. Dredging operations harvested the aggregates, while men washed and screened the gravel by hand. In 1874, for example, contractors in New York City purchased between fifteen and twenty thousand tons of sand and gravel dredged from the harbors near Port Jefferson, New York. They then utilized the gravel in iron smelting, roofing, glassmaking, and sandpaper manufacturing.[29]

By the turn of the century, however, city streets, sidewalks, and buildings made use of a revolutionary construction material, Portland cement, which galvanized aggregate extraction. Invented in England in 1824, Portland cement was made of kiln-fired limestone, chalk, and clay. This amalgamation of aggregates created clinker, which was then ground into fine powder, "the cement of commerce," as one trade book put it. U.S. construction firms first purchased Portland cement from European plants as early as 1868, but demand remained low. This changed dramatically in the 1880s, when Portland cement imported to the United States went from ninety-two thousand barrels in 1878 to over one million barrels a decade later. It was also about this time that local processing plants began opening for business across the United States.[30]

Mixed with sand and gravel, Portland cement created concrete, which contractors began utilizing in the early 1900s. By the 1920s concrete had become the new construction material of choice.[31] Commenting on concrete's many virtues, one trade publication from 1924 seemed to echo Brigham Young: "Cement means concrete; concrete means stone; and stone spells eternity, so far as our finite minds can comprehend."[32] Nature seemed to have endowed gravel with particular properties that made it especially valuable. "The rounded shapes of the component particles permit gravel to be more closely tamped than broken stone and give less danger of voids from bridging," an early publication from 1908 noted.[33] In response, sand and gravel extraction took off, especially once excavation and dredging methods mechanized—steam shovels, suction dredgers, vibrating screens, crushers, and conveyor belts.[34]

The rise of Utah's sand and gravel industry along the Wasatch Front paralleled national trends, with the hillsides near Ensign Peak among the earliest to be commercially excavated for lime, sand, and gravel. When G. K. Gilbert camped in the area in July 1877, he specifically noted an active limestone quarry nearby and several thriving kilns.[35] The kilns themselves refined the quarried limestone between layers of burning wood, likely lacing the air with sharp sulfuric traces.[36] Further north, James Moroni Thomas's Ogden Canyon limekilns produced approximately three hundred bushels of lime a day. Teamsters hauled the lime to Ogden, where it "sold by the bushel or wagonload to 'whitewashers' or builders."[37] By the early 1900s, Portland cement plants were located north of Salt Lake City in Weber Canyon (near Devil's Slide) and east of Salt Lake City in Parley's Canyon.[38] In 1910, the *Salt Lake Tribune* proudly stated that the Hotel Utah would be constructed almost entirely of local materials: "The

Ogden Portland Cement Company will furnish the cement from its plant at Devil's Slide in Weber Canyon. All the sand and gravel will be furnished from the nearby hills by local men employing home men and home teams to haul them."[39]

Beck Street's sand and gravel deposits, on the other hand, remained intimately connected with railroad traffic, which continued to define land use along the industrial corridor. For some time the Oregon Shortline Railroad Company utilized Beck Street's sand and gravel for "fill material . . . and to supply gravel ballast." In 1909 the Mellen Company started extracting sand and gravel near the Oregon Railroad property.[40] Ten years later, Salt Lake County floated a $750,000 bond to improve the city's streets, which encouraged a number of construction startups, including Utah Sand and Gravel, owned and operated by William E. and Eric W. Ryberg, second generation Mormon converts from Sweden.[41] In 1928, the Rybergs acquired a significant amount of property along Beck Street, absorbing already existing sand and gravel operations.[42] Beck Street's Swede Town, just south of the gravel pits, supplied the labor.[43]

Incidentally, the Rybergs' business successes elevated them within the larger Mormon community. In 1937, the LDS Church appointed William to serve as trustee of the newly created Security Plan, an early version of the LDS welfare program. From 1938 until his death in 1951, Eric likewise headed up Deseret Industries, a welfare program subsidiary. When church leaders decided to construct their first underground parking structure near Temple Square in 1939, the Rybergs handily won the contract. William's sudden death in 1950 produced an outpouring of sympathy from church hierarchy. David O. McKay, then serving as a counselor in the First Presidency, eulogized, "He was able to contribute much to the common good of mankind. . . . He was loyal, respectful and reverential." Apostle Henry Moyle called William a "contractor among contractors" and praised William's devotion to his community and church.[44] Such remarks are telling. While not explicitly stated, William and Eric Ryberg's prominence as business leaders was intimately linked to their spiritual standing and provided them with the material means and status to more effectively build the kingdom of God in highly visibly ways—especially given the fact that they obtained their wealth by excavating away the mountainside.

Modernity came with its attendant costs, however. In 1908, gravel pits up Salt Lake City's City Creek Canyon raised citizen ire, perhaps because of new housing developments going in on the east edge of town.[45] Although the Beck Street quarries were not directly implicated in the complaints, aggregate extraction did not sit well with the pristine alpine aspirations some had for Ensign Peak. In 1908 Lon J. Haddock, secretary of the local Manufacturers and Merchants' Association, put forward a plan to beautify Ensign Peak by planting trees and flowers and constructing a road up to the top of the hill, capturing nascent automobile tourist traffic.[46] An article in the *Deseret News* endorsed these plans, stating, "It is believed also that arrangements could be made for a large school picnic in Ensign peak park on Arbor day next year, when the school children of Salt Lake could plant hundreds of trees and make the beginning of the forest which, by the time they reach manhood and womanhood, would become one of the interesting possessions of the city." The only stipulation park investors had was that the surrounding area "be protected from the encroachments of the sand and gravel diggers who are now defacing the landscape back of Salt Lake."[47] Haddock's plans for a tourist trap never came to fruition. Salt Lake residents wanted a woodland retreat like the kind their pioneer parents and grandparents first encountered, but they also desired paved streets and other modern amenities the "gravel diggers" provided.

Despite the Rybergs' and others' quarrying activities, Ensign Peak apparently retained its historical significance for the larger Mormon community. On July 26, 1916, a Boy Scout troop from Salt Lake City's Ensign Stake hiked to the top of the peak, commemorating Brigham Young's original ascent and exchanging campfire signals with a troop at the Point of the Mountain in Draper.[48] The next year participation extended to members of the larger stake and was marked by patriotic devotional programs reenacting the ensign's original unfurling. (What this ensign exactly was—a U.S. flag, handkerchief, or other pioneer banner—was debated.)[49] In 1919 at least one thousand people alone tramped up the hillside and the annual hikes continued through the late 1930s.[50]

Beginning in the early 1930s the Utah Trails and Landmarks Association and Ensign Stake president Arza Hinckley worked together to erect a permanent monument on top of Ensign Peak. The winning proposal—others suggested a neon sign reading "This is the Place"—was to be an 18.47-feet-high stone monument.[51] The height, of course, symbolized the year the Mormon pioneers arrived in the Salt Lake Valley. Furthermore, each stone carried additional historic significance: one came from Big

Cottonwood Canyon, where Brigham Young had a paper mill, and the granite base originated from the quarry site for the Salt Lake Temple up Little Cottonwood Canyon.[52] No one said where the cement mortar holding the rocks came from, but it was likely from the very mountains the monument was supposed to commemorate.

Indeed, the Rybergs had continued to expand their quarrying activities along Beck Street. Through the 1930s, Utah Sand and Gravel's Beck Street pits continued to provide railroad ballast. In addition the company acquired several public works contracts, including one in 1936 to provide sand and gravel for additional concrete runways and aprons at the Salt Lake City airport.[53] With the onset of World War II, the Beck Street plant experienced unprecedented expansion. In 1942 the plant produced a record 564,000 tons of sand and gravel and for the next twenty years the Beck Street pit remained Utah Sand and Gravel's "basic production unit . . . in the Salt Lake metropolitan area."[54]

Following the war, Beck Street gravel operations shifted to ready-mix concrete and, most importantly, road base for the expanding highway infrastructure.[55] The expansion of the sand and gravel industry both responded to and precipitated postwar America's love affair with cars. Along the Wasatch Front, car dealerships quickly multiplied. Car advertisements even graced the pages of the LDS church magazine, *The Improvement Era*.[56] In 1960, for example, American Fork's population numbered just over six thousand, yet the town boasted at least three car dealerships.[57] The increased number of cars on the road strained the region's infrastructure. "The large number of war plants indicates a definite need for a wide road, free from local traffic, which will provide fast transportation between plants," one report indicated.[58] Plans for Utah's Interstate 15 south of Davis County were made public in late 1956.[59] A few years later as construction commenced, the Utah Highway Progress newsletter in 1959 "estimated that the paint purchased for Utah roads this year, would cover all the houses in the city of Ogden. . . . Asphalt also comes into the picture to add to the big kitty for Utah's business. The Finance Commission has already awarded bids for over six million gallons of asphalt to be used by the State Road Commission."[60]

Interstate road construction, automobile sales, and gravel pits reflected the Wasatch Front's postwar economic boom. What may be less apparent is that these activities also indirectly affected how Mormons memorialized their past. In the 1950s and 1960s, for instance, no one seemed to know what to do with a peak inaccessible to automobile traffic. Strapped for cash, in 1952 Salt Lake City toyed with the idea of selling 290 acres surrounding the peak. The Sons of the Utah Pioneers protested the move, encouraging the city to create a park instead. One newspaper editorial responded in favor of abandoning Ensign Peak, pushing the city instead to develop City Creek Canyon, a spot equally rich in pioneer history. "As a recreational facility, a developed picnic area at hand would seem to be worth far more than a proposed park in the bushes."[61] In 1959 the Utah National Guard offered to construct an automobile road up to Ensign Peak in the hopes of attracting tourist traffic, and a few years later Salt Lake City parks commissioner L. C. Romney and businessman Nicholas G. Morgan talked about creating a scenic lookout point, but nothing came of either plan.[62] Sometime in the 1960s, the monument's bronze marker went missing, later to be found in a West Jordan chicken coop in 1992.[63] Over the next few decades off-road vehicles scarred the south face of Ensign Peak, destabilizing the soil with deep ruts and gouges.[64]

In contrast, Emigration Canyon's "This is the Place Monument," located southeast of Ensign Peak, experienced unprecedented visitation. Constructed as part of the 1947 centennial celebrations, the monument commemorated the supposed spot where Brigham Young, overlooking the Salt Lake Valley, pronounced it fit for Mormon settlement. Nestled within Salt Lake City's east bench neighborhoods, the monument's parking lot and easy accessibility to suburban roadways made it a popular stopover. In 1951 an estimated four hundred thousand visitors flocked to the site. An editorial for the *Deseret News* observed, "'This is the Place Monument' is rapidly becoming one of the major attractions of the state. With the development of the long-range program it is expected that the area at the mouth of Emigration Canyon will provide one of the outstanding tourist attractions of the entire West."[65] The plans in question called for an expansion of the monument property and included "terraced landscaping, parking areas, formal gardens, observation shelters and a new information center . . . and eventually a pioneer village."[66] The monument occupied a prime location with vast, untapped potential. "Tied in with the Hogle Zoo and perhaps a lovely garden area along Emigration creek, this could become one of the nation's outstanding tourist centers."[67] Another editorial added, "'How would you like to go up to the mouth of Emigration Canyon and take another look at "This is the Place Monument"?' is a question every Latter-day Saint father and mother might well ask their children."[68] No one was making that kind of suggestion about Ensign Peak—perhaps in large part because it had become more tied to sand and gravel extraction, replacing its earlier value as sacred space.[69]

Indeed as Beck Street's gravel pits unmasked the hillside's striated sediments, clashing with the subdued, low-lying hillside slopes, other pits threatened bench land real estate values. Bountiful resident Rita Keetch best expressed homeowners' concerns in her 1971 letter to the editor. "People have built beautiful homes in these areas only to find that their back yards are being turned into gravel pits. Must we sacrifice everything for the sake of bigger, better high ways?"[70] In a 1974 editorial that discussed a coal strip-mining bill then being debated in Washington, the *Deseret News* came down hard on gravel pits. "Coal is not the only problem in strip-mining," the newspaper emphasized. "Gravel pits are another example of an eye-sore caused by strip-mining."[71] One letter to the editor called the pits "visual pollution."[72] Despite the complaints and attempted legislative action, little was done to limit aggregate extraction along Beck Street. In 1996, the Salt Lake International Airport noted that 80 percent of the sand and gravel they used for various construction projects that year alone came from Beck Street. The State Road Commission also acquired Beck Street property, scattering its test pits across the area. As of 1999 the Utah Department of Transportation continued to possess the property, choosing to save it for a rainy day when Beck Street's private pits closed.[73]

During the 1990s, the Beck Street gravel pits became the focus of conflict between Salt Lake City and Bountiful. Increasing development along the Bountiful bench meant Salt Lake commuters had to descend their hillside perch every morning and wend their way through city traffic before speeding onto the interstate onramp. To ameliorate the problem, residents hoped to extend Bountiful Boulevard, the city's main bench land thoroughfare, to Salt Lake. Salt Lake City would have none of it, noting that the city's infrastructure could not handle the increase in commuter congestion. In the middle of the debate, Beck Street gravel pit companies approached Salt Lake City for permission to expand excavation eastward into the mountainside. The proposal could not have come at a better time. With the Salt Lake salient's east face carved away, there would be no room for Bountiful Boulevard. By strategically utilizing Beck Street's gravel pits as a front, Salt Lake City effectively obstructed the roadway extension, thereby maintaining the city's ability to spatially manipulate and control its northern industrial corridor.[74]

In response to these and similar types of land use concerns and complaints, residents, geologists, gravel pit owners, and city planners worked to create the Beck Street Reclamation Framework and Foothill Area Plan. Both North Salt Lake and Salt Lake City hoped to come to an agreement regarding the future of Beck Street's gravel pits as well as address larger land use concerns. An added impetus may well have been the 2002 Winter Olympics. "The narrow Beck Street corridor, with the Great Salt Lake on the West and Ensign Peak on the east, has been decried as an eyesore in recent decades. . . . When athletes . . . arrive with the 2002 Winter Olympics, they too could get a glimpse of the area."[75]

Housecleaning aside, the plan recommended that the Beck Street bench be preserved as open space, exempt from sand and gravel excavation and future housing developments. Genevieve Atwood, former head of the Utah Geological Survey, explained that the Salt Lake salient has much to teach about the region's seismic action and risk. Here the Lake Bonneville story can be extended four million years or more, encompassing the global glacial and interglacial climate patterns. As important layers of lake sediment, sand and gravel deposits serve as a kind of time map geologists can use to understand wind patterns and lake levels. Atwood wondered if it might be possible to establish monitoring stations at current gravel pits—at least then valuable information might be recorded even if it is ultimately destroyed.[76] As University of Utah geology professor John Bowman explained, "This piece of Bonneville Shoreline is one of the very last undeveloped pieces of shoreline in the entire Salt Lake Valley."[77] Gravel pit owners went along with the plan "as long as there was a general recognition of their economic concerns."[78]

In the intervening years Ensign Peak likewise gradually reentered public consciousness and began to reclaim its place as sacred space. In 1980 Ensign Downs Inc., a real estate agency, planned to develop 299 lots adjacent to Ensign Peak and City Creek Canyon, near the former Salt Lake Gun Range. Salt Lake City mayor Ted Wilson offered Ensign Downs the gun range property in exchange for city access to Ensign Peak and to limit development on the steep slopes near the peak. This move helped satisfy the Capitol Hill Neighborhood Council, which had long been actively petitioning the city for increased open space.[79] In 1989 efforts to preserve the peak itself increased. That year the Ensign Peak Foundation formed to raise money for a city park and nature trail up to Ensign Peak.[80] The foundation timed their efforts to coincide with Utah's 1996 statehood centennial and the pioneer sesquicentennial a year later. Beginning in 1992, the foundation revived the 1916 tradition, holding a devotional program and family hike to the peak every July to help raise money to rehabilitate the peak. They sold the hikers t-shirts that featured Brigham Young and the slogan, "Brother Brigham hiked it . . . I did too." With yearly speakers like Gordon B. Hinckley, the event was bound to be a success.[81]

The foundation's plans finally came to fruition in 1996. The completed park featured an entrance plaza, symbolically forty-seven feet from the street. Even the concrete carried symbolism. "Rough aggregate . . . suggests the rock formations of the peak," R. Scott Lloyd wrote for the *Church News.* Repaired and carefully graded, the nature trail included signage identifying native plant species. In his dedicatory prayer, the recently ordained prophet and president of the church, Gordon B. Hinckley, observed, "Through the efforts of many good people, the monument and its surroundings are beautified." He added, "We pray that through the years to come, many thousands of people of all faiths and all denominations, people of this nation and of other nations, may come here to reflect on the history and the efforts of those who pioneered this area. May this be a place of pondering, a place of remembrance, a place of thoughtful gratitude, a place of purposeful resolution."[82] A year later the church dedicated its own small memorial garden near the Ensign Peak trailhead. "In an overwhelmingly colorful display, youth from the Salt Lake, Salt Lake Parleys, and Salt Lake Monument Park stakes carried flags of many nations from the meeting ground to the peak, while the audience sang 'High on the Mountain top,' the familiar hymn written with Ensign Peak in mind," the *Church News* reported.[83]

In 1999, three years after the Ensign Peak trail and park dedication, the Beck Street Reclamation Framework and Foothill Area Plan was finalized. The plan brought together a number of perspectives regarding sand and gravel, the mountains, and development. Geologists and other citizens saw to it that they retained at least a slice of the Lake Bonneville shoreline, creating valuable open space; yet, the plan did not include how to erase or mask the massive quarries that would be left behind when the last trucks eventually rumbled away, leaving a visible history of Salt Lake City's development and growth.[84] It recognized that sand and gravel excavation had exposed the hillside's striated geological underbelly, a feature some wanted to preserve.[85] As one man stated in a letter to the editor, "The different colors of rock and dirt in the various veins in the mountain are very interesting to look at. They are certainly more interesting than just a hill covered with weeds and sagebrush."[86] Salt Lake City planner Craig Hinckley at the time compared the shorn hillside to "a slice of Bryce Canyon."[87] Artist Stephen Goldsmith suggested threading a geological trail through the abandoned gravel pits up to the Bonneville shoreline. Goldsmith asserted the earthen sculpture was not designed to look natural but was to be "more of a shrine—a celebration. We're saying, yes, man has been here. It's more of an honest approach."[88]

It is difficult to assess whether there is anything particularly Mormon in how mid-twentieth century NIMBY (not in my backyard) activists, environmental preservationists, or sand and gravel operators viewed the Beck Street pits and the adjacent Ensign Peak. Does the erasure of the Salt Lake salient imply a Mormon fall from environmental grace? On the other hand, do decisions to preserve Ensign Peak as a heritage site mean renewed Mormon sensitivity to land and memory? Perhaps both places, intertwined as they are, reveal that Mormon approaches to land use, memory, and the environment generally occur within a larger constellation of competing ethics and values. Lake Bonneville's sand and gravel deposits proved to be an invaluable resource in constructing and developing the Salt Lake Valley. For early Mormon settlers, the mountains' overabundant utility enhanced their sacred significance. Wood, lime, and gravel for adobe were extracted to help build and beautify the kingdom of God. Ensign Peak and the nearby quarries help illustrate these seemingly contradictory perspectives. With the post–World War II emphasis on infrastructure, sand and gravel extraction increased, accelerating geologic processes as mountainside aggregates were chemically mixed and repurposed, creating a concrete interface that permeates and coats the valley floor.[89] As aggregate extraction became intimately associated with road and interstate construction, it encouraged the attendant growth of America's car culture. For a time, how Mormons remembered their past was based on how well memorials accommodated the family station wagon. In Salt Lake City, focus shifted away from Ensign Peak and its associated gravel pits to the "This is the Place" Monument, conveniently located at the mouth of Emigration Canyon, whose valley view remained unblemished by unsightly sand and gravel excavations—even as cars parked on its asphalt lot.

Yet Ensign Peak, with its proximity to the Beck Street pits, may be the more authentic memorial. When Wilford Woodruff and Brigham Young surveyed the Salt Lake Valley for the first time on July 24, 1847, Young beheld the valley as if in vision. What did Brigham see as he overlooked the valley? There are a few scattered clues that he did not necessarily see Interstate 15 or the Belt Route that slithers around and through the valley.[90] But maybe Young did see the gravel pits. After all, as he observed in his October 1859 General Conference address, using adobe tempered with gravel, the Saints might create buildings and structures that would last through the millennium, leaving behind human-made canyons and striated landforms.

EPILOGUE

On the Moral Lessons of Mormon Environmental History

George B. Handley

JUST A YEAR OR SO AFTER I arrived at Brigham Young University in 1998, my students in an environmental humanities seminar and I organized a campus-wide panel discussion about Mormonism and the environment. We created a poster in an effort to pique interest, and it asked such questions as: Are Mormonism and environmentalism compatible? What kinds of answers might the restored gospel offer in the face of the environmental crisis? Walking by the poster one day, I noticed that someone had taken a black marker and written in angry capital letters across the face of the poster, "THERE IS NO ENVIRONMENTAL CRISIS!" Notwithstanding the anonymous author's zeal, we threw caution to the wind, went ahead with the panel discussion, and had a wonderful experience.

I was surprised at first by such vehemence, but I have since come to understand the roots of such denial. I don't say this because I am convinced that there is little reason for concern but because of the particular *kinds* of problems environmental degradation presents. As theologian Willis Jenkins has suggested, the more globalized and collective dimensions of environmental degradation present unprecedented levels of complexity that make identifying individual accountability especially difficult, and thus they present special challenges to our ability to define our ethical responsibilities, especially within religious communities. In the face of either unprecedented or unanticipated circumstances such as those we face today with alarming rates of species extinction, global climate change, often invisible and toxic pollution, degraded watersheds, and depletion of topsoil, it is not uncommon for people to respond with denial as a way to protect the integrity of their traditions and values. The only problem, of course, is that the values they cling to are not only inadequate to the problems they face but they may very well turn out to be the result of a poor or misguided interpretation of tradition. What is called for is not reinforced defensiveness in the face of the unprecedented or unanticipated but creative reformulations of tradition in light of an honest assessment of new circumstances. As Jenkins explains, people "can learn new moral competencies . . . by participating in projects *that use their inheritances to create new responsibilities for unexpected problems*."[1] This is more likely to happen if environmental problems are presented in such a way that does not expect or demand an abandonment of one's most deeply held beliefs but instead encourages creative and faithful responses. Surely it is largely such creativity—not inflexible denial or negative protectiveness—that has already sustained religious traditions and allowed them to survive and remain relevant across centuries of dramatic change.

It is hard to see our own moment in historical retrospect, but I suspect we will look back at the last two decades as a remarkable moment of religious awakening to the environmental crisis, both across world religions and within Mormonism. Although I wish I could say that that small incident with the poster was the last time I have encountered outright denial about the problems we face in the twenty-first century, in the almost twenty years I have lectured and written on the relationship between Mormonism and environmental stewardship things have only improved with each passing year. My experience tells me that Mormons are more aware, more concerned, and more willing to listen and act on their concerns. This is not only because Mormonism is now a global church with a much wider variety of views on the environment across its many cultures, but also because the younger generations are asking new and vital questions that are inspiring the kind of creative use of inheritance that Jenkins advocates. The younger generation of church membership, justifiably, has increasing concerns about the environment, and these concerns often leave them feeling isolated in a church context. The saddest conversations I have with students are about the denial that has been passed on to them from parents, extended family, or their fellow ward members, people whose faith they love and admire but who have been profoundly mistaken about the reality and extent of the problems that surround us. In a time when cynicism and disillusionment with religion are current, some opt to leave, but many have forged ahead in faith, honoring the past and using their inheritance to create new responsibilities with which to face the future. They should be a source of great inspiration for the Mormon tradition.

This volume of essays is itself the fruit of this awakening. Indeed, an environmental awakening has revolutionized the study of history. With the advent of environmental history, we have seen a redefinition of the field of history itself that places the relationship between nature and culture at the center of what drives and shapes our story. There are many new questions and insights that such history provokes, as evidenced by these essays. Consider how differently the Mormon past looks when we examine the sheer number of cattle and other domesticated animals Mormon pioneer settlers brought with them into the Great Basin. Such a fact shaped the land, their relationship to the Native Americans, their relationship to predators, and their interpretations of their own theology. Or consider the ubiquitous gravel pit: innocent enough, perhaps, in its beginning and yet a growing symbol of a dependence on the automobile and of the costs of paved roads and urban sprawl built to conform to the conveniences of an automobile society. Surrounding ourselves with built environments has profoundly influenced our overall relationship to air, water, and land. The artificial climates of our homes and our cars, the energy for which comes from fossil fuels, have as much to do with our contributions to contemporary air pollution as they do with our inability to see them as ours. Environmental history, perhaps like most history, tells stories of error but more specifically error with regard to a fundamental relationship to nature that we have pretended in the modern era either does not exist or does not matter. At least by highlighting that relationship, these stories allow us to begin to imagine a story of redemption.

But we cannot establish such hope without a clear-eyed reckoning of how an accelerated modern, globalized economy challenges our capacity to anticipate the environmental degradations we cause. Early histories of resettlement, such as we read about here, tend to show that there is a long and slow accretion of an environmental ethos, certainly longer and slower than we need in order to keep pace with the problems we instigate. Sadly, stronger scientific understanding often comes only after significant trial and error, to say nothing of the challenge of convincing the public of the legitimacy of that science. While these essays raise awareness and increase our sense of urgency about environmental degradation, we also feel compassion for forebears who might have mistakenly believed that they could aggressively plant exotic species, extirpate predators, or dam and straighten rivers with impunity, since it was only the extension of such errors over time that finally taught us their costs.

But precisely because these stories expose the accretion of consequences and the rising stakes of our environmental choices, they also warn us against passivity and complacency. Environmental history helps us to see not how we should have known better then, but why we should know better now. I say this because these authors trace stories of success from which to take inspiration, but they also tell stories of error that go beyond the naming of individuals and their misuse of agency in deciding our present. Instead they outline the accumulated effects of collective choices that are societal and structural and thus expose our modern and globalized reality: individual accountability matters as much as it ever did, but the subject of history is also collectively human, rather than merely individual, and the collective human acts and is acted upon in the large unfolding of history between peoples and places. The contemporary environmental crisis asks us not only to write environmental histories that reveal a complexity in our relationship to nature that is often missed by traditional historiography, but also to redefine ethics in ways that move us beyond the model of a radically autonomous individual who pretends to act independently of others or of the earth.

Environmental history, then, gives account of the sometimes mystical and often denied relationship between our choices and environmental degradation that make us more accountable for their causal links. We might think of it as a bit like how Alcoholics Anonymous makes use of frank testimonials about choices and their empirical consequences to awake the conscience of the individual alcoholic. And my analogy to addiction here is intentional because the portrait of the human species that emerges from environmental history shows us to be inconsistent at best and compulsively reckless at worst about protecting our own best interests. We are, after all, the only species that seems willing to foul our own nest repeatedly. So I would say that this environmental history is inherently sobering, in the sense that it gives us the chance to shake off our dream of innocence and learn to act with greater caution and self-distrust so as to minimize our destructive tendencies. Unlike our forebears, we live in an age of unprecedented ecological understanding, an understanding that informs the stories these authors have unearthed. Thus, as compassionate as we might feel toward those who erred in the past, we simply no longer have excuses for ignoring accountability for our choices. The growing urgency around recovering and promoting environmental stewardship that we have seen in both world religions and Mormonism, in other words, is an exercise in repentance, and repentance, as any Christian will tell you, is far easier and more meaningful when we have a clearer picture of the damage as well as of the good we are capable of doing. And thanks to these essays, we see more clearly the moral dimensions of our relationship to nature that was previously unseen or unacknowledged.

Change is hard and it requires moral urgency, which is no doubt why we have seen so much interest and engagement among religious communities in seeking to define and promote environmental values. From the Pope's encyclical on the climate in 2015 to the Patriarch Bartholomew's decades-long effort to preach environmental stewardship among Orthodox Christians to the Dalai Lama's strong pronouncements to care for the earth throughout his career, faith-based environmentalism over the past quarter of a century has grown at a remarkable rate.[2] And again, these are not efforts to teach a new religion but to reengage. Indeed, evidence suggests that from its nineteenth-century origins, environmentalism—at least in the United States—was profoundly influenced by a Christian ethos of care and reverence for Creation that got lost in the age of heightened partisanship and the cultural wars of the 1980s.[3] The hope recently is that by framing environmental problems around the quest to show proper reverence and respect for the sanctity of the Creation, to care for the poor and the most vulnerable, and to live lives of humility, modesty, and moderation, religion can promote a stronger ethic and point the way to meaningful change far more effectively than any political or educational campaign. This is the reason for the Harvard biologist E. O. Wilson's appeal to religious ministers. He writes: "You are well prepared to present the theological and moral arguments for saving the Creation," something science alone cannot do.[4] Although the research is somewhat undecided on the exact nature of the influence of religious belief on environmental behavior, there is no question, as these essays demonstrate, that ideas and values about environmental issues—such as how we define sacred space, how we conceptualize wildlife management, what we understand to be the relationship between consumption of resources and the plight of the poor, or why and when a gardening ethic waxes or wanes—have real consequences in the world and on the environment.

An environmental history of a religious and cultural community, then, is not some secularization or debunking of their story that is meant to drag their ethereal ideas down into the mud, but rather a way of respecting and testing its potency as a living tradition. It is an honest history, to be sure, since it doesn't allow ideals or rhetoric to take the place of actual impact, but it is a complex history, too, since it is never precisely clear what the relationship is between belief and action. Rebecca Andersen speaks perhaps for most of these essays when she says that "Mormon approaches to land use, memory, and the environment generally occur within a larger constellation of competing ethics and values." It is fair to say, then, that pointing fingers of blame when environmental damage results is neither fruitful nor accurate. This larger constellation of values implies that it isn't always clear what accounts for the tradition's failures as well as its environmental successes. Successes deserve more research so as to understand the delicate interplay between environmental values as inherited from the tradition and those that are shaped by or are reactions to currents in contemporary political culture. We can only hope that religious values might prevail over ideology.

Because political ideology has such powerful sway, especially these days, it is especially important to mine the tradition and promote the values we need to guide us into the future. This history makes clear that the ubiquity and clarity of Mormon environmental values and the strength of a Mormon environmental ethos cannot be assumed or taken for granted. Indeed, the history laid out in these essays suggests that the very value of stewardship itself is often fragile and endangered, vulnerable to distortion or loss but also capable of being refurbished and renewed. It is fair to say that stewardship within Mormonism has had varying and sometimes opposing meanings throughout Mormon history, which isn't to say that it is an arbitrary value. It is, however, to say that stewardship is an important value to the extent that it can be shown to motivate the most effective responses to a contemporary crisis. For this reason, Elder Marcus B. Nash's remarkable discourse in this volume's appendix stands out as a particularly significant achievement. Although not about Mormon history, as are the other essays in this collection, it is no doubt a response to our historical moment and stands a good chance, if sufficiently advanced by the church, of inspiring adequate responses among believers to the challenges we face in the way that Willis Jenkins envisions. It represents the first discourse of its kind by any LDS church leader, despite the fact that it doesn't introduce any new ideas or doctrines. It is inaccurate to assume that church leaders have not preached against environmental waste or degradation before. Indeed, it is hard to find major leaders of the church who haven't—at least at some point—mentioned important stewardship principles. But Mormon teachings about the plan of salvation, the Creation and the Fall, and the principles of stewardship and consecration have never been interpreted before by a church leader exclusively for their bearing on the question of our relationship to and use of natural resources.

Elder Nash's essay is also found on two new church websites on stewardship and conservation that represent further mining and refurbishing of the tradition in light of contemporary challenges.[5] They, along with a significant development in scholarship on Mormon ecotheology over the past twenty years, make evident a treasury of doctrines and teachings about stewardship of the earth that make Mormonism unique in the Judeo-Christian tradition for their clarity, prominence, and profundity.[6] Research in the field of ecotheology has convinced me, at any rate, that there are scarcely any religious traditions with so many explicit revelations and teachings about the centrality of our careful stewardship of material resources. I don't say that out of any particular Mormon pride or religious chauvinism. In fact, if anything, the existence of so many rich and useful teachings only suggests reasons for embarrassment for their neglect. These are no doubt the teachings that have inspired generations of Mormons to embrace their stewardship of the earth, as this research shows, because they saw those teachings as integral to their fundamental Christian duties. But these essays also suggest at least some reasons why these teachings have not stuck.

This collection does the vital spade work of addressing some key questions that will no doubt require further exploration. We still need to understand better the reasons why, as a culture, Mormons are not more widely recognized for their explicit and strong commitment to earth stewardship in the twenty-first century, as well as when and why there has been a decreased emphasis on stewardship in Mormon culture. Was this the result of benign neglect or proactive suppression of an environmental ethos that was perceived to be motivated by Malthusian designs on population control? A lack of a pronounced and clear commitment to earth stewardship is especially curious given the fact that the church websites point to many exemplary environmental practices in their architectural design and in other arenas, even though scarcely little has been written about them to understand their origin or logic. Precisely because local autonomy over the stewardship of physical facilities has declined so dramatically over the last generation, the membership need and deserve to know this history.

Research on the roots of contemporary attitudes and practices might have ramifications too for understanding how and why Mormons eat as they do, how they conceive of city planning and land use, why and when they have been willing to support public transportation or public lands, what their changing views of air pollution have been, and what their views are of science and its relationship to policy. Two Mormon candidates recently ran for the Republican ticket for president and they were the only candidates among a slate of climate change deniers with at least moderate concern about climate change, and this raises the question: Were there theological grounds for that difference? Why, by contrast, has Utah produced political leaders—all of them Mormon—with such pronounced and formally antagonistic attitudes toward environmental regulations? What are the reasons why conservative leanings predominate in American Mormon culture today, and why does partisan affiliation today determine, far more than any other factor, attitudes about environmental issues such as climate change? What kind of religious culture could mitigate that influence?

Contemporary environmental attitudes in the Intermountain West need further study, as do the impact and understanding of Mormonism's unique and powerful teachings about stewardship throughout the worldwide church. Such research might help point us to the regional, rather than doctrinal, roots of environmental apathy or antagonism and help us understand more clearly the differences, as well as the common ground between, say, a Swedish Mormon's environmental views and those of an Idaho Mormon. Even locally, Mormonism has produced, after all, both the Bundy family and Terry Tempest Williams. What might explain these radical differences? And what, for that matter, can ensure greater stability and clarity about the importance of stewardship and how to implement it among the worldwide membership? Answers to such questions will go a long way in celebrating the values of self-restraint in our consumption of natural resources, regard for the sanctity and beauty of the earth, and proactive interest in addressing the vulnerabilities of the poor and of future generations, thus providing a blueprint for a more sustainable future. It would seem that in all of the confusion about the difficult decisions we face, Mormonism's unique and powerful beliefs and values about environmental stewardship ought to prevail and make common ground a lot easier to identify—and hopefully make endangered species of divisiveness and denialism.

APPENDIX

Righteous Dominion and Compassion for the Earth

Marcus B. Nash

Members of the Church of Jesus Christ of Latter-day Saints (LDS) share with many of the great faith traditions of the world a common concern for the environment, as well as a common desire to draw others unto the Creator of heaven and earth. Prior to my call as a general authority for the church, I practiced law in the city of Seattle, Washington, where I was born and raised, a place where nature's splendor is abundantly displayed.

One of the first clients I represented as a young lawyer was a berry farmer near Mount Vernon, Washington. This farmer raised and sold certified nursery strawberry stock, which could be sold as such only if it was not infested with nematodes, small ground worms that can damage the productivity of the plant. The proven way (at the time) to rid the plant of this pest and to achieve "certified" status was to fumigate the soil.

According to the fumigant manufacturer, the fumigant agent would gasify upon application and evaporate through the soil into the air; the nematodes would be killed and the soil and groundwater would be unaffected. However, neighbors began to notice an odor in their water and had tests performed. It seems that not all of the fumigant evaporated into the air, but some portion had leached into the groundwater, rendering it unusable.

This story shows the inherent complexity of human interaction with the environment: the farmer recognized that the only way to produce nursery stock not infested with nematodes was to fumigate, but by so doing he unwittingly polluted the neighbors' groundwater. Although the farmer was eventually dismissed from the case, he had to endure significant stress and some economic repercussions; and the fumigant manufacturers (whom we did not represent) dealt with some financial consequences. And, what of future generations who must grapple with groundwater contamination of an unknown duration? It is interesting, if not mildly ironic, that the very families who sued my client likely had at some point during the litigation periodically enjoyed fresh, delicious strawberries grown from certified nursery stock.

These complexities show that our approach to the environment must be prudent, realistic, balanced, and consistent with the needs of the earth and of current and future generations. In an effort to go to the root of the issue, most of us agree that: (1) we depend upon this earth to sustain life; and (2) the quality of the earth and its environment will directly affect the quality of our life—as well as that of future generations.

Despite what I believe to be almost universal agreement on these postulates, they have been (by many) ignored, unappreciated, and/or simply seen as too costly or inconvenient. Yet if we understand who we are, the purpose of our existence, and the reason the earth was created—and keep these things in mind—our conduct will rise to a higher, nobler level. Religion and faith play a major role in shaping our understanding and conduct with respect to the environment.

With that understanding, this essay seeks to explore my interpretation of the doctrine of the LDS Church pertaining to this earth and all life on it. It contains eight different sections, each organized around a verse accepted and canonized by the LDS Church as scripture. The LDS Church defines "scripture" as "a writing recognized by the Church as sacred and inspired."[1] The term "canon" is "used to denote the authoritative collection of the sacred books used by the true believers in Christ."[2] The LDS Church recognizes both ancient and modern scripture in its official canon.[3]

This essay draws upon the full array of the scriptural canon to explain my understanding of the doctrine of the LDS Church as it pertains to the environment. The first three sections of this essay provide a broad doctrinal context for the subsequent five sections, which in turn address more directly what the church's doctrine teaches about the environment.

"What is man that thou are mindful of him?" (Psalms 8:4).

In this psalm, David of the Old Testament considers the majestic creations of God and wonders aloud why—amongst such wonders—God is mindful of man. David concludes that the fact that God cares and gives humankind a dominant role on this earth is evidence that humankind is special, "a little lower than the angels."[4] What we understand of who we are and why we are on this earth can (and should) have a profound effect upon how we choose to relate to the earth and its life.

Why was the earth created, according to LDS doctrine? One of the prophets we revere, along with much of Christianity, is Moses of the Old Testament. According to the Pearl of Great Price (a book of scripture unique to the LDS Church), after seeing in a vision all of God's creations pertaining to this earth in a limitless expanse, Moses declared: "Now, for this cause I know that man is nothing, which thing I never had supposed."[5] In Moses's humility before the magnificence of God's creation he failed to comprehend a great truth that the Lord wanted him to understand. Thus, according to the scriptural account, the Lord returns to Moses, shows him again the earth and all pertaining thereto, speaks of His vast and limitless creation, and then enlarges Moses's perspective by declaring that he—God—has made these creations "for [His] own purpose," which is "to bring to pass the immortality and eternal life of man."[6] In short, according to LDS doctrine, God desires that humankind (i.e., His children[7]) progress, improve, and receive eternal life in his presence—and He created this earth for that purpose.

"And we will prove them herewith" (Abraham 3:24).

Other passages in the Pearl of Great Price teach that before this world was created God explained in practical terms the purpose behind its creation: "we will make an earth whereon these [meaning us] may dwell; and we will prove them herewith, to see if they will do all things whatsoever the Lord their God shall command them."[8] The same account adds that those who choose to follow and serve God in this life upon this earth will "have glory added upon their heads for ever and ever."[9] In short, the creation of this earth affords us the opportunity to choose to seek and someday receive all that God offers. However, one cannot, and should not, be forced to follow God; indeed, real growth and development in an individual occurs only when one has the opportunity to choose for him or herself. As a passage in the Book of Mormon—also accepted by church members as scripture—teaches, "the Lord God gave unto man that he should act for himself."[10]

The earth, then, was created to provide a place for the children of God to be tested, to learn and gain necessary experience in a place where they would have the opportunity to choose whether they will or will not do all things that the Lord commands. Once the process of the creation of this earth was completed, God was pleased, and He stated: "Behold, all things that I had made were very good."[11] He was pleased, for He saw that this earth would serve His purpose for us, His children.

According to the biblical account, once the earth was created with plant and animal life, the stage was set and Adam and Eve, the first man and woman, were placed upon this earth. Latter-day Saints believe God's plan was—and is—that His children come to this earth through marriage between a man and a woman, who are to procreate, form a family, and teach children to choose the good part in a world with real moral choices and consequences.[12] According to a Joseph Smith revelation, the Almighty designed these purposes so that "the earth might answer the end of its creation; and that it might be filled with the measure of man according to His creation before the world was made."[13] Consistent with this, a prophet in the Book of Mormon stated that "the Lord hath created the earth that it should be inhabited; and he hath created his children that they should possess it."[14] According to LDS doctrine, men and women are not mere interlopers or a sideshow on this earth; rather, they and the children they bring into this world are central to its purpose.

"For I, the Lord God, created all things . . . spiritually, before they were naturally upon the earth" (Moses 3:5).

Latter-day Saints believe that not only did God create a beautiful world of mountains, valleys, rivers, streams, seas, sunsets, and sunrises, He also adorned it with plant and animal life. According to LDS scripture, each form of plant and animal life has a spirit: "I, the Lord God, made the heaven and the earth, And every plant of the field before it was in the earth, and every herb of the field before it grew. For I, the Lord God, created all things . . . spiritually, before they were naturally upon the earth."[15] Further, He declared: "Out of the ground made I, the Lord God, to grow every tree, naturally, that is pleasant to the sight of man. . . . And it became also a living soul. For it was spiritual in the day that I created it."[16] Then with regard to animal life we read: "Out of the ground I, the Lord God, formed every beast of the field, and every fowl of the air . . . and they were also living souls."[17]

Since both plant and animal life are living souls, they are capable of experiencing happiness as they fulfill the measure of their creation.[18] As church president Joseph Fielding Smith, taught, "the Lord gave life to every creature . . . [and] commanded [them] to be fruitful and multiply and fill the earth. It was intended that all creatures should be happy in their several elements."[19]

Not only is animal life capable of happiness, but it is also included within the scope of His redeeming power, as taught in this LDS scripture: "And the end shall come, and the heaven and the earth shall be consumed and pass away, and there shall be a new heaven and a new earth. For all old things shall pass away, and all things shall become new, even the heaven and the earth, and all the fullness thereof, both men and beasts, the fowls of the air, and the fishes of the sea; And not one hair, neither mote, shall be lost, for it is the workmanship of mine hand."[20] Plainly, all forms of life identified in this verse have great value in the eyes of God, for they are the workmanship of His hand, and will be blessed by His redeeming power. This doctrine leads one to view plant and animal life differently, as living souls created by God.

"Ordained for the use of man" (Doctrine and Covenants 49:19–21).

According to LDS doctrine, this earth, as well as its plant and animal life, were provided for the use of man. However, we believe that God has commanded that the earth and its resources be utilized responsibly to abundantly sustain the human family. Joseph Smith dictated a revelation in 1831 that declared: "Behold, the beasts of the field and the fowls of the air, and that which cometh of the earth, is ordained for the use of man for food and for raiment, and that he might have in abundance."[21]

Nevertheless, LDS doctrine is clear: all humankind are stewards over this earth and its bounty—not owners—and will be accountable to God for what we do with regard to His creation. In an 1834 revelation dictated by Joseph Smith, the Lord rightfully asserts His ownership over this earth and its resources: "For it is expedient that I, the Lord, should make every man accountable, as a steward over earthly blessings, which I have made and prepared for my creatures. I, the Lord, stretched out the heavens, and built the earth, my very handiwork; and all things therein are mine."[22]

How we care for the earth, how we utilize and share in its bounty, and how we treat all life that has been provided for our benefit and use is part of our test in mortality. Thus, when God gave unto man "dominion over the fish of the sea, and over fowl of the air, and over the cattle, and over all the earth, and over every creeping thing that creepeth upon the earth," it was not without boundaries or limits.[23] He intends man's dominion to be a righteous dominion, meaning one that is guided, curbed, and enlightened by the doctrine of His gospel—a gospel defined by God's love for us and our love for Him and his works. The unbridled, voracious consumer is not consistent with God's plan of happiness, which calls for humility, gratitude, and mutual respect.

In other words, as stewards over the earth and its life, we are to gratefully make use of that which the Lord has provided, avoid wasting life and resources, and use the bounty of the earth to care for the poor. Another revelation dictated by Joseph Smith states: "But it is not given that one man should possess that which is above another, wherefore the world lieth in sin. And wo be unto man that sheddeth blood or that wasteth flesh and hath no need."[24] The 1834 revelation continues, "For the earth is full, and there is enough and to spare; yea, I prepared all things, and have given unto the children of men to be agents unto themselves. Therefore, if any man shall take of the abundance which I have made, and impart not his portion, according to the law of my gospel, unto the poor and the needy, he shall, with the wicked, lift up his eyes in hell, being in torment."[25]

The Lord gave to men and women agency, or the capacity to choose; however, because He cares deeply for all life—and especially for His children—He will hold us accountable for what we choose to do (or not do) with the bounties of His creation.

"It pleaseth God that he hath given all these things unto man" (Doctrine and Covenants 59:20).

In an 1831 revelation, humankind is promised that if we choose to follow the Lord and judiciously use the resources of the earth with thanksgiving and respect,

> the fulness of the earth is yours, the beasts of the field and the fowls of the air, and that which climbeth upon the trees and walketh upon the earth; Yea, and the herb, and the good things which come of the earth, whether for food or for raiment, or for houses, or for barns, or for orchards, or for gardens, or for vineyards; Yea, all things which come of the earth, in the season thereof, are made for the benefit and the use of man, both to please the eye and to gladden the heart; Yea for food and for raiment, for taste and for smell, to strengthen the body and to enliven the soul. And it pleaseth God that he hath given all these things unto man; for unto this end were they made to be used, with judgment, not to excess, neither by extortion.[26]

According to this passage, we have been provided with a beautiful and bountiful world, teeming with life and resources to bless, strengthen, and enliven mankind, and we are to use these resources joyfully—but we must do so as careful, grateful stewards over God's handiwork. We are to use the resources with judgment, gratitude, prudence, and with an eye to bless our fellowmen and women and those of future generations, and in that way help Him accomplish His purpose to help humankind progress, improve, and receive His blessings in time and eternity.

"The God of heaven looked upon the residue of the people, and he wept" (Moses 7:28).

Because of the principle of agency, Latter-day Saints believe that earth is a place where individuals may choose to reject God and treat His creation with disdain. When this occurs, God and creation are pained. Passages in the Pearl of Great Price relate a vision in which Enoch of Old Testament times saw that the God of heaven wept on account of the poor choices and suffocating selfishness of humankind—His children.[27] Similarly, the Book of Mormon prophesied that in the latter-days there would be "fires, tempests, and vapors of smoke . . . and . . . great pollutions upon the face of the earth," and that such conditions would be coupled with "murders, and robbing, and lying, and deceiving, and whoredoms, and all manner of abominations; when there shall be many who will say, Do this, or do that, and it mattereth not."[28] According to LDS scripture, there is a corollary between the selfish, materialistic man out to hoard money and material possessions—and/or the man with irreverence for life—and pollutions (spiritual or temporal) upon the face of the earth. As former church president Ezra Taft Benson stated: "Irreverence for God, of life, and for our fellowmen takes the form of things like littering, heedless strip-mining, [and] pollution of water and air. But these are, after all, outward expressions of the inner man."[29] Gordon B. Hinckley, another former president of the church, added: "This earth is his creation. When we make it ugly, we offend him."[30]

According to LDS scripture, when man pollutes this world spiritually or temporally, not only God but nature also suffers. The same Pearl of Great Price passages discussing Enoch's vision of humankind's wickedness state that "Enoch looked upon the earth; and he heard a voice from the bowels thereof, saying: Wo, wo is me, the mother of men; I am pained, I am weary, because of the wickedness of my children. When shall I rest, and be cleansed from the filthiness which is gone forth out of me? When will my Creator sanctify me, that I may rest, and righteousness for a season abide upon my face?" After hearing the earth mourn (which may be metaphorical), Enoch "wept and cried unto the Lord, saying: O Lord, wilt thou not have compassion upon the earth?"[31] Prevalent today are the spiritual and temporal pollutions, scars, and damage wrought by man upon this earth. May we all ask ourselves as Enoch did: will we not have compassion upon the earth? Or are we too caught up in our personal pursuits and desires?

"Every man seeking the interest of his neighbor, and doing all things with an eye single to the glory of God" (Doctrine and Covenants 82:19).

In an 1832 revelation dictated by Joseph Smith, the Lord stated that His aim in organizing His church included creating a society in which every man "sought the interest of his neighbor, and to do all things with an eye single to the glory of God."[32] Faith and religion should have the capacity to stretch, enlarge, and change the human soul beyond self, to inspire love of God and His creations, to think of others, and to consider the needs of future generations, even to the point of sacrificing personal desires. We need that soul-stretching, for the state of the human soul will directly impact the condition and health of the environment—which, in turn, affects our quality of life. Thus, the late Neal A. Maxwell (a member of the Quorum of Twelve Apostles in the LDS Church) invited followers of Christ to live lives of moral integrity: "True disciples [of Christ] . . . would be consistent environmentalists—caring both about maintaining the spiritual health of a marriage and preserving a rain forest; caring about preserving the nutrient capacity of a family as well as providing a healthy supply of air and water."[33]

Faith in the Lord Jesus Christ teaches us to live lives of internal consistency, true to God, true to his present and yet-to-be born children, and true to the purpose of his creations. To the degree that it enlarges our understanding of who we are, our knowledge of why this earth was created, and inspires us to both respect this earth as the handiwork of God and think of others (including future generations), religion can change how we will treat the earth and its resources.

Ezra Taft Benson, who was president of the church from 1985 to 1994, expressed it this way: "The Lord works from the inside out. The world works from the outside in. . . . The world would mold men by changing their environment. Christ changes men, who then change their environment."[34] As the human soul is thus changed, the environment is better cared for. The doctrine and commandments of God lead us beyond the suffocating, self-limiting weight of selfishness, the blinding press of self-gratification or aggrandizement. The gospel of Jesus Christ helps us think beyond ourselves, to think of the earth and all life given by God, including others now and in future generations, rather than pursue the immediate vindication of our personal desires or avowed rights. If I pursue a selfish, irreverent course, I pursue a course that gives license to despoil the earth, for pollution, damage, and waste are frequently the product of selfishness or irreverence. To the degree that religion teaches reverence for God, for His creations, for life, and for our fellowmen, it will teach us to care for the environment. In short, the state of the human soul and the environment are interconnected; each affects and influences the other.

Brigham Young, who led church members from the eastern states to the Salt Lake Valley in the mid-1800s, reportedly stated: "In the mind of God there is no such a thing as dividing spiritual from temporal, or temporal from spiritual; for they are one in the Lord."[35] When the early pioneers arrived, this valley was mostly uncultivated desert, but with fertile soil and water from mountain runoffs. So, they went to work to carve a civilization out of the wilderness. Young understood the doctrine of which I have spoken, and reportedly told those early pioneers, "Keep your valley pure, keep your towns as pure as you possibly can, keep your hearts pure."[36] He added the need to study and reverence the Lord's creations: "Fields and mountains, trees and flowers, and all that fly, swim or move upon the ground are lessons for study in the great school [of] our heavenly Father . . . in the great laboratory of nature."[37] He enjoined those pioneers to care and not waste nature and its bounty, stating that "it is not our privilege to waste the Lord's substance."[38]

The LDS Church continues to seek to care for this earth and judiciously utilize its resources. In so doing, the church makes real effort to conduct itself by what it should do, not just what is legally required. Speaking some years ago to members of the legal profession, James E. Faust, who served in the church's governing First Presidency from 1995 to 2007, stated:

> In our own standards of personal conduct we must remember that the laws of men are the lesser law. I cite to you that the laws of many jurisdictions do not require or encourage being a Good Samaritan. There is a great risk in justifying what we do individually and professionally on the basis of what is "legal" rather than what is "right." In so doing, we put our very souls at risk. The philosophy that what is legal is also right will rob us of what is highest and best in our nature. What conduct is actually legal is, in many instances, way below the standards of a civilized society and light years below the teachings of the Christ. If you accept what is legal as your standard of personal or professional conduct, you will deny yourself of that which is truly noble in your personal dignity and worth.[39]

The Church of Jesus Christ of Latter-day Saints desires to do what is right in its temporal affairs, including environmental practices—even if it is something more than the requirements of the law.

For example, the LDS Church wants to be good stewards of the environment, conserve energy and water resources, minimize pollution, create and maintain cost-effective properties, and be a good neighbor. It thus believes in sustainable design strategies for its facilities. The church has built several projects that are certified under the U.S. Leadership in Energy and Environmental Design (LEED) green building program. Some of these facilities include the City Creek Center development, the Church History Library, and several meetinghouses. The church is also developing its own green building program for meetinghouses based on LEED models plus the International Green Construction Code (IGCC). This self-enacted and self-enforced green building program will create sustainable properties and achieve green building objectives at a reduced cost.

The LDS Church employs sustainable agriculture best practices on farms and ranches that it operates, such as Deseret Cattle and Citrus, a 295,000 acre farm and ranch in Florida.[40] Examples include:

Integrated pest management (IPM) for pesticide application.

Use of pheromone traps at orchards as a means of pest mating disruption.

No-till dry farming to reduce erosion and retain moisture.

Good range management practices to avoid overgrazing.

Community gardens are sponsored to give people an opportunity to grow nutritious foods and reduce costs.

Significant recycling efforts are employed at forty-two Deseret Industries thrift stores. Over 66 million pounds of materials were recycled in 2012, including clothing, shoes, household goods, books, metal, and electronics. Recycling programs for materials such as cardboard, paper, aluminum, and plastic are practiced at major locations such as the church headquarters campus, printing operations, distribution centers, and apparel manufacturing locations.

"All things which come of the earth . . . are made for the benefit . . . of man, both to please the eye and to gladden the heart . . . and to enliven the soul" (Doctrine and Covenants 59:18–19).

The earth and its life are much more than items to be consumed and/or conserved; some parts and portions of it are also to be preserved. As we nurture and appreciate nature, we will become better acquainted with our God, for unspoiled nature is designed to inspire and uplift humankind. Nature in its pristine state brings us closer to God, clears the mind and heart of the noise and distractions of materialism, lifts us to a higher, more exalted sphere, and helps us to better know our God. In an 1832 revelation from Joseph Smith, the Lord declared: "The earth rolls upon her wings, and the sun giveth his light by day, and the moon giveth her light by night, and the stars also give their light, as they roll upon their wings in their glory, and in the midst of the power of God . . . Behold . . . any man who hath seen any or the least of these hath seen God moving in his majesty and power."[41]

I am an avid hiker; I grew up hiking the North Central Cascades in Washington State, and have hiked in other various parts of the country and world. In my childhood I loved to be in the woods, to sense and feel the silent, eloquent witness the towering evergreen trees bore of the Creator. As I grew older, my wanderings in wilderness took me beyond the woods, to climb the magnificent granite rocks and peaks rising above the timberline, where the only sound is the wind moving through rock and some scrub trees fighting to survive in a harsh alpine environment. These high peaks, humble in their magnificence before the one who designed and made this earth, touch the blue vault of the heavens. Although silent, they speak of the power and majesty of God—and of His matchless genius for beauty.[42]

There is a spirit among the trees—are they not living souls? This is even more true for me above the timberline, amongst the mountaintops, where I feel a closeness to God. I love to sit or stand under the sky where heaven and earth meet, the high alpine peaks around me, and to gaze at the stars at night, trying—always unsuccessfully—to wrap my mind around the eternity within my gaze, an eternity of both time and space. (Imagine, for example, the hundreds or millions of lightyears it took for some of the light of the stars to reach this earth.) Yet, I always marvel at the quiet knowledge that settles upon me in those solitary moments of tranquility: that, despite the vastness of the cosmos, the Lord of the universe knows puny me. And He knows you, and each of His children. This earth, every aspect of it, was created for the purpose of giving each of us the opportunity to be blessed now and in eternity. This planet witnesses of the Creator, and if we preserve these special places in their unspoiled state, they will silently, eloquently witness of our God and inspire us onward and upward.

This earth is provided to help each of us return to Him, having grown through testing and experience to become more like Him, and enjoy eternal felicity with Him. Our test on this earth is whether we will choose wisely and follow God, treat His creations with respect, and use them to bless our fellowmen and women. The better we care for this world and all in it, the better it will sustain, inspire, strengthen, enliven, and gladden our hearts and spirits—and prepare us to dwell with our Heavenly Father and with our families in a Celestial sphere, which members of the LDS Church believe will be the very earth upon which we stand today, but in a glorified state.[43]

May we care for this earth—our present and potentially future home—well.

ACKNOWLEDGMENTS

THIS VOLUME began as a session on Mormon environmental history at the Mormon History Association Conference in 2013. We wish to thank John Alley, editor-in-chief at the University of Utah Press, for his unwavering support, as well as each of the contributors for enduring multiple rounds of drafts. We owe a debt of gratitude to Brian Cannon and Brenden Rensink of the Charles Redd Center for Western American Studies at Brigham Young University for hosting a working seminar for the editors and contributors of the volume on November 5–6, 2015. This gathering gave contributors an opportunity to hone their arguments through discussions with other scholars as well as to think broadly about the intersection of Mormonism and environmental history. A special thanks to Brenden for reading and commenting on a number of the papers.

Thanks go to Jared Farmer, Kevin Marsh, and an anonymous reviewer for a careful reading of and thoughtful comments on the manuscript.

NOTES

Introduction

1. Since Hugh Nibley's work on early Mormon environmental thought, a handful of writers and historians have examined the intersection of Mormonism and the environment. See the essay "History, Nature, and Mormon Historiography" in this volume for mention of some of these authors. The invisibility of environmental history in Mormon studies is reflected in the criticism leveled by Jared Farmer that his award-winning book *On Zion's Mount: Mormons, Indians, and the American Landscape* (Harvard University Press, 2008) was not reviewed in the *Journal of Mormon History*. See Farmer, "Crossroads of the West," *Journal of Mormon History* (Winter 2015): 171–72.
2. See Donald Worster, "Transformations of the Earth: Toward an Agroecological Perspective in History," *Journal of American History* 76 (March 1990): 1087–1106, esp. 1089. Other articles in the same issue of the *Journal of American History* that discuss this topic include William Cronon, "Modes of Prophecy and Production: Placing Nature in History," 1122–31; Alfred W. Crosby, "An Enthusiastic Second," 1107–10; Carolyn Merchant, "Gender and Environmental History," 1117–21; Stephen J. Pyne, "Firestick History," 1132–41; Richard White, "Environmental History, Ecology, and Meaning," 1111–16; and Donald Worster, "Seeing beyond Culture," 1142–47.
3. Paul S. Sutter, "The World with Us: The State of American Environmental History," *Journal of American History* 100 (June 2013): 96.
4. Donald Worster, "The Vulnerable Earth: Toward a Planetary History," *Environmental Review* 11 (Summer 1987): 8–14.
5. Christian Schwägerl, *The Anthropocene: The Human Era and How It Shapes Our Planet* (Synergetic Press, 2014); Jeremy Davies, *The Birth of the Anthropocene* (Oakland: University of California Press, 2016).

6. Lynn White Jr., "The Historical Roots of Our Ecological Crisis," *Science* 155 (March 1967): 1203–7.
7. Mark R. Stoll, *Inherit the Holy Mountain: Religion and the Rise of American Environmentalism* (Oxford: Oxford University Press, 2015), 2.
8. Roger S. Gottlieb, "Introduction: Religion in an Age of Environmental Crisis," in *This Sacred Earth: Religion, Nature, Environment*, ed. Roger S. Gottlieb, 2nd ed. (New York City: Routledge, 2004), 8–12.
9. See, for example, the *Journal for the Study of Religion, Nature and Culture*, published by Equinox Press.
10. Mark Fiege, *The Republic of Nature: An Environmental History of the United States* (Seattle: University of Washington Press, 2012).

History, Nature, and Mormon Historiography

1. Pomeroy, "Toward a Reorientation of Western History: Continuity and Environment," *Mississippi Valley Historical Review* 41 (March 1955): 579–80. For Turner's thesis, see "The Significance of the Frontier in American History," *Annual Report of the American Historical Association for the Year 1893* (Washington, D.C.: GPO and American Historical Association, 1894), as reprinted in Richard W. Etulain, ed., *Does the Frontier Experience Make America Exceptional?* (Boston: Bedford/St. Martin's Press, 1999).
2. Webb is best known for *The Great Plains: A Study in Institutions and Environment* (Waltham, MA: Ginn and Co., 1931). For a good collection of Malin's writings and thought, see *History and Ecology: Studies of the Grassland*, edited by Robert P. Swierenga (Lincoln: University of Nebraska Press, 1984). It should be noted that both men were regionalists interested primarily in the Great Plains, and their characterizations little applied to the complicated mountain topography of the Intermountain West, Southwest, or the West Coast.
3. See, e.g., Leland H. Creer, *The Founding of an Empire: The Exploration and Colonization of Utah, 1776–1856* (Salt Lake City: Bookcraft, 1947); Milton R. Hunter, *The Mormons and the American Frontier* (Salt Lake City: LDS Department of Education, 1940); Andrew Love Neff, *History of Utah, 1847–1869*, edited by Leland H. Creer (Salt Lake City: Deseret News, 1940); Thomas C. Romney, *The Story of Deseret* (Independence, MO: Zion's Printing & Publishing, 1948).
4. Strahorn, *To the Rockies and Beyond: Or a Summer on the Union Pacific Railway and Branches* (Omaha: Omaha Republican Print, 1878), 94.
5. Smythe, *The Conquest of Arid America* (New York: The MacMillan Company, 1905).
6. The best general works on Mormon village studies are Bahr's *Saints Observed: Studies of Mormon Village Life, 1850–2005* and his edited collection *Four Classic Mormon Village Studies*, both published by the University of Utah Press in 2014. Twentieth-century Mormon village studies include

Nels Andersen, *Desert Saints: The Mormon Frontier in Utah* (Chicago: University of Chicago Press, 1942); Lowry Nelson, *The Mormon Village: A Pattern and Technique of Land Settlement* (Salt Lake City: University of Utah Press, 1952); Thomas F. O'Dea, "The Mormon Village," *American Journal of Sociology* 59 (July 1953); Joseph Earle Spencer, "The Development of Agricultural Villages in Southern Utah," *Journal of Agricultural History* 14, no. 4 (1940).

7. Stegner, *Mormon Country* (Lincoln: University of Nebraska Press, 1942), 21, 23–24.
8. Ibid., 112.
9. See, e.g., *Beyond the Hundredth Meridian: John Wesley Powell and the Second Opening of the West* (Boston: Houghton Mifflin Company, 1954) and (edited) *This Is Dinosaur: Echo Park Country and Its Magic Rivers* (New York: Knopf, 1955).
10. See selections from DeVoto, *The Western Paradox: A Conservation Reader*, edited by Douglas Brinkley and Patricia Nelson Limerick (New Haven: Yale University Press, 2001), esp. 449; Interview with Abbey in Doug Biggers, "From Abbey's Tower," *Tucson's Mountain Newsreal*, September 1979, 6–7.
11. William Cronon, "Landscapes of Abundance and Scarcity," in *The Oxford History of the American West*, edited by Clyde A. Milner II, Carol A. O'Connor, and Martha A. Sandweiss (New York: Oxford University Press, 1994), 604.
12. DeVoto, *Western Paradox*, 272.
13. Farmer, *On Zion's Mount: Mormons, Indians, and the American Landscape* (Cambridge, MA: Harvard University Press, 2008), 105–38.
14. *Utah: A Centennial History*, edited by Wain Sutton, vol. II (New York: Lewis Historical Publishing Co., Inc., 1949), 551–52; qt. from Bolton, "The Mormons in the Opening of the Great West," *Utah Genealogical and Historical Magazine* 44 (1926), 69.
15. Ricks, *Forms and Methods of Early Mormon Settlement in Utah and the Surrounding Region, 1847 to 1877* (Logan: Utah State University Press, 1964), 13.
16. Ibid., 32.
17. Ibid., 75–76, 125, 127.
18. Arrington, *Great Basin Kingdom: An Economic History of the Latter-day Saints* (Cambridge, MA: Harvard University Press, 1958), ix.
19. Arrington, May, and Fox, *Building the City of God: Community and Cooperation among the Mormons* (Salt Lake City: Deseret Book, 1976).
20. For an insightful critique of Arrington's writings, see Gary Topping, *Leonard J. Arrington: A Historian's Life* (Norman, OK: Arthur H. Clark Company, 2008).
21. Meinig, "The Mormon Culture Region: Strategies and Patterns in the Geography of the American West, 1847–1964," *Annals of the Association*

of American Geographers 55 (June 1965): 217; also published as "The Mormon Culture Region," in *Cultural Geography: Selected Reading*, (New York: Thomas Crowell, 1967). Another early essay from a geographer, though less known, is William W. Speth, "Environment, Culture and the Mormon in Early Utah: A Study in Cultural Adaptation," *Yearbook of the Association of Pacific Coast Geographers* 29 (1967): 53–67.

22. Francaviglia, *The Mormon Landscape: Existence, Creation and Perception of a Unique Image in the American West* (New York: AMS Press, 1978).
23. Jackson, "Great Salt Lake and Great Salt Lake City: American Curiosities," *Utah Historical Quarterly* 56 (Spring 1988): 130.
24. Jackson, "Righteousness and Environmental Change: The Mormons and the Environment," in *Essays on the American West, 1973–74*, Charles Redd Monographs in Western History, no. 5, edited by Thomas G. Alexander (Provo, UT: Brigham Young University Press, 1975), 21–43. For other works, see Jackson, "Mormon Perception and Settlement," *Annals of the Association of American Geographers* 68, no. 3 (1978): 317–34; Jackson, "Utah's Harsh Lands, Hearth of Greatness," *Utah Historical Quarterly* 49 (Winter 1981): 4–25; Jackson and R. Henrie, "Perceptions of Sacred Space," *Journal of Cultural Geography* 3 (Spring/Summer 1983): 94–107; Jackson, "The Mormon Experience: The Plains as Sinai, the Great Salt Lake as the Dead Sea, and the Great Basin as Desert-cum-Promised Land," *Journal of Historical Geography* 18 (January 1992): 41–58; Jackson, "Geography and Settlement in the Intermountain West: Creating an American Mecca," *Journal of the West* 33 (July 1994): 22–34.
25. Francaviglia, *Mapping and Imagination in the Great Basin: A Cartographic History* (Reno: University of Nevada Press, 2005), 5–6. See also Francaviglia, *Believing in Place: A Spiritual Geography of the Great Basin* (Reno: University of Nevada Press, 2003).
26. Francaviglia, *The Mapmakers of New Zion: A Cartographic History of Mormonism* (Salt Lake City: University of Utah Press, 2015), 55.
27. Carter, *Building Zion: The Material World of Mormon Settlement* (Minneapolis: University of Minnesota Press, 2015), xxiii–xxiv.
28. Mark Fiege, *Irrigated Eden: The Making of an Agricultural Landscape in the American West* (Seattle: University of Washington Press, 1999), 207.
29. Lynn White, Jr., "The Historical Roots of Our Ecological Crisis," *Science* 155 (March 1967): 1203–7; Roderick Nash, *Wilderness and the American Mind*, 3rd ed. (New Haven: Yale University Press, 1982), 13–20. An analysis of White's argument is found in Roderick Nash, *The Rights of Nature: A History of Environmental Ethics* (Madison: University of Wisconsin Press, 1989), 88–97.
30. Flores, *The Natural West: Environmental History in the Great Plains and Rocky Mountains* (Norman: University of Oklahoma Press, 2001); Oelschlaeger, *Caring for Creation: An Ecumenical Approach to the*

Environmental Crisis (New Haven: Yale University Press, 1994), 204; Wright, *Rocky Mountain Divide: Selling and Saving the West* (Austin: University of Texas Press, 1993).

31. Flores, *The Natural West*, 135, 136–37, 141.
32. Kristen Rogers, "Steward of the Earth," *This People* (Spring 1990): 12–16. See also Alexander, "Mormon Prophets and the Environment: Creation, Sin, the Fall, Redemption, and the Millennium," in *Dreams, Myths & Reality: Utah and the American West*, edited by William Thomas Allison and Susan J. Matt (Salt Lake City: Signature Books, 2008), 85–103; Alexander, "Stewardship and Enterprises: The LDS Church and the Wasatch Oasis Environment, 1847–1930," *Western Historical Quarterly* 25 (Autumn 1994): 341–66; Hugh Nibley, "Brigham Young on the Environment," in *To the Glory of God: Mormon Essays on Great Issues—Environment, Commitment, Love, Peace, Youth, Man*, edited by Truman Madsen and Charles D. Tate (Salt Lake City: Deseret Book, 1972), 3–29.
33. Williams, Smart, and Smith, *New Genesis: A Mormon Reader on Land and Community* (Layton, UT: Gibbs Smith, 1999), ix.
34. Handley, Ball, and Peck, *Stewardship and the Creation: LDS Perspectives on the Environment* (Provo, UT: BYU Religious Studies Center, 2006).
35. For more, see Nash, *The Rights of Nature*, 110–11.
36. See Thomas G. Alexander's essay herein for his belief that the de-emphasis of an environmental ethic in the modern church is due to historical amnesia.
37. This idea comes from Cronon, "Ecological Prophecies."
38. Dunlap, *Faith in Nature: Environmentalism as Religious Quest* (Seattle: University of Washington Press, 2001); Ellingson, *To Care for Creation: The Emergence of the Religious Environmental Movement* (Chicago: The University of Chicago Press, 2016); Gottlieb, ed., *This Sacred Earth: Religion, Nature, Environment*, 2nd ed. (New York City: Routledge, 2004); Stoll, *Inherit the Holy Mountain: Religion and the Rise of American Environmentalism* (New York: Oxford University Press, 2015).
39. Abbey, *Desert Solitaire: A Season in the Wilderness* (New York: McGraw-Hill, 1968); Geary, *The Proper Edge of the Sky: High Plateau Country of Utah* (Salt Lake City: University of Utah Press, 2002); Trimble, *Bargaining for Eden: The Fight for the Last Open Spaces in America* (Berkeley: University of California Press, 2008); Wilkinson, *Fire on the Plateau: Conflict and Endurance in the American Southwest* (Washington, D.C.: Island Press, 1999); Williams, *Red: Passion and Patience in the Desert* (New York: Pantheon Books, 2001).
40. Trimble, *Bargaining for Eden*, 6.
41. Rogers, *Roads in the Wilderness: Conflict in Canyon Country* (Salt Lake City: University of Utah Press, 2013).
42. Brehm and Eisenhauer, "Environmental Concern in the Mormon Culture Region," *Society and Natural Resources* 19 (May–June 2006): 393–410.

43. Cronon, *Nature's Metropolis: Chicago and the Great West* (New York: Norton, 1992).
44. See "Sylvester Q. Cannon and the Revival of Environmental Consciousness in the Mormon Community," *Environmental History* 3 (October 1998): 488–507, and "Cooperation, Conflict, and Compromise: Women, Men, and the Environment in Salt Lake City, 1890–1930," *BYU Studies* 35, no. 1 (1995): 6–39.
45. Worster, "Transformations of the Earth: Toward an Agroecological Perspective in History," *Journal of American History* 76 (March 1990): 1111–12.
46. Hall, *Earth Repair: A Transatlantic History of Environmental Restoration* (Charlottesville: University of Virginia Press, 2005).
47. Coleman, *Vicious: Wolves and Men in America* (New Haven: Yale University Press, 2004).
48. Cheney, *Plain but Wholesome: Foodways of the Mormon Pioneers* (Salt Lake City: University of Utah Press, 2012), 10, 91.
49. Farmer, *On Zion's Mount.*
50. Handley, *Home Waters: A Year of Recompense on the Provo River* (Salt Lake City: University of Utah Press, 2011), 19.
51. See especially Limerick, *The Legacy of Conquest: The Unbroken Past of the American West* (New York: Norton, 1987); Worster, *Rivers of Empire: Water, Aridity, and the Growth of the American West* (New York: Pantheon Press, 1985).
52. See, for example, the periodical *Ecotheology* and its successor, the *Journal for the Study of Religion, Nature and Culture,* which have published interdisciplinary articles examining "the role of theistic religion in nature-human relationships." Bron Taylor, "Exploring Religion, Nature and Culture—Introducing the *Journal for the Study of Religion, Nature and Culture,*" *Journal for the Study of Religion, Nature and Culture* 1, no. 1 (2007): 5–9.
53. As an example, a Keres Puebloan stated that "the land is not really a place separate from ourselves, where we act out the drama of our isolate destinies... [but] *is* being, as all creatures are also being: aware, palpable, intelligent, alive." Paula Gunn Allen, *The Sacred Hoop: Recovering the Feminine in American Indian Traditions* (Beacon Press, 1992).

The "Lion of the Lord" and the Land

Portions of this essay first appeared in "Field Notes: Brigham Young's 'All the People' Quote Quandary," *Western Historical Quarterly* 46 (Summer 2015): 219–23; also Sara Dant, *Losing Eden: An Environmental History of the American West* (Hoboken, NJ: John Wiley & Sons, 2017).

1. John C. Frémont, *Report of the Exploring Expedition to the Rocky Mountains in the Year 1842 and to Oregon and North California in the Years 1843–44*

(Washington, DC: United States Senate, 1845), 271. See also Alexander L. Baugh, "John C. Frémont's 1843–44 Western Expedition and Its Influence on Mormon Settlement in Utah," *Utah Historical Quarterly* 83 (Fall 2015): 254–69.

2. The Church of Jesus Christ of Latter Day Saints, *The Doctrine and Covenants of the Church of Jesus Christ of Latter Day Saints* (D&C hereafter), 104:13.
3. Jules Rémy, *The Westminster Review*, vol. 20, July and October 1861 (London: George Manwaring, 1861), 397.
4. *Journal of Discourses* (hereafter *JD*), 9:243. Note: Sermons in the *Journal of Discourses* may or may not accurately reflect the contents of a discourse and should be used with scholarly discretion.
5. *JD* 2:308. For a thorough discussion of Young's thoughts on the relationship between humans and nature, see Hugh Nibley, "Brigham Young on the Environment," in *Brother Brigham Challenges the Saints* (Salt Lake City, UT: Deseret Book, 1994).
6. Leonard J. Arrington, *Great Basin Kingdom: An Economic History of the Latter-Day Saints 1830–1900* (Salt Lake City: University of Utah Press, 1993), 52.
7. Ibid., 437 (note 54).
8. B. H. Roberts, *A Comprehensive History of The Church of Jesus Christ of Latter-day Saints Century I* (Salt Lake City, 1930), 3:269.
9. *Journal History of the Church of Jesus Christ of Latter-day Saints* (*JH* hereafter), Accession No. 1146, Special Collections and Archives, University of Utah, J. Willard Marriott Library, Salt Lake City, Utah. See also *Journal History of the Church*, LDS Church History Library, Salt Lake City, Utah (hereafter CHL), accessed December 2–4, 2012, and January 1–3, 2013, eadview.lds.org/findingaid/CR%20100%20137; Joseph Fielding Smith, *Essentials in Church History* (Salt Lake City: Deseret News Press, 1922); *JD*, journalofdiscourses.com/; Wilford Woodruff, *Wilford Woodruff's Journal, 1833–1898* (Midvale, UT: Signature Books, 1983), Special Collections, Weber State University, Stewart Library, Ogden, Utah.
10. Sources checked at the CHL include: Brigham Young Collection (CR 1234 1); *Wilford Woodruff's Journal* (M270.1 W893wiL v. 1–9 1833–1898); *Comprehensive History of the Church* (M272 R643c 1912 no. 2); *JD* (M230 J86 1966); Salt Lake Stake Minutes (LR 604 109, LR 604 107); Robert G. Dunbar, *Forging New Rights in Western Waters* (346.7304 D899f 1983); John B. Wright, *Rocky Mountain Divide: Selling and Saving the West* (333.73 W951r 1993); *Levi Jackman's Journal* (M270.1 J123j 18–?); *Thomas Bullock's Journal* (MS 9469); Thomas Bullock's Papers (MS 12475); *Pioneering the West: 1846–1878: Major Howard Egan's Diary* (MS 12475); *The Mormon Vanguard Brigade of 1847: Norton Jacob's Record* (M270.1 J15j 2005); *William Clayton's Journal* (M270.1 C622c 1921); *State of Deseret Papers* (MS

2918); *Laws and Constitution of the State of Deseret, 1849–1851* (MS 8375); Richard S. Van Wagoner, ed., *The Complete Discourses of Brigham Young: Volume 1, 1832–1852* (Salt Lake City: The Smith-Pettit Foundation, 2009), April 13, 1851.

11. Leonard J. Arrington, *Brigham Young: American Moses* (New York: Knopf, 1985), 431. "The quote" is also notably missing from his coauthored article with Dean May, "'A Different Mode of Life': Irrigation and Society in Nineteenth-Century Utah," *Agricultural History* 49 (January 1975): 3–20.
12. Gordon B. Hinckley, *What of the Mormons?: A Brief Study of the Church of Jesus Christ of Latter Day Saints* (Salt Lake City, 1947), 175. The *Journal History* for this date contains no quotes from Young regarding ownership of the streams nor do the next four weeks' worth of entries leading up to Young's departure for Winter Quarters.
13. Richard T. Ely, "Economic Aspects of Mormonism," *Harper's Monthly Magazine* 106 (April 1903): 669.
14. John Bennion, "Water Law on the Eve of Statehood: Israel Bennion and a Conflict in Vernon, 1893–1986," *Utah Historical Quarterly* 82 (Fall 2014): 289–305.
15. D&C 59:18; *The Pearl of Great Price* (Salt Lake City: LDS Printing and Publishing, 1878), 6; *JD* 11:136; Leonard J. Arrington and Ronald K. Esplin, "Building a Commonwealth: The Secular Leadership of Brigham Young," *Utah Historical Quarterly* 45 (Summer 1977): 219.
16. *JD* 1:272–73.
17. D&C 104:13–14.
18. *JD* 19:46.
19. Brigham Young, "Management of the Canyons—Paying Debts—Keeping Stores—Material for the Temple," *JD* 1:209.
20. Dale L. Morgan, "The State of Deseret," *Utah Historical Quarterly* 8 (April, July, October, 1940): 73.
21. Section 39, Territory of Utah Legislative Assembly, *Resolutions and Memorials Passed by the First Annual, and Special Sessions, of the Legislative Assembly of the Territory of Utah, Begun and Held at Great Salt Lake City, on the 22nd Day of September, A.D. 1851 also the Constitution of the United States, and the Act Organizing the Territory of Utah* (Salt Lake, 1852), 46.
22. See also Jeanne Kay and Craig J. Brown, "Mormon Beliefs about Land and Natural Resources, 1847–1877," *Journal of Historical Geography* 11 (July 1985): 253–67; Morgan, "The State of Deseret," 195, 200, 207, 235.
23. Thomas G. Alexander, "Brigham Young and the Transformation of Utah Wilderness, 1847–58," *Journal of Mormon History* 41 (January 2015); Van Wagoner, *Complete Discourses of Brigham Young*, 774.
24. *Journal History*, January 1, 1849, 2; *JH*, March 5, 1849; see also Victor Sorensen, "The Wasters and Destroyers: Community-Sponsored Predator Control in Early Utah Territory," *Utah Historical Quarterly* 62 (Winter

1994): 26–41. This predator toll is somewhat misleading as the high numbers for wolves and foxes probably included numerous coyotes, and "mink" likely included ermine and weasels.

25. Sorensen, "The Wasters and Destroyers," 33–36.
26. Morgan, "The State of Deseret," 175–76; Sorensen, "The Wasters and Destroyers," 37; *Kane County Standard*, August 22, 1912; *Kane County Standard*, September 12, 1912; Michael J. Robinson, *Predatory Bureaucracy: The Extermination of Wolves and the Transformation of the West* (Boulder: University Press of Colorado, 2005), 40.
27. Garrett Hardin, "The Tragedy of the Commons," *Science* 162 (December 13, 1968): 1243–48.
28. Milton R. Hunter, *Brigham Young: The Colonizer* (Salt Lake City, UT: Deseret News Press, 1940), 155.
29. Charles S. Peterson, "Imprint of Agricultural Systems on the Utah Landscape," in *The Mormon Role in the Settlement of the West*, edited by Richard H. Jackson (Provo, UT: Brigham Young University Press, 1978), 92, 102; Morgan, "The State of Deseret," 73.
30. Hunter, *Brigham Young*, 15.
31. Howard A. Christy, "Open Hand and Mailed Fist: Mormon-Indian Relations in Utah, 1847–52," *Utah Historical Quarterly* 46 (Summer 1978): 216–35.
32. *JD* 9:71, "Prosperity of Zion, Etc." March 10, 1861.
33. Frederick Jackson Turner, "The Significance of the Frontier in American History," in *Does the Frontier Experience Make America Exceptional?*, edited by Richard W. Etulain (Boston: Bedford/St. Martin's, 1999), 19.
34. Marshall Sahlins, *Stone Age Economics* (New York: Routledge, 1972), 1–2.
35. Ibid., 39.
36. See, for example, Mark Stoll, *Inherit the Holy Mountain: Religion and the Rise of American Environmentalism* (New York: Oxford University Press, 2015), 230–32.
37. John Winthrop, *The Journal of John Winthrop: 1630–1649*, Abridged, edited by Richard S. Dunn and Laetitia Yeandle (Cambridge, MA: Harvard University Press, 1996), 10.
38. Arrington and Esplin, "Building a Commonwealth," 224.
39. *JH*, March 28, 1858; see also Brigham Young, "Governor's Message to the General Assembly of the State of Deseret," January 22, 1866, in *The Latter-Day Saints' Millennial Star* 28 (March 17, 1866).
40. Gregory Umbach, "Learning to Shop in Zion: The Consumer Revolution in Great Basin Mormon Culture, 1847–1910," *Journal of Social History* 38 (Fall 2004): 31.
41. 43 US Code 321, accessed November 4, 2016, law.cornell.edu/uscode/text/43/32; Bennion, "Water Law on the Eve of Statehood," 293.
42. See, for example, Utah Department of Administrative Services, Division of

Archives and Records, "Original Land Titles in Utah Territory," accessed March 9, 2016, archives.utah.gov/research/guides/land-original-title.htm.

43. Thomas Alexander, "An Ambiguous Heritage," in *Dialogue: A Journal of Mormon Thought* 2 (Autumn 1967): 129; Dan Flores, "Zion in Eden: Phases of the Environmental History of Utah," in *The Natural West: Environmental History in the Great Plains and Rocky Mountains* (Norman: University of Oklahoma Press, 2001), 133; Donald Worster, *The Wealth of Nature: Environmental History and the Ecological Imagination* (New York: Oxford University Press, 1993), 115.

44. Dan Flores, "Zion in Eden: Phases of the Environmental History of Utah," *Environmental Review* 4 (Winter 1983): 331. See also Flores's expanded version of this article in *The Natural West* and Thomas Alexander, "Stewardship and Enterprise," *Western Historical Quarterly* 25 (Autumn 1994): 340–64.

45. Albert F. Potter, "Proposed Reserves in the State of Utah, 1903," Accession No. R4-1680-2006-0204, U.S. Forest Service Region 4 History Collection, Ogden, Utah; Albert F. Potter, "Diary of Albert F. Potter," 10, accessed May 11, 2013, forestry.usu.edu/htm/rural-forests/forest-history/the-potter-diaries.

46. Albert F. Potter, *Diary of Albert Potter*, July–November 22, 1902, U.S. Forest Service, Department of Agriculture, Ogden, UT; Transcription of R4 History Collection, R4-1680-2006-0204, 1903, USU Special Collections and Archives: forestry.usu.edu/files/uploads/PotterDiaries.pdf#page=1.

47. For a thorough discussion of Potter's survey, see Charles S. Peterson, "Albert F. Potter's Wasatch Survey, 1902: A Beginning for Public Management of Natural Resources in Utah," *Utah Historical Quarterly* 39 (Summer 1971): 243–55.

48. Flores, "Zion in Eden," 333–34; see also Andrew Honker, "'Been Grazed Almost to Extinction': The Environment, Human Action, and Utah Flooding, 1900–1940," *Utah Historical Quarterly* 67 (Winter 1999): 23–47.

49. For a thoughtful discussion of current LDS thought on environmental stewardship, see Marcus B. Nash, "Righteous Dominion and Compassion for the Earth," in this volume's appendix. See also the Church of Jesus Christ of Latter Day Saints, "Environmental Stewardship and Conservation" website, accessed March 9, 2016, lds.org/topics/environmental-stewardship-and-conservation?lang=eng.

50. For another examination of problematic historical quotes, see Richard F. Hamilton, "McKinley's Backbone," *Presidential Studies Quarterly* 36 (September 2006): 482–92, which argues that Theodore Roosevelt likely never said President William McKinley had "no more backbone than a chocolate éclair."

51. James Wilson to The Forester [historians believe Gifford Pinchot authored the letter for Wilson's signature], February 1, 1905, Forest History Society,

foresthistory.org/ASPNET/Policy/Agency_Organization/Wilson_Letter.aspx. See also, for example, Richard H. Jackson, "Righteousness and Environmental Change: The Mormons and the Environment," in *Charles Redd Monographs in Western History: Essays on the American West 1973–1974*, no. 5, edited by Thomas G. Alexander (Provo: Brigham Young University Press, 1975), 31–34.

52. James Wilson to The Forester.
53. Mitt Romney, "Mitt Romney Speaks about Public Lands to the *Reno Gazette-Journal* Editorial Board," February 2, 2012, accessed May 11, 2013, rgj.com/VideoNetwork/1450884666001/Mitt-Romney-speaks-about-public-lands-to-the-Reno-Gazette-Journal-editorial-board.
54. Van Wagoner, *Complete Discourses*, 425.
55. Wells A. Hutchins, *Mutual Irrigation Companies in Utah*, Bulletin 199, May 1927, accessed May 11, 2013, digitalcommons.usu.edu/uaes_bulletins/164/.
56. *Adams v. Portage Irrigation Co.*, 72 P.2d 648, 652–53 (Utah 1937). Several other Utah cases also confirm these principles.
57. Donald Worster, "John Muir and the Roots of American Environmentalism," in *The Wealth of Nature: Environmental History and the Ecological Imagination* (New York: Oxford University Press, 1993), 185. See also Roderick Nash, "The Greening of Religion," in *This Sacred Earth: Religion, Nature, Environment*, edited by Roger S. Gottlieb (New York: Routledge, 1996), 194–229.
58. John Muir, *My First Summer in the Sierra* (Mineola, NY: Dover Publications, Inc., 2004), 139.
59. Rachel Carson, *Silent Spring* (Boston: Houghton Mifflin Company, 2002), 75.
60. Richard A. Baer, Jr., "Higher Education, the Church, and Environmental Values," *Natural Resources Journal* 17 (July 1977): 48.
61. Richard A. Baer, Jr., "The Church and Man's Relationship to His Natural Environment," *Quaker Life* 12 (January 1970): 421.
62. Worster, "John Muir," 195; Van Wagoner, *Complete Discourses*.
63. Greenpeace, "About," accessed May 15, 2015, greenpeace.org/usa/about/.
64. Vincent Rossi, "The Eleventh Commandment: A Christian Deep Ecology," *The Eleventh Commandment Newsletter* (1985), 3.

LOST MEMORY AND ENVIRONMENTALISM

This essay began as an article: "Stewardship and Enterprise: The LDS Church and the Wasatch Oasis Environment, 1847–1930" in *Western Historical Quarterly* 25 (Autumn 1994): 341–64. The portions from that original article included in this publication are reprinted with permission from *Western Historical Quarterly.* I thank David R. Hall and Sharon S. Carver for their assistance with the research on the original article and Olga De La Roca and Kris Nelson for their secretarial assistance. I appreciate also the comments and criticisms of Charles S. Peterson and Clyde A. Milner II, as well as the editorial assistance and critique of Howard A. Christy. And thanks go to Christina Diel for the map and Susan Whetstone for the photos. Thanks also to Jedediah Rogers for his comments and for the comments of participants in a symposium on environmentalism at Brigham Young University in November 2015. Since writing the article, others and I have done considerable research on this topic, and I have tried to incorporate much of that research in this article.

1. See Lynn White Jr., "The Historical Roots of Our Ecologic Crisis," *Science* (March 10, 1967): 1203–7.
2. Andrew Greeley, "Religion and Attitudes toward the Environment," *Journal for the Scientific Study of Religion* 32 (March 1993): 19–28; Roderick Nash, *Wilderness and the American Mind*, 3rd ed. (New Haven: Yale University Press, 1982), 13–22; Donald Worster, *Dust Bowl: The Southern Plains in the 1930s* (New York: Oxford University Press, 1979); Donald Worster, *Rivers of Empire: Water, Aridity, and the Growth of the American West* (New York: Pantheon Books, 1985), esp. 48–60, 216–17; Donald Worster, "Transformations of the Earth: Toward an Agroecological Perspective in History," *Journal of American History* 76 (March 1990): 1087–1106, esp. 1092–93 and the literature cited in note 11 on 1093. See also Max Oelschlaeger, *Caring for Creation: An Ecumenical Approach to the Environmental Crisis* (New Haven: Yale University Press, 1994) for a mixed picture.
3. Wendell Berry, "Religion and the Environment," in *American Environmentalism: Readings in Conservation History*, 3rd ed., edited by Roderick Frazier Nash (New York: McGraw Hill Humanities/Social Sciences, 1990), 275–79. Berry argued that the experience of Israel entering the Promised Land provides a better model than the book of Genesis. Both Callicott and Santmire argue that the question is really one of interpretation or hermeneutics, since the way people read the Bible or tradition affects the environmental message they see in it. J. Baird Callicott, "Genesis Revisited: Muirian Musings on the Lynn White, Jr. Debate," *Environmental History Review* 14 (Spring/Summer 1990): 65–90. Santmire wrote, "an ecological hermeneutic of history … allow[s] us—if [it does] not indubitably require us—to develop [a positive] ecological reading of biblical faith." H. Paul Santmire, *The Travail of Nature: The Ambiguous Ecological Promise of Christian Theology* (Philadelphia: Fortress Press, 1985), 189; see also 189–218. Susan Power Bratton, "Christian Ecotheology and the Old Testament," in *Religion and Environmental Crisis*, edited by Eugene C. Hargrove (Athens, GA, 1986), 53–75, draws on the Genesis accounts of the roles of God and Adam in the creation. Mark Stoll, in *Protestantism, Capitalism, and Nature in America* (Albuquerque: University of New Mexico Press, 1997), sees a mixed picture.
4. In this connection, the debate over climax theory is also relevant. For

an example, see Daniel B. Botkin, *Discordant Harmonies: A New Ecology for the Twenty-first Century* (New York: Oxford University Press, 1990), 60–62. For a discussion of chaos theory, see James Gleick, *Chaos: Making a New Science* (New York: Penguin Books, 1988). I realize that a number of scientists and historians still cling to Frederick E. Clements's climax vegetation theories, but frankly, these seem untenable to me in the face of the chaos theorists' arguments. For another discussion of Worster's views, see "The Ecology of Order and Chaos," *Environmental History Review* 14 (Spring/Summer 1990): 1–18.

5. See Richard L. Bushman, *From Puritan to Yankee: Character and the Social Order in Connecticut*, 1690–1765 (Cambridge, MA: Harvard University Press, 1967), esp. 267–88; John Opie, "Renaissance Origins of the Environmental Crisis," *Environmental Review* 11 (Spring 1987): 2–17; and Max Weber, *The Protestant Ethic and the Spirit of Capitalism*, translated by Talcott Parsons (New York, 1930).
6. Thomas G. Alexander, "Stewardship and Enterprise: The LDS Church and the Wasatch Oasis Environment, 1847–1930," *Western Historical Quarterly* 25 (Autumn 1994): 341–64.
7. On collective memory see: James Fentress and Chris Wickham, *Social Memory* (Cambridge, MA: Blackwell Publishers, 1992); Rauf Gargagozov, *Collective Memory: How Collective Representations About the Past Are Created, Preserved, and Reproduced* (New York: Nova Publications, 2015 online edition); Maurice Halbwachs, *On Collective Memory* (Chicago: University of Chicago Press, 1992); and James V. Wertsch, *Voices of Collective Remembering* (New York: Cambridge University Press, 2002). The quoted passage is from Fentress and Wickham, *Social Memory*, 122. Halbwachs argues that remembering is to a great extent determined by social groups. Gargagozov's book is about memory in breakaway nations of the former Soviet Union, but his insights are valuable for understanding the force of collective memory.
8. For examples of progress replacing worker consciousness, see Fentress and Wickham, *Social Memory*, 122–25.
9. Dan L. Flores, "Agriculture, Mountain Ecology, and the Land Ethic: Phases of the Environmental History of Utah," in *Working on the Range: Essays on the History of Western Land Management and the Environment*, edited by John R. Wunder (Westport, CT: Greenwood Press, 1985), 164, 165, 168, 174; Max Oelschlaeger, *Caring for Creation: An Ecumenical Approach to the Environmental Crisis* (New Haven: Yale University Press, 1994), 204; John B. Wright, *Rocky Mountain Divide: Selling and Saving the West* (Austin: University of Texas Press, 1993), 243–45.
10. Charles S. Peterson, "Small Holding Land Patterns in Utah and the Problem of Forest Watershed Management," *Forest History* 17 (July 1973): 5–13.
11. "Environmental Stewardship and Conservation," accessed May 17, 2016,

https://www.lds.org/topics/environmental-stewardship-and-conservation?lang=eng.

12. "Millennium," accessed May 17, 2016, lds.org/topics/millennium?lang=eng.

13. Robert F. Berkhofer Jr., *A Behavioral Approach to Historical Analysis* (New York: Free Press, 1969), 99–100, 106–15. For examples, see George B. Handley, "LDS Belief and the Environment," accessed May 18, 2015, saveourcanyons.org/campaigns/resources/latter-day-saints-and-the-environment.

14. Hugh W. Nibley, "Subduing the Earth," in *Nibley on the Timely and Timeless: Classic Essays of Hugh W. Nibley*, Religious Studies Monograph Series (Provo, UT: Religious Studies Center, Brigham Young University, 1978), 85–99; and Hugh W. Nibley, "Brigham Young on the Environment," in *To the Glory of God: Mormon Essays on Great Issues—Environment, Commitment, Love, Peace, Youth, Man*, edited by Truman G. Madsen and Charles D. Tate Jr. (Salt Lake City: Deseret Book Company, 1972), 3–29. See also George B. Handley, Terry B. Ball, and Steven L. Peck, *Stewardship and the Creation: LDS Perspectives on the Environment* (Provo, UT: BYU Religious Studies Center, 2006); Richard H. Jackson, "Righteousness and Environmental Change: The Mormons and the Environment," in *Essays on the American West, 1973–1974*, edited by Thomas G. Alexander, Charles Redd Monographs in Western History, no. 5 (Provo, UT: Brigham Young University Press, 1975); Jeanne Kay and Craig J. Brown, "Mormon Beliefs about Land and Natural Resources, 1847–1877," *Journal of Historical Geography 11* (July 1985): 253–67; Peterson, "Small Holding Land Patterns in Utah," 5–13; Terry Tempest Williams, William B. Smart, and Gibbs M. Smith, eds., *New Genesis: A Mormon Reader on Land and Community* (Salt Lake City: Gibbs Smith, 1998). Such individuals include *Deseret News* editor William Smart; musician and poet Emma Lou Thayne; LDS general authorities Vaughn J. Featherstone, Hugh W. Pinnock, and Steven E. Snow; publisher Gibbs Smith; and scholars Terry Tempest Williams, Ted Wilson, Marilyn Arnold, P. Jane Hafen, Ardean Watts, Eugene England, Clayton White, Sam Rushforth, George Handley, Terry Ball, Steven Peck, Paul Cox, Andrew Hedges, Donald Adolphson, Aaron Kelson, Clark Monson, Larry Rupp, Roger Kjelgren, and Jack Brotherson.

15. Thomas G. Alexander, "Mormon Prophets and the Environment: Creation, Sin, the Fall, Redemption, and the Millennium," in *Dreams, Myths, and Reality: Utah and the American West*, the Critchlow Lectures at Weber State University, edited by William Allison and Susan J. Matt (Salt Lake City: Signature Books, 2008), 85–103.

16. In a revelation of March 1832, Smith wrote, "that which is spiritual being in the likeness of that which is temporal; and that which is temporal in the

likeness of that which is spiritual; the spirit of man in the likeness of his person as also the spirit of beast and every other creature which God has created." In a sermon on April 8, 1843, Smith taught that beasts have souls and that they would be saved in heaven. Joseph Smith, *History of the Church*, 2nd edition, edited by B. H. Roberts (Salt Lake City, 1967–69), 5:343–44. For the Native American point of view, see Catherine L. Albanese, *Nature Religion from the Algonkian Indians to the New Age* (Chicago: University of Chicago Press, 1990), esp. 19–33. See also William Cronon, *Changes in the Land: Indians, Colonists, and the Ecology of New England* (New York: Hill and Wang, 1983), for Native American attitudes toward resource use and their confrontation with Euro-Americans. For the pre-eighteenth-century attitudes and their breakdown, see Botkin, *Discordant Harmonies*, 91–99, 103–5. On the idea of Gaia, see J. E. Lovelock, *Gaia:* A *New Look at Life on Earth* (Oxford: Oxford University Press, 1979).

17. D&C 77:2–3. See also Section 88 for a more complete elaboration. Moses 7:48. For a more thorough discussion of these matters, see Nibley, "Subduing the Earth" and "Brigham Young on the Environment."
18. D&C 29:34.
19. Wilford Woodruff, *Wilford Woodruff's Journal, 1833–1898*, edited by Scott G. Kenney (Midvale, UT: Signature Books, 1983–85), 3:178.
20. Heber C. Kimball, "Discourse Delivered in Salt Lake City, August 2, 1857," in *JD* 5:137–38; Orson Pratt, "Sermon Delivered at Mount Pleasant, 12 November 1879," in *JD* 21:200–201; Woodruff, *Wilford Woodruff's Journal*, 3:203.
21. See Ball and Brotherson, "Environmental Lessons"; Nibley, "Subduing the Earth" and "Brigham Young on the Environment"; Jackson, "Righteousness and Environmental Change"; Kay and Brown, "Mormon Beliefs".
22. Richard S. Van Wagoner, *The Complete Discourses of Brigham Young*, 5 vols. (Salt Lake City: Smith-Pettit Foundation, 2009), 1:235, July 28, 1847.
23. Van Wagoner, *Complete Discourses*, 1:238, July 30, 1847.
24. John Locke, *Second Treatise on Civil Government*; The Project Gutenberg E-book of Second Treatise of Government: E-book 7370, 2010), Chapter 5, Sections 25, 26, 27. The quotation is from Section 27.
25. Van Wagoner, *Complete Discourses*, 1:403, September 22, 1850.
26. Thomas G. Alexander, "Conflict and Fraud: Utah Public Land Surveys in the 1850s, the Subsequent Investigation, and Problems with the Land Disposal System," *Utah Historical Quarterly* 80 (Spring 2012): 111.
27. Van Wagoner, *Complete Discourses*, 1:443, July 14, 1851; Hosea Stout, *On the Mormon Frontier: The Diary of Hosea Stout*, 2 vols, edited by Juanita Brooks (Salt Lake City: University of Utah Press, 1964), 2:434, April 6, 1852.
28. William Clayton, *An Intimate Chronicle: The Journals of William Clayton*, edited by George D. Smith (Salt Lake City: Signature Books, 1991),

375; and Orson Hyde, "Instructions Concerning Things Temporal and Spiritual...October 7, 1865," in *JD* 11:147–51; Woodruff, *Wilford Woodruff's Journal,* 3:236.

29. Thomas G. Alexander, *The Rise of Multiple-Use Management in the Mountain West: A History of Region 4 of the Forest Service* (Washington, DC, 1987), 1–32; Charles S. Peterson, "Small Holding Land Patterns in Utah and the Problem of Forest Watershed Management," *Forest History* 17 (July 1973): 5–13.
30. Thomas G. Alexander, *Things in Heaven and Earth: The Life and Times of Wilford Woodruff, A Mormon Prophet* (Salt Lake City: Signature Books, 1991), 140; Brigham Young, "Remarks Made by Brigham Young, March 4, 1860," in *JD* 8:8; Brigham Young, "Remarks Made by Brigham Young, May 17, 1868," in *JD* 12:218.
31. Van Wagoner, *Complete Discourses,* 1:403, September 22, 1850.
32. Ibid., 1:330, April 29, 1849.
33. Ibid., 2:759, February 12, 1854.
34. Ibid., 2:1015, 1016, October 8, 1855.
35. Ibid., 5:2836, May 14, 1871.
36. Leonard J. Arrington, *Great Basin Kingdom: An Economic History of the Latter-day* Saints, 1830–1900 (Cambridge, MA, 1958), 48 says that someone asked Charles Crismon to build the mill. Young did not agree. Van Wagoner, *Complete Discourses,* 1:294, September 24, 1848.
37. Van Wagoner, *Complete Discourses,* 1:425, April 13, 1851. See also Stout, Brooks, ed., 2:343, February 19, 1849, and note 15.
38. Van Wagoner, *Complete Discourses,* 2:1094, April 20, 1956
39. Ibid., 2:1143, July 23, 1856.
40. Ibid., 1:597–600, 602–3, October 9, 1852.
41. Ibid.
42. See Garrett Hardin, "The Tragedy of the Commons," *Science* 162 (1968): 1243–48, cited in Ball and Brotherson, "Environmental Lessons," 79, 82.
43. Stout, *On the Mormon Frontier,* 2:507–8, March 2, 1854.
44. See "Laws," January 14, 1857, Journal History of the Church of Jesus Christ of Latter-day Saints, Salt Lake City; see also Journal History, January 8, 18, 1858.
45. Arrington, *Great Basin Kingdom,* 150, esp. note 73; Stout, *On the Mormon Frontier,* 2:577, 79.
46. David H. Burr to Jeremiah Black, delivered March 20, 1857, by John F. Kinney, "Records Relating to the Appointment of Federal Judges, Attorneys and Marshals for the Territory and State of Utah, 1852–1901," Pierce's Administration, 1853–57, RG 60, National Archives, Microfilm 680, Reel 1; see also Alexander, "Conflict and Fraud."
47. For a discussion of Max Weber's ideas in connection with Salt Lake City, see Thomas G. Alexander and James B. Allen, *Mormons and Gentiles: A*

History of Salt Lake City (Boulder, CO, 1984), 3–5. For a general discussion of settlement patterns and development, see Arrington, *Great Basin Kingdom*, esp. 39–63.

48. For a discussion of irrigation development, see George Thomas, *The Development of Institutions under Irrigation with Special Reference to Early Utah Conditions*, Rural Science Series, edited by L. H. Bailey (New York: The Macmillan Company, 1920). For one of the early reservoir projects, see Leonard J. Arrington and Thomas C. Anderson, "The 'First' Irrigation Reservoir in the United States: The Newton, Utah, Project," *Utah Historical Quarterly* 39 (Summer 1971): 207–23; for a discussion of irrigation development in the Salt Lake Valley, see Alexander, "Brigham Young and the Transformation of Utah Wilderness, 1847–58," *Journal of Mormon History* 41 (January 2015): 114, 118–19; and Alexander, "Irrigating the Mormon Heartland: The Operation of the Irrigation Companies in Wasatch Oasis Communities, 1847–1880" *Agricultural History* 76 (Spring, 2002): 172–87.
49. G. K. Gilbert, "Salt Lake Drainage System," in *Report on the Lands of the Arid Region of the United States, with a More Detailed Account of the Lands of Utah*, edited by John Wesley Powell (1878; reprint, Washington, DC, 1879), 115–26; Elwood Mead, *Report of Irrigation Investigations in Utah* (Washington, DC, 1904).
50. Alexander, *Things in Heaven and Earth*, 162; Richard F. Burton, *The City of the Saints and Across the Rocky Mountains to California*, edited by Fawn M. Brodie (1861; reprint, New York: Knopf, 1963), 211, 315; Woodruff, *Wilford Woodruff's Journal*, 4:287, 5:51; "Diary of Albert F. Potter, 1 July 1902–22 November 1902," typed manuscript, historical files, supervisor's office, Fishlake National Forest, Richfield, Utah. For an assessment of Potter's work, see Charles S. Peterson, "Albert F. Potter's Wasatch Survey, 1902: A Beginning for Public Management of National Resources in Utah," *Utah Historical Quarterly* 39 (Summer 1971): 238–53, esp. 243–46.
51. On the Walker brothers see Jonathan Bliss, *Merchants and Miners in Utah: The Walker Brothers and Their Bank* (Salt Lake City, 1983). On the Godbeites see Ronald W. Walker, *Wayward Saints: The Godbeites and Brigham Young* (Urbana: University of Illinois Press, 1998).
52. Franklin B. Hough, *Report upon Forestry* (Washington, DC: 1878–1880, 1882), 1:594–95.
53. Ibid., 3:5.
54. In 1880, Utah produced 25.8 million board feet of lumber from 107 mills. In 1900, Utah had only 81 mills and produced only 12.1 million board feet. Bureau of the Census, *Compendium of the Tenth Census, 1880*, Part 2 (Washington, DC: 1883), 1162–63; Bureau of the Census, *Twelfth Census of the United States, 1900*, Vol. 9, Part 3 (Washington, DC: 1902), 808–10, 817.

55. Hough, *Report upon Forestry*, 1:389; L. F. Kneipp, "Utah's Forest Resources: Their Administration, Development and Use," in *Third Report of the State Bureau of Immigration, Labor, and Statistics,* 1915–1916 (Salt Lake City: 1917), 176–77.
56. Thomas G. Alexander, "Wilford Woodruff, Intellectual Progress, and the Growth of an Amateur Scientific and Technological Tradition in Early Territorial Utah," *Utah Historical* Quarterly 59 (Spring 1991): 164–88; Oliphant Autobiography, 10–11, 12–14; "Excerpts from the Diary of William F. Rigby," in *Our Pioneer Heritage*, edited by Kate B. Carter (Salt Lake City: Daughters of Utah Pioneers, 1961), 4:257. For one example, see Woodruff, *Wilford Woodruff's Journal*, 4:440, 5:41.
57. David B. Madsen and Brigham D. Madsen, "One Man's Meat is Another Man's Poison: A Revisionist View of the Seagull 'Miracle,'" *Nevada Historical Society Quarterly* 30 (Fall 1987): 165–81.
58. Arrington, *Great Basin Kingdom*, 48–49; "Autobiographies of Pioneers," entry for Barbara Gowans Bowen, in *Our Pioneer Heritage*, edited by Carter, 9:399–400; "Autobiographies," entry for Charles Sperry, in *Our Pioneer Heritage*, edited by Kate B. Carter, 9:445.
59. Davis Bitton and Linda P. Wilcox, "Pestiferous Ironclads: The Grasshopper Problem in Pioneer Utah," *Utah Historical Quarterly* 46 (Fall 1978): 351. Bitton and Wilcox believe that Young's comments about feeding the crickets were tongue in cheek, but given his attitude about the sanctity of life there may have been at least a little seriousness behind it. At the same time, most Mormon leaders recognized the devastation of the crops and tried to eradicate the crickets and grasshoppers.
60. Thomas G. Alexander, "Brigham Young and the Transformation of Utah Wilderness," *Journal of Mormon History* 41 (January 2015): *passim* and table 2, 153.
61. John D. Lee, *A Mormon Chronicle: The Diaries of John D. Lee, 1848–1876*, 2 vols., edited by Robert Glass Cleland and Juanita Brooks (Salt Lake City: University of Utah Press, 1983), 1:80, 82, 99–100; Stout, *On the Mormon Frontier*, 2:337, 338.
62. For the statistical information, see "Measures of Economic Changes in Utah, 1847–1947," *Utah Economic and Business Review* 7 (December 1947): 49.
63. For a study of these companies, see Leonard J. Arrington, *Beet Sugar in the West: A History of the Utah-Idaho Sugar Company, 1891–1966* (Seattle: University of Washington Press, 1966); J. R. Bachman, *Story of the Amalgamated Sugar Company, 1897–1961* (Caldwell, ID: Caxton Printers, 1962); and Matthew C. Godfrey, *Religion, Politics and Sugar: The Mormon Church, the Federal Government, and the Utah-Idaho Sugar Company, 1907–1921* (Logan: Utah State University Press, 2007). For statistical information on the growth of beet sugar production, see "Measures of Economic Changes in Utah," 49.

64. See also Flores, "Agriculture, Mountain Ecology, and the Land Ethic," 172; and Flores, "Measures of Economic Changes in Utah," 47. The arithmetic mean increased to 212 acres, but this was skewed by some relatively large farms.
65. Letter to Editor, Outlook, 17 March 1915, 652–53. See also George W. Snow, "Smoke Elimination in Salt Lake City," *American City*, September 1915, 196–97.
66. "Measures of Economic Changes in Utah," 51.
67. Charles E. Hughes, "An Investigation of Smelting in the Salt Lake Valley Prior to 1900," a packet prepared for the law firm of Parsons, Behle, and Latimer by Timpanogos Research Associates, 18 January 1990, copy in author's possession.
68. Charles H. Fulton, *Metallurgical Smoke*, Department of the Interior, Bureau of Mines, Bulletin 84 (Washington, DC, 1915), 82–83, reproduced in Hughes, "Investigation of Smelting."
69. For a detailed discussion of this process, see Alexander, *Things in Heaven and Earth*, 235–87, 307–32.
70. For a discussion of the changes, see Thomas G. Alexander, *Mormonism in Transition: A History of the Latter-day Saints, 1890–1930*, 3rd edition (Salt Lake City: Greg Kofford Books, 2012).
71. See the discussion in George Thomas, *The Development of Institutions under Irrigation: With Special Reference to Early Utah Conditions* (New York: Macmillan, 1920), 55, 170.
72. For a discussion of this concept, see Sherry H. Olson, *The Depletion Myth: A History of Railroad Use of Timber* (Cambridge, MA: Harvard University Press, 1971); and Harold K. Steen, *The U.S. Forest Service: A History* (Seattle, 1976).
73. Thomas G. Alexander, "Senator Reed Smoot and Western Land Policy, 1905–1920," *Arizona and the West* 13 (Autumn 1971): 246.
74. Joseph F. Smith, *Gospel Doctrine: Sermons and Writings of Joseph F. Smith*, 13th edition, edited by John A. Widtsoe, et al. (Salt Lake City, 1963), 265–66.
75. Robert M. Crunden, *Ministers of Reform: The Progressive's Achievement in American Civilization, 1889–1920* (New York: Basic Books, 1982).
76. For a more thorough treatment of activities of these women and men in Salt Lake City, see Thomas G. Alexander, "Cooperation, Conflict, and Compromise: Women, Men, and the Environment in Salt Lake City, 1890–1930," *BYU Studies* 35 (1995): 6–46.
77. Alexander, "Senator Reed Smoot and Western Land Policy," 246. See also Alexander, "Red Rock and Grey Stone: Reed Smoot, the Establishment of Zion and Bryce National Parks and the Rebuilding of Downtown Washington, D.C." *Pacific Historical Review* 72 (Spring 2003): 1–38. After inspections such as those conducted by Albert Potter, the federal government set aside a number of small forest reserves, later consolidating them

in the Wasatch oasis into the Cache, Wasatch, and Uinta National Forests. Alexander, *Rise of Multiple-Use Management*, 18–20, 22–23.

78. Alexander, *Rise of Multiple-Use Management*, 24–25, 41–45, 79–99.
79. For a discussion of this matter, see Stanford J. Layton, *To No Privileged Class: The Rationalization of Homesteading and Rural Life in the Early Twentieth Century American West* (Provo, UT: Brigham Young University, Charles Redd Center for Western Studies, 1988), 84–87.
80. Thomas G. Alexander, "Latter-day Saints, Utahns, and the Environment: A Personal Perspective," edited by Williams, et al., *New Genesis*, 204–10.
81. For a general treatment of the air pollution problem, see Walter E. Pittman Jr., "The Smoke Abatement Campaign in Salt Lake City, 1890–1925," *Locus* 2 (Fall 1989): 69–78.
82. Alexander and Allen, *Mormons and Gentiles*, 180–82.
83. Ibid., 108–9.
84. "General Conference 28-Page Special Section," *Salt Lake Tribune*, September 29, 2015, 4.
85. By contrast, farmers in Montana were not so fortunate, since the courts there ruled in favor of polluting smelters instead of protecting the environment. See Gordon Morris Bakken, "Was There Arsenic in the Air?: Anaconda Versus the Farmers of Deer Lodge Valley," *Montana: The Magazine of Western History* 41 (Summer 1991): 30–41.
86. See, for instance, Émile Durkheim, *Durkheim on Religion: A Selection of Readings with Bibliographies*, edited by W. S. E Pickering, translated by Jacqueline Redding and W. S. E Pickering (London, 1975); Max Weber, *From Max Weber: Essays in Sociology*, translated and edited by H. H. Gerth and C. Wright Mills (New York, 1946), 303–13.
87. Stan L. Albrecht and Tim B. Heaton, "Secularization, Higher Education, and Religiosity," *Review of Religious Research* 26 (September 1984): 43–58.
88. Halbwachs, *On Collective Memory*, 24, 50, 51.

THE NATURAL WORLD AND THE ESTABLISHMENT OF ZION, 1831–1833

1. See, for example, Mark Ashurst-McGee, "Zion Rising: Joseph Smith's Early Social and Political Thought," Ph.D. diss. (Arizona State University, 2008); Mario S. DePillis, "Christ Comes to Jackson County: The Mormon City of Zion and Its Consequences," *John Whitmer Historical Association Journal* 23 (2003): 21–44.
2. See, for example, Martha Sonntag Bradley, "Creating the Sacred Space of Zion," *Journal of Mormon History* 31 (Spring 2005): 1–30; Steve L. Olsen, *The Mormon Ideology of Place: Cosmic Symbolism of the City of Zion, 1830–1846* (Provo, Utah: Joseph Fielding Smith Institute for Latter-day Saint History and BYU Studies, 2002).
3. Mark Fiege has recently argued that examining history through the lens of nature can bring great insights into events. "To recover the nature of

familiar historical subjects," he declared, "is to come to terms with nature in its fullest sense and with its centrality to the human experience." Fiege, *The Republic of Nature: An Environmental History of the United States* (Seattle: University of Washington Press, 2012), 9.

4. When Smith organized the church on April 6, 1830, in Fayette, New York, he dictated a revelation that designated him as "a seer & Translater & Prophet an Apostle of Jesus Christ an Elder of the Church" and instructed church members to "give heed unto all his words & commandments which he Shall give unto you." Revelation, April 6, 1830 [D&C 21:1, 4], in Michael Hubbard MacKay, Gerrit J. Dirkmaat, Grant Underwood, Robert J. Woodford, and William G. Hartley, eds., vol. 1 of the Documents series of *The Joseph Smith Papers*, general editors Dean C. Jessee, Ronald K. Esplin, Richard Lyman Bushman, and Matthew J. Grow (Salt Lake City: The Church Historian's Press, 2013), 129 (hereafter referred to as *JSP*, D1).
5. *The Book of Mormon: An Account Written by the Hand of Mormon, upon Plates Taken from the Plates of Nephi* (Palmyra, NY: E. B. Grandin, 1830), 1–4.
6. 3 Nephi 21:22–25.
7. Revelation, circa March 7, 1831 [D&C 45:66–67], in *JSP*, D1:280.
8. Revelation, September 1830 [D&C 28:9], in *JSP*, D1:185–86.
9. Ibid.
10. See Covenant of Oliver Cowdery and Others, October 17, 1830, in *JSP*, D1:204.
11. Richard Cummins, Delaware and Shawnee Agency, to William Clark, St. Louis, Missouri, February 15, 1831, Correspondence of the U.S. Office of Indian Affairs, vol. 6, 113–14, Kansas State Historical Society, Topeka, Kansas; Oliver Cowdery to Joseph Smith and others, January 29, 1831, in Letter to Hyrum Smith, March 3–4, 1831, in *JSP*, D1:272–73.
12. Letter from Oliver Cowdery, April 8, 1831, in *JSP*, D1:291–94.
13. Revelation, June 6, 1831 [D&C 52:3–4], in *JSP*, D1:328.
14. Revelation, July 20, 1831 [D&C 57:1–5], in Matthew C. Godfrey, Mark Ashurst-McGee, Grant Underwood, Robert J. Woodford, and William G. Hartley, eds., vol. 2 of the Documents series of *The Joseph Smith Papers*, general editors Dean C. Jessee, Ronald K. Esplin, Richard Lyman Bushman, and Matthew J. Grow (Salt Lake City: The Church Historian's Press, 2013), 7–11 (hereafter referred to as *JSP*, D2; spelling retained from original).
15. Hilary Davidson, "Restoring the Prairie, Missouri's Endangered Habitat," *St. Louis Beacon*, September 30, 2013; Richard Manning, *Grassland: The History, Biology, Politics, and Promise of the American Prairie* (New York: Viking, 1995), xi, 2–3. For an excellent history of the grasslands region in the United States, see James E. Sherow, *The Grasslands of the United States: An Environmental History* (Santa Barbara, CA: ABC–CLIO, 2007).

16. Walter A. Schroeder, "The Presettlement Prairie in the Kansas City Region (Jackson County, Missouri)," *Missouri Prairie Journal* 7, no. 2 (1985): 11.
17. Ibid., 9.
18. *The History of Jackson County, Missouri, Containing a History of the County, Its Cities, Towns, Etc.* (Kansas City, MO: Union Historical Company, 1881), 71–73.
19. William E. Foley, *The Genesis of Missouri: From Wilderness Outpost to Statehood* (Columbia: University of Missouri Press, 1989), 7–13.
20. *The History of Jackson County, Missouri*, 101–2.
21. "Mormonism—No. VI," *The Ohio Star*, November 17, 1831.
22. Washington Irving to Mrs. Paris, September 26, 1832, in *The Life and Letters of Washington Irving* (New York: G. P. Putnam, 1868), 3:38.
23. See Zachary McLeod Hutchins, *Inventing Eden: Primitivism, Millennialism, and the Making of New England* (Oxford: Oxford University Press, 2014), 53–64.
24. Roderick Frazier Nash, *Wilderness and the American Mind*, 4th edition (New Haven, CT: Yale University Press, 2001), 24.
25. Genesis 1:28; Nash, *Wilderness and the American Mind*, 30–32.
26. Daniel Walker Howe, *What Hath God Wrought: The Transformation of America, 1815–1848* (New York: Oxford University Press, 2007), 188.
27. Joseph Smith 1838 Manuscript History, vol. A-1, 127, LDS Church History Library, Salt Lake City [hereafter referred to as CHL].
28. Nash, *Wilderness and the American Mind*, 35–37. For more on John Winthrop and the Puritans' efforts to construct a city on a hill in the 1600s, see Edmund Morgan, *The Puritan Dilemma: The Story of John Winthrop* (Glenview, IL: Scott, Foresman, and Company, 1958).
29. *The Book of Mormon*, 412–13 [Helaman 3:3–9].
30. Hutchins, *Inventing Eden*, 35–67.
31. William Cronon, *Nature's Metropolis: Chicago and the Great West* (New York: W. W. Norton & Company, 1991), 35–37.
32. Revelation, August 1, 1831 [D&C 58:50], in *JSP*, D2:19.
33. Sidney Rigdon to the Churches, August 31, 1831, CHL. This was Rigdon's second draft; his first draft was deemed overly enthusiastic and embellished, leading to a revelation telling him that "his writing is not exceptable [*sic*] unto the Lord" and to compose a more restrained description. Revelation, August 30, 1831 [D&C 63:56] in *JSP*, D2:54; Jedediah M. Grant, *A Collection of Facts, Relative to the Course Taken by Elder Sidney Rigdon, in the States of Ohio, Missouri, Illinois and Pennsylvania* (Philadelphia, PA: Brown, Bicking, and Guilbert, 1844), 7.
34. Joseph Smith 1838 Manuscript History, vol. A-1, 137–38, CHL.
35. Eliza Marsh to Lewis and Ann Abbott, ca. September 1831, CHL (spelling and grammar retained from original).
36. W. W. Phelps to the *Ontario Phoenix*, July 23, 1831, in *Ontario Phoenix*, September 7, 1831.

37. "The Elders in the Land of Zion to the Church of Christ Scattered Abroad," *Evening and the Morning Star* 1 (July 1832): 5.
38. "The Far West," *Evening and the Morning Star* 1 (October 1832): 5; *The Book of Mormon*, 412–13 [Helaman 3:4–6].
39. Charles J. Latrobe, *The Alpenstock: Or, Sketches of Swiss Scenery and Manners* (London: Seeley and Burnside, 1829); Charles J. Latrobe, *The Pedestrian: A Summer's Ramble in the Tyrol* (London: Seeley and Burnside, 1832).
40. Charles Joseph Latrobe, *The Rambler in North America* (New York: Harper & Brothers, 1835), 102.
41. Washington Irving to Mrs. Paris, September 26, 1832, in *The Life and Letters of Washington Irving* (New York: G. P. Putnam, 1868), 3:38.
42. Rufus B. Sage to Mrs. J. Sage, May 30, 1841, in *Rufus B. Sage: His Letters and Papers, 1836–1847*, edited by LeRoy R. Hafen and Ann W. Hafen, (Glendale, CA: The Arthur H. Clark Company, 1956), 4:85.
43. "The Mormons," *Daily Missouri Republican*, June 2, 1834.
44. Revelation, July 20, 1831 [D&C 57:7], in *JSP*, D2:11.
45. Edward Partridge to Lydia Partridge, August 5–7, 1831, CHL.
46. "Mormonism—No. VI," *The Ohio Star*, November 17, 1831.
47. Edward Partridge to Lydia Partridge, August 5–7, 1831, CHL.
48. Letter from Oliver Cowdery, January 28, 1832, in *JSP*, D2:178.
49. "The Elders Stationed in Zion to the Churches Abroad, In Love, Greeting," *Evening and the Morning Star* 2 (July 1833): 110.
50. Letter from Oliver Cowdery, January 28, 1832, in *JSP*, D2:167 (grammar and punctuation retained from original).
51. Likewise, Thomas G. Alexander argued that when church members moved to the Great Basin in the 1840s, "they expected to use science and technology to refashion the arid west both as a fit place for Christ's Second Coming and as an earthly home, like the familiar humid region they had fled." Thomas G. Alexander, "Stewardship and Enterprise: The LDS Church and the Wasatch Oasis Environment, 1847–1930," *Western Historical Quarterly* 25 (Autumn 1994): 344.
52. Plat of the City of Zion, ca. early June–25 June 1833, in Gerrit J. Dirkmaat, Brent M. Rogers, Grant Underwood, Robert J. Woodford, and William G. Hartley, eds., vol. 3 of the Documents series of *The Joseph Smith Papers*, general editors Ronald K. Esplin and Matthew J. Grow (Salt Lake City: The Church Historian's Press, 2014), 121–30 (hereafter referred to as *JSP*, D3).
53. Richard L. Bushman, *The Refinement of America: Persons, Houses, Cities* (New York: Alfred A. Knopf, 1992), 354.
54. Revelation, August 12, 1831 [D&C 61:17], in *JSP*, D2:41.
55. Revelation, August 7, 1831 [D&C 59:16–17], in *JSP*, D2:33.
56. "The Elders Stationed in Zion," 6–7.
57. Note from Partridge in Letter from Oliver Cowdery, January 28, 1832, in *JSP*, D2:176.

58. "The Crops," *Missouri Intelligencer and Boon's Lick Advertiser*, July 30, 1831; "Wheat," *Missouri Intelligencer and Boon's Lick Advertiser*, August 13, 1831; News Item, *Missouri Intelligencer and Boon's Lick Advertiser*, October 13, 1832.
59. Note from Partridge in Letter from Oliver Cowdery, January 28, 1832, in *JSP*, D2:177.
60. Edward Partridge to Lydia Partridge, August 5–7, 1831, CHL.
61. James H. Rollins, "Reminiscences," 1896, draft typescript, 2, CHL; Emily Dow Partridge Young, "What I Remember," 5, 1884, typescript, CHL. Gilbert apparently obtained a merchant's license in the name of "Gilbert & Whitney" from Jackson County sometime prior to February 6, 1832. On February 20, 1832, "Gilbert & Whitney" paid $371 for the former log courthouse in Independence, which was used as the store. Jackson Co., Missouri, Deed Records, 1827–1909, vol. B, 32–33, February 20, 1832, microfilm 1,017,978, U.S. and Canada Record Collection, LDS Family History Library, Salt Lake City, Utah; Joanne C. and O. B. Eakin, *Jackson County Court Minutes, Book 1, 1827–1833* (Independence: J. C. Eakin, 1988), 127, 143–44; Max H Parkin, *Sacred Places: Missouri*, edited by LaMar C. Berrett (Salt Lake City: Deseret Book, 2004), 4:47–48, 58.
62. Note from Partridge in Letter from Oliver Cowdery, January 28, 1832, in *JSP*, D2:176.
63. "The Gathering," *Evening and the Morning Star* 1 (November 1832): 5–6.
64. Ibid.; "The Elders Stationed in Zion to the Churches Abroad, in Love, Greeting," *Evening and the Morning Star* 2 (July 1833): 6.
65. Letter to William W. Phelps, July 31, 1832, in *JSP*, D2:264. For more information on the conflicts between Kirtland and Independence leaders, see Matthew C. Godfrey, "'Seeking after Monarchal Power and Authority': Joseph Smith and Leadership in the Church of Christ, 1831–1832," *Mormon Historical Studies* 13 (Spring/Fall 2012): 15–37.
66. Revelation, September 22–23, 1832 [D&C 84:54–59], in *JSP*, D2:298.
67. Letter to Edward Partridge and Others, January 14, 1833, in *JSP*, D2:376–77; Letter to William W. Phelps, January 11, 1833, in *JSP*, D2:367.
68. For an explanation of the expulsion from Missouri, see Edward Partridge, "'A History of the Persecution, of the Church of Jesus Christ, of Latter Day Saints in Missouri,' December 1839–October 1840," in Karen Lynn Davidson, Richard L. Jensen, and David J. Whittaker, eds., vol. 2 of the Histories series of *The Joseph Smith Papers*, general editors Dean C. Jessee, Ronald K. Esplin, and Richard Lyman Bushman (Salt Lake City: The Church Historian's Press, 2012), 206–23 (hereafter referred to as *JSP*, H2).
69. Letter from John Whitmer, July 29, 1833, in *JSP*, D3:196. Church members believed that a lack of obedience to God's commandments had a deleterious effect on the earth itself. Not only would the earth not provide its

fullness to the Saints, it would also suffer pain because of the Saints' sins. This idea was present in Joseph Smith's new translation of the Bible, which he commenced in 1830, whereby he revised the text of the Bible according to what he believed was inspiration from God. As part of his revision of the book of Genesis, Smith produced a lengthy account of the biblical prophet Enoch and the Zion community he developed in Old Testament days. In this account, Enoch sees a vision of the history of "all the inhabitants of the earth," in which he perceives "the children of men" descending into wickedness and sin. With wickedness running rampant over the earth, Enoch hears "a voice from the bowels" of the earth, crying, "Wo, wo is me the mother of men? [*sic*] I am pained: I am weary because of the wickedness of my children? [*sic*] When shall I rest, and be cleansed from the filthiness which has gone forth out of me?" "Extract from the Prophecy of Enoch," *Evening and the Morning Star* 1 (August 1832): 2. See also Craig D. Galli, "Enoch's Vision and Gaia: An LDS Perspective on Environmental Stewardship," *Dialogue: A Journal of Mormon Thought* 44 (Summer 2011): 36–56.

70. See Numbers 33:51–54.
71. Revelation, January 2, 1831 [D&C 38:18–20], in *JSP*, D1:231–32.
72. Revelation, July 20, 1831 [D&C 57:7], in *JSP*, D2:11.
73. Letter to Edward Partridge, December 5, 1833, in *JSP*, D3:374–94.
74. Letter to William W. Phelps, November 27, 1832, in *JSP*, D2:316–21. This letter was later canonized in the church's Doctrine and Covenants.
75. Revelation, February 9, 1831 [D&C 42], in *JSP*, D1:251–52; Lyndon W. Cook, *Joseph Smith and the Law of Consecration* (Provo, UT: Grandin Book Company, 1985), 7–21.
76. Revelation, August 1, 1831 [D&C 58:52–53], in *JSP*, D2:20.
77. Revelation, February 9, 1831 [D&C 42:34–35], in *JSP*, D1:252.
78. "The Progress of the Church of Christ," *Evening and the Morning Star* 2 (June 1833): 100.
79. For more on American notions about land, see Donald Worster, *The Wealth of Nature: Environmental History and the Ecological Imagination* (New York: Oxford University Press, 1993), 95–111.
80. Partridge, "A History of the Persecution, of the Church of Jesus Christ, of Latter Day Saints in Missouri," in *JSP*, H2:209.
81. Letter to Church Leaders in Jackson County, Missouri, August 18, 1833, in *JSP*, D3:268.
82. Letter to Edward Partridge, December 5, 1833, in *JSP*, D3:374.
83. Letter to Edward Partridge and Others, December 10, 1833, in *JSP*, D3:378–79.
84. Revelation, December 16–17, 1833 [D&C 101:99], in *JSP*, D3:397.
85. This included petitioning Missouri governor Daniel Dunklin for a militia escort back to Jackson County, the marching of an expedition led by Joseph

Smith to Missouri to protect church members from future mob attacks, and negotiations with a committee of Jackson County citizens to purchase their holdings in the county. Although Smith did lead Zion's Camp, as it was later called, to Missouri, Dunklin did not call out the militia, negotiations with the Jackson County committee broke down, and the Saints remained barred from the county. John Whitmer, Edward Partridge, John Corrill, and A. S. Gilbert to Daniel Dunklin, December 6, 1833, CHL; Sidney Rigdon and Oliver Cowdery to Dear Brethren, May 10, 1834, CHL; Proposition of the Jackson Committee to the Mormons and their answer, June 16, 1834, CHL; W. W. Phelps, W. E. McLellin, A. S. Gilbert, John Corrill, and Isaac Morley to S. C. Owens, June 21, 1834, CHL; A. S. Gilbert, W. W. Phelps, and John Corrill to Daniel Dunklin, June 26, 1834, CHL; S. C. Owens to Mr. Amos Rees, June 26, 1834, CHL.

86. See, for example, Thomas G. Alexander, "Stewardship and Enterprise: The LDS Church and the Wasatch Oasis Environment, 1847–1930," *Western Historical Quarterly* 25 (Autumn 1994): 340–64; Dan L. Flores, "Agriculture, Mountain Ecology, and the Land Ethic: Phases of the Environmental History of Utah," in *Working on the Range: Essays on the History of Western Land Management and the Environment*, edited by John R. Wunder (Westport, CT: Greenwood Press, 1985), 157–86; Richard H. Jackson, "Righteousness and Environmental Change: The Mormons and the Environment," in *Essays on the American West, 1973–1974*, edited by Thomas G. Alexander, Charles Redd Monographs in Western History, no. 5 (Provo, UT: Brigham Young University Press, 1975), 21–42; Jeanne Kay and Craig J. Brown, "Mormon Beliefs about Land and Natural Resources, 1847–1877," *Journal of Historical Geography* 11 (July 1985): 253–67.

"We Seldom Find Either Garden, Cow, or Pig"

1. Wilford Woodruff, Journal, January 11, 1840, in Scott G. Kenney, ed., *Wilford Woodruff's Journal*, 9 vols. (Midvale, Utah: Signature Books, 1985), 1:403.
2. Woodruff, January 12, 1840, *Wilford Woodruff's Journal*, 1:404.
3. Woodruff, January 14, 1840, *Wilford Woodruff's Journal*, 1:405.
4. Alfred W. Crosby, *The Columbian Exchange: Biological and Cultural Consequences of 1492* (Westport, CT: Praeger, 2003).
5. Daniel Walker Howe, *What Hath God Wrought: The Transformation of America, 1815–1848* (New York: Oxford University Press, 2007), 672–73, 714–19; Anne Hyde, *Empires, Nations, & Families: A History of the North American West, 1800–1860* (Lincoln: University of Nebraska Press, 2011), 104–5, 375.
6. As evidence for her belief in an impending war, Richards cited not only the tense international relations, but a recently published revelation of Joseph Smith, which declared that peace would "soon be taken from the earth" and that a "very fierce and very terrible war is near at hand, even at

your doors." Hepzibah Richards to William Richards, January 22, 1838, Richards Family Papers, MS 1558, box 1, folder 1, page 222, LDS Church History Library, Salt Lake City, Utah (hereafter CHL); Hepzibah Richards to Rhoda Richards, January 28, 1838, Richards Family Papers, MS 1558, box 1, folder 1, page 226, CHL; Joseph Smith, Revelation circa November 1837, in Joseph Smith, Untitled Editorial, *Elders' Journal of the Church of Latter Day Saints* (Kirtland, Ohio), November 1837, 28.

7. John Smith to George A. Smith, June 17, 1840, George A. Smith Papers, MS 1322, box 9, folder 2, CHL.
8. Revelation, August 12, 1831, Matthew C. Godfrey, Mark Ashurst-McGee, Grant Underwood, Robert J. Woodford, and William G. Hartley, eds., vol. 2 of the Documents series of *The Joseph Smith Papers*, general editors Dean C. Jessee, Ronald K. Esplin, Richard Lyman Bushman, and Matthew J. Grow (Salt Lake City: The Church Historian's Press, 2013), 37–44 [D&C 61] (hereafter cited as *JSP* D2).
9. Brigham Young to Mary Ann Angell Young, April 6, 1840, George W. Thatcher Blair Collection, MS 15616, box 1, folder 6, CHL.
10. Franklin D. Richards, Diary, October 14, 1846, Richards Family Collection, MS 1215, reel 1, box 1, volume 6, CHL.
11. Woodruff, November 28, 1840, *Wilford Woodruff's Journal*, 1:553–54.
12. George A. Smith, Memoir, July 20, 1840, History of George A. Smith, MS 8839, typescript, 150–51, CHL.
13. Woodruff, January 20, 1840, *Wilford Woodruff's Journal*, 1:409.
14. John Dizikes, *Opera in America: A Cultural History* (New Haven: Yale University Press, 1993), 4. On visiting Philadelphia in 1842, Charles Dickens described the city as possessing a "cold, cheerless air" resulting from widespread financial chaos. Charles Dickens, quoted in Alasdair Roberts, *America's First Great Depression: Economic Crisis and Political Disorder after the Panic of 1837* (Ithaca, NY: Cornell University Press, 2012), 14.
15. Peter A. Coates, "The Strange Stillness of the Past: Toward an Environmental History of Sound and Noise," *Environmental History* 10 (October 2005): 639; Richard Cullen Rath, *How Early America Sounded* (Ithaca, NY: Cornell University Press, 2003), 11.
16. Woodruff, August 19, 1840, *Wilford Woodruff's Journal*, 1:495.
17. Brigham Young to Mary Ann Angell Young, March 13, 1840, Thatcher Blair Collection, MS 15616, box 1, folder 10, CHL.
18. Brigham Young to Mary Ann Angell Young, May 26, 1840, Thatcher Blair Collection, MS 15616, box 1, folder 7, CHL.
19. Young to Young, April 6, 1840, Thatcher Blair Collection, CHL; Brigham Young to Joseph Smith, May 7, 1840, Joseph Smith Letterbook, http://josephsmithpapers.org/paperSummary/letter-from-brigham-young-7-may-1840 (accessed September 7, 2014).
20. Wilford Woodruff, August 8, 1839, *Wilford Woodruff's Journal*, 1:350;

Wilford Woodruff to Phebe Carter Woodruff, circa April 6, 1840, postscript to Young to Young, April 6, 1840, CHL.

21. Brigham Young to Mary Ann Angell Young, February 11, 1841, George W. Thatcher Blair Collection, MS 15616, box 1, folder 8, CHL; Brigham Young to Mary Ann Angell Young, March 1, 1841, George W. Thatcher Blair Collection, MS 15616, box 1, folder 8, CHL.
22. John Smith to George A. Smith, July 18, 1840, George A. Smith Papers, MS 1322, box 9, folder 2, CHL; Wilford Woodruff to Phebe Carter Woodruff, January 29, 1840, Wilford Woodruff Collection, MS 19535, folder 14, CHL.
23. Reuben Hedlock to Brigham Young, January 4, 1841, George W. Thatcher Blair Collection, MS 15616, box 1, folder 11, CHL.
24. Woodruff to Woodruff, January 29, 1840, Woodruff Collection, CHL.
25. William Cronon, in *Nature's Metropolis: Chicago and the Great West* (New York: W. W. Norton, 1991), described the connections between the urban and natural environments in the nineteenth century.
26. Kenneth Pomeranz has argued that coal not only fueled British industrialization, but gave it a vital spot in the international marketplace during the eighteenth and nineteenth centuries. Other historians take issue with Pomeranz's postulations, but still agree on the importance of coal in British industrialization. Thomas Andrews notes that coal played a similar role for the United States during the late nineteenth and early twentieth centuries, with similar environmental outcomes. Thomas G. Andrews, *Killing for Coal: America's Deadliest Labor War* (Cambridge, MA: Harvard University Press, 2010); Kenneth Pomeranz, *The Great Divergence: China, Europe, and the Making of the Modern World Economy* (Princeton, NJ: Princeton University Press, 2009); P. H. H. Vries, "Are Coal and Colonies Really Crucial? Kenneth Pomeranz and the Great Divergence," *Journal of World History* 12 (Fall 2001): 407–46.
27. Ian Douglas, quoted in Joel A. Tarr, "The Material Basis of Urban Environmental History," *Environmental History* 10 (October 2005): 744–45.
28. Marcus Hall, *Earth Repair: A Transatlantic History of Environmental Restoration* (Charlottesville: University of Virginia Press, 2005), 196; Frederick Turner, "The Invented Landscape," in *Beyond Preservation: Restoring and Inventing Landscapes*, edited by A. Dwight Baldwin Jr., Judith de Luce, and Carl Pletsch (Minneapolis: University of Minnesota Press, 1994), 35–66.
29. Richards, Diary, October 14, 1846, CHL; Brigham Young to Mary Ann Angell Young, October 1, 1839, George W. Thatcher Blair Collection, MS 15616, box 1, folder 2, CHL; Brigham Young, History, October 12–November 22, 1839, "History of Brigham Young," *Deseret News*, February 24, 1858, page 2, column 2; George A. Smith to John and Clarissa Lyman

Smith, November 22, 1839, John Smith Papers, MS 1326, box 1, folder 5, CHL.

30. Smith, Memoir, April 6, 1840, History of George A. Smith, typescript, 136–37, CHL.
31. Young to Young, April 6, 1840, Thatcher Blair Collection, CHL.
32. Woodruff, July 4, 1840, *Wilford Woodruff's Journal*, 1:479.
33. Richard L. Bushman, *The Refinement of America: Persons, Houses, Cities* (Alfred A. Knopf, 1992), 127–31.
34. Young to Young, May 26, 1840, Thatcher Blair Collection, CHL.
35. Brigham Young, Discourse, June 13, 1860, in Richard S. Van Wagoner, ed., *The Complete Discourses of Brigham Young*, 5 vols. (Salt Lake City: Smith-Pettit Foundation, 2009), 3:1607; Brigham Young, *Complete Discourses*, 4:2007.
36. Brigham Young and Willard Richards to First Presidency, September 5, 1840, Joseph Smith Collection, MS 155, box 3, folder 1, CHL.
37. Woodruff, July 3, 1840, *Wilford Woodruff's Journal*, 1:479.
38. Woodruff, September 4, 1840, *Wilford Woodruff's Journal*, 1:507.
39. Willard Richards to William Richards, March 5, 1840, Richards Family Papers, MS 21558, box 1, folder 2, 11–13, CHL.
40. Young and Richards to First Presidency, September 5, 1840, CHL.
41. Richards to Richards, March 5, 1840, Richards Family Papers, CHL.
42. Lorenzo Barnes to Elijah Malin and Edward Hunter, June 8, 1842, MS 7191, CHL.
43. Original spelling maintained wherever possible in this quotation. Young to Young, April 6, 1840, Thatcher Blair Collection, CHL.
44. Heber C. Kimball to Joseph Smith Jr., July 9, 1840, *Times and Seasons* 6 (April 1, 1845): 862.
45. Kimball to Smith, July 9, 1840, *Times and Seasons* 6 (April 1, 1845): 863.
46. Woodruff, June 26, 1840, *Wilford Woodruff's Journal*, 1:471–72.
47. George D. Watt to Brigham Young, June 29, 1841, George W. Thatcher Blair Collection, MS 15616, box 1, folder 11, CHL; Exodus 1:11, 13–14.
48. Johann Georg Khol, *The British Isles and Their Inhabitants* (1844), 131.
49. Friedrich Engels, *The Condition of the Working Class in England in 1844; With a Preface Written in 1892*, translated by Florence Kelley Wischnewetzky (London, England: George Allen & Unwin Ltd., 1892), 26.
50. Brigham Young and Willard Richards to First Presidency, September 5, 1840, Joseph Smith Collection, MS 155, reel 2, box 3, folder 1, CHL.
51. Brigham Young to Joseph Smith, May 7, 1840, Joseph Smith Collection, MS 155, reel 2, box 2, folder 2, CHL.
52. Roderick Nash, *Wilderness and the American Mind*, revised edition (New Haven, CT: Yale University Press, 1974), 44–83.
53. Joseph Smith to Quorum of the Twelve, December 15, 1840, Joseph Smith Collection, MS 155, box 2, folder 4, CHL.

54. There were small outcroppings of coal just south of Nauvoo in Hancock County, Illinois, and across the river in Lee County, Iowa, but both were surface deposits, not significant enough to power any long-term industrialization in the area. James Hall and J. D. Whitney, *Report of the Geological Survey of the State of Iowa: Embracing the Results of Investigations Made during Portions of the Years 1855, 56 & 57*, 2 vols. (Des Moines, IA: Charles Van Benthuysen, 1858), 1:190; A. H. Worthen, J. D. Whitney, Leo Lesquereux, and Henry Engelmann, *Geological Survey of Illinois: Volume I, Geology* (Springfield, IL: State Journal Steam Press, 1866), 331–32.
55. Cronon, *Nature's Metropolis*, 47–48.
56. Pomeranz, *The Great Divergence*.
57. Smith to Quorum of the Twelve, December 15, 1840, CHL.
58. Charles Dickens to [unknown], March 13, 1842, in John Forster, *The Life of Charles Dickens: Vol. I. 1812–1842*, 2 vols. (Philadelphia: J. B. Lippincott & Co., 1872), 1:351.
59. Charles Dickens, *American Notes for General Circulation* (Paris, France: A. and W. Galignani and Co., 1842), 216.
60. David and Della Miller, *Nauvoo: The City of Joseph* (Santa Barbara, CA: Peregrine Smith, 1974), 5.
61. "Mormons Arrived from England," *Sangamo Journal* (Springfield, IL), December 15, 1840, page 2, column 2; "Temperance among the Mormons," *North American and Daily Advertiser* (Philadelphia, PA), March 19, 1841, page 1, columns 7–8; "Compendium," *Boston Investigator*, March 31, 1841, page 3, column 5.
62. Joseph Fielding to Parley P. Pratt, January 1842, in "Joseph Fielding's Letter," *Latter-day Saints' Millennial Star* 3 (August 1842): 76.
63. Edward Tolton to J. F. Tolton, September 25, 1885, MSS 409, L. Tom Perry Special Collections, Harold B. Lee Library, Brigham Young University, Provo, Utah.
64. William Ellis Jones, Journal, February 2, 1842, MS 8620, reel 12, item 13, page 3, CHL.
65. William Adams, Autobiography of William Adams, MS 8039, page 2, CHL.
66. William Clayton, Diary, October 11, 1840, in James B. Allen and Thomas G. Alexander, eds., *Manchester Mormons: The Journal of William Clayton, 1840 to 1842* (Santa Barbara, CA: Peregrine Smith, 1974), 183–84.
67. Clayton, November 21, 1840, *Manchester Mormons*, 199.
68. Clayton, October 29, 1840, *Manchester Mormons*, 194.
69. George W. Gee to George A. Smith, January 7, 1841, George A. Smith Papers, MS 1322, box 9, folder 2, CHL.
70. Aspects of this thinking were also a part of American thought. Conquest over nature had motivated projects like the construction of the Erie Canal.

See Carol Sheriff, *The Artificial River: The Erie Canal and the Paradox of Progress, 1817–1862* (New York: Hill and Wang, 1997).

71. William Clayton to Edward Martin, November 29 1840, MS 3682, CHL.
72. Clayton to the Saints in England, December 10, 1840, CHL.
73. Joseph Smith, Revelation, January 19, 1841, http://www.josephsmithpapers.org/paperSummary/revelation-19-january-1841-dc-124#!/paperSummary/revelation-19-january-1841-dc-124&p=3 (accessed May 10, 2016).
74. Clayton, November 24, 1840, *Manchester Mormons*, 201.
75. Clayton to Martin, November 29, 1840, CHL.
76. Clayton to the Saints in England, December 10, 1840, CHL.
77. Clayton to the Saints in England, December 10, 1840, CHL; Clayton to Martin, November 29, 1840, CHL.
78. Kimball to Smith, July 9, 1840, *Times and Seasons*, 6:862.
79. "The Mormons," *Warsaw Signal* (Warsaw, IL), May 29, 1841, page 2, columns 3–4.
80. Untiled *Daily Missouri Republican* (St. Louis, MO), November 25, 1841, page 2, column 2.
81. "Extract of a Letter from Sister Melling Who Lately Emigrated from Preston, England, to Nauvoo, United States," *Latter-day Saints' Millennial Star* 2 (October 1841): 96.
82. Joseph Fielding to Parley P. Pratt, January 1842, in "Joseph Fielding's Letter," *Latter-day Saints' Millennial Star* 3 (August 1842): 77.
83. For a discussion of the dual perceptions of the American frontier during the antebellum era, see Peter J. Kastor, *William Clark's World: Describing America in an Age of Unknowns* (New Haven: Yale University Press, 2011).
84. "The 'Latter-day Saint' Swindle," *Preston Chronicle* (Preston, England), September 18, 1841, page 4, column 4.
85. "The Mormons," *Warsaw Signal* (Warsaw, IL), May 29, 1841, page 2, columns 3–4.
86. Francis Moon to Parley P. Pratt, November 4, 1840, in "Important from America," *Latter-day Saints' Millennial Star* 1 (February 1841): 255.
87. Entry for June 13, 1842, Manuscript History Draft Notes, CR 100 92, box 1, folder 4, CHL.
88. Hyrum Smith to Parley P. Pratt, June 12, 1842, MS 17572, CHL.
89. George and Ellen Douglas to Father and Mother, June 2, 1842, MS 5639, CHL.
90. Thomas Jefferson to George Rogers Clark, December 25, 1780, in Julian P. Boyd, et al., eds., *The Papers of Thomas Jefferson*, 33 vols. (Princeton, New Jersey: Princeton University Press, 1951), 4:237–38; Thomas Jefferson to James Madison, April 27, 1809, edited by J. Jefferson J. Looney, *The Papers of Thomas Jefferson: Retirement Series*, 10 vols. (Princeton, NJ: Princeton University Press, 2005): 1:169.
91. Frederick Jackson Turner, "The Significance of the Frontier in American

History," in John Mack Faragher, comp., *Rereading Frederick Jackson Turner: "The Significance of the Frontier in American History" and Other Essays* (New Haven, CT: Yale University Press, 1994), 31, 33, 53. Turner's work has been sharply criticized by historians, but two recent historians have demonstrated how Turner accurately reflected the mindset of nineteenth-century Americans. Joy S. Kasson, *Buffalo Bill's Wild West: Celebrity, Memory, and Popular History* (New York: Hill and Wang, 2000); David Wrobell, *The End of American Exceptionalism: Frontier Anxiety from the Old West to the New Deal* (Lawrence: University of Kansas Press, 1993).

92. Howe, *What Hath God Wrought*, 525–32; Gordon S. Wood, *Empire of Liberty: A History of the Early Republic, 1789–1815* (New York: Oxford University Press, 2009), 626–28.
93. "Church History," *Times and Seasons* 3 (March 1, 1842): 709.

MAPPING DESERET

This paper began as a presentation titled "Practical and Beautiful: The Aesthetics of Mormon Mapmaking," which was made at the symposium titled "Pictures from an Expedition: Aesthetics of Cartographic Exploration in the Americas." The symposium was held at the Newberry Library in Chicago on June 20–21, 2013. The author wishes to thank both the Newberry Library and the Terra Foundation for American Art, which sponsored that symposium, as well as the presenters and attendees who offered so much encouragement.

1. See D. W. Meinig, "The Mormon Culture Region: Strategies and Patterns in the Geography of the American West, 1847–1964," *Annals of the Association of American Geographers* 55 (June 1965): 191–220; Wallace Stegner, *Mormon Country* (New York: Duell, Sloan and Pierce, 1942).
2. Richard Francaviglia, *The Mormon Landscape: Existence, Creation and Perception of a Unique Image in the American West* (New York: AMS Press, 1978).
3. Richard Francaviglia, "The City of Zion in the Mountain West," *The Improvement Era*, December 1969.
4. Lowell C. "Ben" Bennion, "Mormondom's Deseret Homeland," in *Homelands: A Geography of Culture and Place across America* (Baltimore: The Johns Hopkins University Press, 2001), 184–209.
5. Based in part on the author's contact with historian Kevin Folkman, as described in Richard Francaviglia, *The Mapmakers of New Zion: A Cartographic History of Mormonism* (Salt Lake City: University of Utah Press, 2015), 120–23.
6. Richard Francaviglia, "Geography and Mormon Identity," entry in *The Oxford Handbook on Mormonism*, edited by Philip Barlow and Terryl Givens (Oxford University Press, 2015).
7. Howard Stansbury, *Exploration and Survey of the Valley of the Great Salt*

Lake of Utah, Including a Reconnaissance of a New Route through the Rocky Mountains (Philadelphia: Lippincott, Grambo, 1852).

8. Richard Francaviglia, *The Mapmakers of New Zion*, 81.
9. See Leonard J. Arrington, *Great Basin Kingdom: An Economic History of the Latter-day Saints, 1830–1900* (Cambridge, MA: Harvard University Press, 1958).
10. Orson Pratt, *New Jerusalem; Or, the Fulfilment of Modern Prophecy*, no. 7 (Liverpool: R. James Printer, 1849), 18–19.
11. See Dale Morgan, *The State of Deseret* (Logan and Salt Lake City: Utah State University Press and the Utah Historical Society, 1987).
12. Andrew Love Neff, *History of Utah, 1847 to 1869* (Salt Lake City: Deseret News Press, 1940), 118.
13. See William G. Hartley, "Mormons, Crickets, and Gulls: A New Look at an Old Story," in *The New Mormon History: Revisionist Essays on the Past*, edited by D. Michael Quinn (Salt Lake City: Signature Books, 1992).
14. See Preston Nibley, *Brigham Young: The Man and His Work* (Salt Lake City: Deseret News Press, 1937), 181; and Francaviglia, *The Mormon Landscape*, 79–86.
15. Historical Records of Parowan, Utah, 1856–1859, page 3, LDS Church History Library, Salt Lake City (hereafter referred to as CHL).
16. Richard Francaviglia, *Go East, Young Man: Imagining the American West as the Orient* (Logan: Utah State University Press, 2011), esp. chapter 3, "Chosen People, Chosen Land, Utah as the Holy Land," 87–125.
17. John Davis, *The Landscape of Belief: Encountering the Holy Land in Nineteenth-Century American Art and Culture* (Princeton: Princeton University Press, 1996), 21.
18. Entry for February 9, 1857, page 45, Conferences et al., "Meetings [of the] Nephi Branch, September 23rd 1855 to March 9th, 1862," (Historical Record & Record of Members), CHL.
19. For more on this subject, see the chapter titled "Chosen People, Chosen Land: Utah as the Holy Land," in Francaviglia, *Go East, Young Man*, 87–125.
20. Richard V. Francaviglia, "'Like the Hajis of Meccah and Jerusalem': Orientalism and the Mormon Experience," *Leonard J. Arrington Mormon History Lecture Series*, No. 17 (Logan: Utah State University Press, 2011), 1–44.
21. Juanita Brooks, *Journal of the Southern Indian Mission*, 46–47.
22. Ibid.
23. Kerry William Bate, "John Steele: Medicine Man, Magician, Mormon Patriarch," *Utah Historical Quarterly* 62 (Winter 1994): 72.
24. See Noel Carmack, "Running the Line: John H. Martineau's Survey in Northern Utah, 1860–1862," *Utah Historical Quarterly* 68 (Fall 2000): 292–312.

25. Clifford Stott, *Search for Sanctuary: Brigham Young and the White Mountain Expedition* (Salt Lake City: University of Utah Press, 1984).
26. James H. Martineau Journal, page 85, box 9, Huntington Library, San Marino, California.
27. Richard Francaviglia, *Over the Range: A History of the Promontory Summit Route of the Pacific Railroad* (Logan: Utah State University Press, 2008), 67–100.

American Zion

1. A collection of articles and commentary on the increase of vitriol in southern Utah (as well as other western public lands) is presented in "Forty Years of Sagebrush Rebellion," *High Country News*, January 4, 2016, Web Exclusive, hcn.org/articles/sagebrush-rebellion.
2. Roderick Nash, *Wilderness and the American Mind*, 5th edition (New Haven, CT: Yale University Press, 2014); Alfred Runte, *National Parks: The American Experience*, 4th edition (Lanham, MD: Taylor Trade Publishing, 2010).
3. William Logan Hebner, *Southern Paiute: A Portrait* (Logan: Utah State University Press, 2010); Richard W. Stoffel and Michael J. Evans, "Kaibab Paiute History: The Early Years" (Fredonia, AZ: Kaibab Paiute Tribe, 1978), 9.
4. There exists a prophecy, though highly controversial, that Joseph Smith foresaw the Mormon people finding homeland in the Rocky Mountain region. *History of the Church*, 5:85; Joseph Smith manuscript history, book D-1, page 1362, LDS Church Library, Salt Lake City, Utah (hereafter CHL).
5. Nathan Waite, "Remembering, Branding, Claiming: How Mukuntuweap National Monument Became Zion National Park," paper delivered at Utah History Conference, Salt Lake City, September 6, 2013; Angus Woodbury, *A History of Southern Utah and Its National Parks*, Utah State Historical Society, vol. 12, nos. 3–4, October 1944, revised in 1950, 186–87.
6. William R. Palmer, "Indian Names in Utah Geography," *Utah Historical Quarterly* 1 (1928): 13; Waite, "Mukuntuweap National Monument"; Woodbury, *History of Southern Utah*, 114.
7. Roderick Nash referred to Henry David Thoreau as one of the most important thinkers in and inspirations to the creation of the American Romantic movement and its regard for the wilderness ideal. Roderick Nash, *Wilderness in the American Mind*, 3rd edition (New Haven, CT: Yale University Press, 1982).
8. Fenton E. Moss, "The Man Who Named Zion Canyon: The Story of Isaac Behunin," 1998, 2, CHL.
9. Clarence E. Dutton, "Vermilion Cliffs and the Valley of the Virgin," in *Tertiary History of the Grand Canyon District* (Washington, DC: Government Printing Office, 1882), 60.

10. Moss, "The Man Who Named Zion Canyon," 2.
11. F. S. Dellenbaugh, "A New Valley of Wonders," *Scribner's Magazine* 35 (January 1904): 2.
12. Ibid, 2.
13. Leonard J. Arrington, "The Mormon Cotton Mission in Southern Utah," *Pacific Historical Review* 25 (August 1956): 221–38.
14. Hebner, *Southern Paiute*; Beverly P. Smaby, "The Mormons and the Indians: Conflicting Ecological Systems in the Great Basin," *American Studies* 16 (Spring 1975): 35–48.
15. Arrington, "The Mormon Cotton Mission in Southern Utah," 221–38.
16. Roderick Frazier Nash, *Wilderness and the American Mind*, 5th edition (New Haven, Yale University Press, 2014).
17. Mark David Spence, *Dispossessing Wilderness: Indian Removal and the Making of the National Park* (Oxford: Oxford University Press, 1999), 5.
18. Horace M. Albright, Interview by J. L. Crawford and Fern Crawford, Sherman Oaks, CA, April 10, 1979, 10, box 8, folder 26, J. L. Crawford Papers, Special Collections, Gerald R. Sherratt Library, Southern Utah University.
19. Robert Shankland, *Stephen Mather of the National Parks*, 1st edition (New York: Knopf, 1951), 145.
20. Karl Jacoby, *Crimes against Nature: Squatters, Poachers, Thieves and the Hidden History of American Conservation* (Berkeley: University of California Press, 2003).
21. Spence, *Dispossessing Wilderness*, 131.
22. Thomas G. Alexander, "Red Rock and Gray Stone: Senator Reed Smoot, the Establishment of Zion and Bryce Canyon National Parks, and the Rebuilding of Downtown Washington, D.C.," *Pacific Historical Review* 72 (February 2003): 1–38; Wayne K. Hinton, "Getting Along: The Significance of Cooperation in the Development of Zion National Park," *Utah Historical Quarterly* 68 (Fall 2000): 313–31.
23. Early terms used to describe park visitors. "Dudes" traveled by train or motor stages and "sagebrushers" traveled in their own vehicles.
24. David Quammen, "The Paradox of the Park," *National Geographic* 229 (May 2016): 55–67.
25. Jedediah Rogers, *Roads in the Wilderness: Conflict in Canyon Country* (Salt Lake City: University of Utah Press, 2013), 3, 6.
26. George Bucknam Dorr, "Two National Monuments: The Desert and the Ocean Front. Zion National Monument," *Sieur de Mont Publications* 14, Department of the Interior, National Park Service, October 2, 1920, 11; Woodbury, *History of Southern Utah*, 163.
27. Howard Means, "Zion Park–Mt. Carmel Road," unpublished typescript, 1947–1948, 5, MSSA17, Utah State Archives, Salt Lake City, Utah.
28. Donald T. Garate, *The Zion Tunnel: From Slickrock to Switchback* (Springdale, UT: Zion Natural History Association, May 2, 1989).

29. Jack Lait, "An Appreciation of Zion National Monument," *United States Railroad Administration*, 1917, 3, PAM 599, Utah State Archive.
30. W. H. Murray, "Zion National Park, Bryce Canyon, Cedar Breaks, Kaibab Forest, North Rim of the Grand Canyon," 1926, 6, Union Pacific System, PAM 599, Utah State Archive, 6.
31. Lola Belle DeMille Bryner, Interview with Fielding H. Harris, May 4, 1968, Voices of Remembrance Foundation, Oral History Collection, Dixie State University, St. George, Utah. 12.
32. Arthur Stevens to Utah State Historical Society, June 4, 1982, 1, box 6, folder 12, Crawford Collection.
33. Maurice Cope, "Living for Ages in 24 Hours: Pre-Americans," 5, unpublished typescript, MSA2594, Utah State Historical Society.
34. Craig S. Smith, "James E. Talmage and the 1895 Deseret Museum Expedition to Southern Utah," *Utah Historical Quarterly* 84 (Spring 2016): 147.
35. This sentiment is more in line with Brigham Young's quote, "It is all good, the air, the water, the gold, the silver; the wheat, the fine flour and the cattle upon the thousand hills are all good." *Journal of Discourses* [hereafter *JD*] 1:272–73; see Sara Dant's essay in this volume.
36. Janice F. DeMille. "Generations, William Louis Crawford," *Color Country Spectrum Outlook*, May 7, 1978, 10, in box 6, folder 7, Crawford Collection.
37. The Paiutes saw Sinawava as a central God. Mormon missionaries upheld Sinwava to would-be converts as Jesus and other members of the tribe believed that Sinwava was "God, Jesus, same thing." Hebner, *Southern Paiute*, 16, 19, 107.
38. William Flanigan, "Excerpts from the Diaries of William Wallace," 8, unpublished typescript, August 20, 1990, Special Collections, Gerald R. Sherratt Library, Southern Utah University.
39. Ibid. The first person to travel The Narrows was actually Grove Karl Gilbert, who accompanied John Wesley Powell on the Wheeler Survey in 1872. Powell surveyed Parunuweap Canyon and the East Fork of the Virgin River.
40. *Beehive Girls Handbook*, The General Board of the Young Woman's Mutual Improvement Association, The Church of the Jesus Christ of Latter-Day-Saints, Salt Lake City Utah, 1935.
41. "Bee Hive Girls Have a Good Time at Zion National Park," *Washington County News*, June 17, 1920.
42. *Beehive Girls Handbook*, 40, 46.
43. This brochure was published the year ZNP was designated, so the guidelines were written prior to the area becoming a national park.
44. King Hendricks, memoir 3, 5.
45. Ibid, 4.

46. Robert Baird, "Welcome, Welcome Sabbath Morning," *Hymns of the Church of Jesus Christ of Latter Day Saints* (Salt Lake City: Church of the Latter Day Saints, 1985), 280.
47. Reed Smoot, Journal, Book 37, July 12–July 18, 1927, 97–100, typescript, Reed Smoot Papers, L. Tom Perry Special Collections, Harold B. Lee Library, Brigham Young University.
48. Ibid, 98.
49. David Quammen, "Is Yellowstone Park in Danger of Being 'Loved to Death?'" *Fresh Air*, April 18, 2016, npr.org/2016/04/18/474658536/is-yellowstone-national-park-in-danger-of-being-loved-to-death.
50. Ibid, 97.
51. J. L. Crawford, interview by Don Graff, September 29, 1989, 3, transcript, Zion National Park Oral History Project, CCC Reunion, MS178, Register of Civilian Conservation Corps Oral History Project, Special Collections, Gerald R. Sherratt Library, Southern Utah University Archive.
52. "Eviend T. Scoyen: An Interview about His Experiences as the First Superintendent of Zion National Park, 1927–1931," interview by Lucy C. Schiefer, January 28, 3, transcript, box 5, Zion 12352, National Park Service, Zion National Park Collection; Wayne K. Hinton, "Getting Along: The Significance of Cooperation in the Development of Zion National Park," *Utah Historical Quarterly* 68 (Fall 2000): 313–31.
53. In 1861, Brigham Young announced the roster of Saints who would settle what would become Dixie during a conference in Salt Lake City. According to the historian H. Lorenzo Reid, "One of the proverbial statements that aptly characterized the early pioneers of the Dixie Mission was to the effect that they were extremely loyal to the call of their church that had they been called to build their home on barren rock they would have done so willingly, and would have remained there until they were released from that call." H. Lorenzo Reid, *Brigham Young's Dixie of the Desert: Exploration and Settlement* (Salt Lake City, UT: Zion Historical Association, 1964), 98. To the Saints, "the call" was the word of God.
54. H. E. Petersen, "Zion Easter Pageant: Outline of the Zion National Park Easter Pageant," Report by Easter Pageant Executive Committee, 1938, 1, Easter History Pageant Box, Special Collections, Gerald R. Sherratt Library, Southern Utah University.
55. For more on the idea of embedding the Great Basin with notions of the Holy Land, see Richard Francaviglia, *Go East, Young Man: Imagining the American West as the Orient* (Logan: Utah State University Press, 2011).
56. "Stage All Set for Zion Easter Pageant," *Garfield News*, April 7, 1938.
57. "Thousands Expected to Witness Fourth Annual Presentation of Zion Easter Pageant," *Iron County Record*, March 21, 1940.
58. Phil Hepworth, Interview by Don Graff, Zion National Park Oral History

Project, CCC Reunion, September 27, 1989, MS 178, Register of Civilian Conservation Corps Oral History Project, 12, Special Collections, Gerald R. Sherratt Library, Southern Utah University.

59. Anon to Mariam Nelson, c/o Gilliam Advertising Agency, Salt Lake City, Utah, February 13, 1938, 1, Easter Pageant Box, Special Collections, Gerald R. Sherratt Library, Southern Utah University.

60. I make the assumption that the author is a Dixie Mormon, as he refers to the people of Utah, "as our own people," a reference to kinship with other Saints ubiquitous in LDS culture. The writer indicated that he/she is directly involved in the production, mentioning in the letter "our Easter Pageant" and "a wish that we had an advertising man down here."

61. David O. McKay and J. Reuben Clark to Stake President Harold B. Snow, March 6, 1940, reprinted in "Famed Zion Easter Pageant to be Discontinued—Objections Voiced by General Church Authorities," *Iron Country News*, February 27, 1941.

62. According to correspondence between the acting and the former ZNP supervisors just prior to the discontinuation of the pageant, the National Park Service was very supportive of the Easter production. P. P. Patraw, to C. Marshall Finnan, March 22, 1940, Zion 14578, box 8, National Park Service, ZNP Collection; Marshall C. Finnan, "Memorandum for Superintendent Patraw," March 30, 1940, box 8, Zion 14578, National Park Service, ZNP Collection.

Before the Boom

1. On the Mormon pioneer migration, see Will Bagley, ed., *The Pioneer Camp of the Saints: The 1846 and 1847 Mormon Trail Journals of Thomas Bullock* (Spokane: Arthur H. Clark, 1997); Richard E. Bennett, *We'll Find the Place: The Mormon Exodus, 1846–1848* (Norman: University of Oklahoma, 2009).

2. Matthew J. Grow and Ronald W. Walker, eds., *The Prophet and the Reformer: The Letters of Brigham Young and Thomas L. Kane* (New York: Oxford University Press, 2015), 17. Gary S. Ford argues that the Mormons' livestock literally represented "temporal salvation": "Cornelius P. Lott and his contribution to the temporal salvation of the Latter-day Saint Pioneers through the care of livestock" (master's thesis, Brigham Young University, 2005).

3. Leonard J. Arrington, *Great Basin Kingdom: An Economic History of the Latter-day Saints, 1830–1900*, new edition (Urbana: University of Illinois, 2005), 18.

4. On the animals a typical family might bring, see John Mack Faragher, *Women and Men on the Overland Trail* (New Haven: Yale University Press, 1979), 44. See also Diana L. Ahmad, *Success Depends on the Animals: Emigrants, Livestock, and Wild Animals on the Overland Trails, 1840–1869* (Reno: University of Nevada Press, 2016).

5. William Cronon describes how keeping livestock constituted one of the key cultural dividing lines between whites and natives; *Changes in the Land: Indians, Colonists, and the Ecology of New England,* twentieth anniversary edition (New York: Hill and Wang, 2003), chapter 7, 127–56. See also Virginia DeJohn Anderson, "King Philip's Herds: Indians, Colonists, and the Problem of Livestock," *William and Mary Quarterly* 49 (1992): 183–209; Edmund Morgan, *American Slavery, American Freedom: The Ordeal of Colonial Virginia* (New York: Norton, 1975), 175, 310.
6. Alfred Crosby, *Ecological Imperialism: The Biological Expansion of Europe, 900–1900* (New York: Cambridge University Press, 1986), 89. The most consequential items in the portmanteau were pathogens that decimated Native populations.
7. Specific examples of these statements are discussed in the essays in this volume by Thomas Alexander and Sara Dant.
8. Ironically, Utah's tradition of many small ranchers, which did not cause excessive damage as long as numbers of animals stayed relatively low, meant many more grazers and evidently more range damage in Utah in the 1920s than in neighboring states with fewer herds. See Charles S. Peterson, "Smallholding Land Patterns in Utah and the Problem of Forest Watershed Management," *Forest History Newsletter* 17 (July 1973): 4–13. On Potter, see Susan Deaver Olberding, "Albert F. Potter: The Arizona Rancher Who Shaped U.S. Forest Service Grazing Policies," *Journal of Arizona History* 50 (Summer 2009): 167–82.
9. Thomas G. Alexander, "Stewardship and Enterprise: The LDS Church and the Wasatch Oasis Environment, 1847–1930," *Western Historical Quarterly* 25 (Autumn 1994): 363. John B. Wright concurs in *Rocky Mountain Divide: Selling and Saving the West* (Austin: University of Texas Press, 1993), 169–70.
10. Ford, "Cornelius P. Lott," 72–82.
11. "Biographical Sketch of the Life of Peter Wilson Conover," typescript, n.d., MS 7720, LDS Church History Library, (hereafter CHL); Jared Farmer, *On Zion's Mount: Mormons, Indians, and the American Landscape* (Cambridge: Harvard University Press, 2008), 59; Ford, "Cornelius P. Lott," 123–31; Grow and Walker, *Prophet and the Reformer*, 28, 46–47.
12. Grow and Walker, *Prophet and the Reformer*, 23–24.
13. "History of Grazing," chapter 2, "Early Background," unpublished Utah Writer's Project manuscript, in "WPA History of Grazing (Utah) Notes," box 1, folder 4, 16, MS B 100, Utah State Historical Society, Salt Lake City.
14. Jedediah S. Rogers, ed., *The Council of Fifty: A Documentary History* (Salt Lake City: Signature Books, 2014), 107, 112. Pratt's and a committee's later recommendations differ slightly.
15. "History of Grazing," chapter 2, "Early Background," 18; Charles S. Peterson, "Livestock Industry," in Allan Kent Powell, ed., *Utah History Encyclopedia* (Salt Lake City: University of Utah Press, 1994).

16. James H. Beckstead, *Cowboying: A Tough Job in a Hard Land* (Salt Lake City: University of Utah Press, 1991), 5.
17. Rick Rasby, "Determining How Much Forage a Beef Cow Consumes Each Day," University of Nebraska-Lincoln, accessed September 8, 2015, beef.unl.edu/cattleproduction/forageconsumed-day. Rasby suggests a present-day animal may eat twenty-four pounds of dry forage in a day; assuming that cattle on the trail in 1847 were smaller but walking miles every day yields a rough estimate of twenty pounds.
18. Elliott West, *The Contested Plains: Indians, Goldseekers, and the Rush to Colorado* (Lawrence: University of Kansas, 1998), 228–35. For the theft of a Mormon cow that spiraled into a pitched battle, the so-called "Grattan massacre," see John D. Unruh, *The Plains Across: The Overland Emigrants and the Trans-Mississippi West, 1840–1870* (Urbana: University of Illinois, 1979), 170–71.
19. Ted J. Warner, ed., *The Domínguez-Escalante Journal*, translated by Fray Angelico Chavez (Provo: Brigham Young University Press, 1976), 58–60.
20. Alexander L. Baugh, "John C. Frémont's 1843–44 Western Expedition and Its Influence on Mormon Settlement in Utah," *Utah Historical Quarterly* 83 (Fall 2015): 254–69, quote on 268. For Frémont's reports, see Donald Jackson and Mary Lee Spence, eds., *The Expeditions of John Charles Frémont*, 3 vols. (Urbana: University of Illinois Press, 1970–1984).
21. For historic elevation graphs, see "Utah Water Science Center," last modified January 10, 2013, accessed September 24, 2015, ut.water.usgs.gov/greatsaltlake/elevations/.
22. Dan L. Flores, "Agriculture, Mountain Ecology, and the Land Ethic: Phases of the Environmental History of Utah," in John R. Wunder, ed., *Working the Range: Essays on the History of Western Land Management and the Environment* (Westport, CT: Greenwood Press, 1985), 162. A standard source is Stanley L. Welsh, et al., eds., *A Utah Flora*, fourth edition, revised (Provo: Brigham Young University, 2008); see especially xvii–xx.
23. Bagley, *Pioneer Camp of the Saints*, 232.
24. *An Intimate Chronicle: The Journals of William Clayton*, edited by George D. Smith (Salt Lake: Signature Books, 1995), July 24, 1847, 364.
25. On wild grazer numbers, see "History of Grazing," chapter 2, "Early Background." On grazing's effect on native grasses, see Richard N. Mack and John N. Thompson, "Evolution in Steppe with Few Large, Hooved Mammals," *The American Naturalist* 119, no. 6 (June, 1982): 757–73; E. W. Tisdale and M. Hironaka, "The Sagebrush-Grass Region: A Review of the Ecological Literature," contribution no. 209 (Moscow, Idaho: University of Idaho Forest, Wildlife and Range Experiment Station): 1–31. See also Alexander, "Stewardship and Enterprise," 347; Terry D. Ball and Jack D. Brotherson, "Environmental Lessons from our Pioneer Heritage," *BYU Studies* 38, no. 3 (1999): 63–82; Farmer, *On Zion's Mount*, 27; James A.

Young and B. Abbott Sparks, *Cattle in the Cold Desert*, expanded edition (Reno and Las Vegas: University of Nevada Press, 2002), 36.

26. See, for example, Neal Salisbury, *Manitou and Providence: Indians, Europeans, and the Making of New England, 1500–1643* (New York and Oxford: Oxford University Press, 1982), 176–77; and Morgan, *American Slavery, American Freedom*, 51–55.
27. Bagley, *Pioneer Camp of the Saints*, 254; Farmer, *On Zion's Mount*, 78, 86, 90.
28. On conflicts between Mormons and Indians, see Howard Christy, "Open Hand and Mailed Fist: Mormon-Indian Relations in Utah, 1847–1852," in John S. McCormick and John R. Sillito, eds., *A World We Thought We Knew: Readings in Utah History* (Salt Lake City: University of Utah Press, 1995), 34–51; Farmer, *On Zion's Mount*, 62–83.
29. Thomas G. Alexander and James B. Allen, *Mormons and Gentiles: A History of Salt Lake City* (Boulder: Pruett Publishing Company, 1984), 29–34. On the "shadow government" of Deseret, see Dale L. Morgan, *The State of Deseret* (Logan: Utah State University Press, 1987).
30. D&C 42:104.
31. Leonard J. Arrington, "Property among the Mormons," *Rural Sociology* 16 (1951): 339–52.
32. In 1854, the law of consecration was restored; an estimated 40 percent of household heads consecrated their property, although the LDS Church did not actually take possession. From 1868–1884, some two hundred enterprises were labeled "cooperative"; Arrington, *Great Basin Kingdom,* 146, 293.
33. See, for example, Brigham Young, April 9, 1852, *Journal of Discourses*, 1:49 (hereafter *JD*). See also Sara Dant's and Thomas Alexander's essays in this volume; and Dan L. Flores, "Zion in Eden: Phases of the Environmental History of Utah," *Environmental Review* 7 (Winter 1983): 325–44.
34. Jeanne Kay and Craig J. Brown, "Mormon Beliefs about Land and Natural Resources, 1847–1877," *Journal of Historical Geography* 11, no. 3 (1985): 260.
35. Arrington, *Great Basin Kingdom,* 51–52. For examples of the land market, see Alexander, "Stewardship and Enterprise," 352–53.
36. Richard H. Jackson, "Geography and Settlement in the Intermountain West: Creating an American Mecca," *Journal of the West* 33, no. 3 (1994): 22–34; Leonard J. Arrington, "Agriculture and Mormonism: The Historical Perspective," *Brigham Young Centennial: College of Biological and Agricultural Sciences Symposium*, March 30, 1976, 3.
37. Charles S. Peterson, "Grazing in Utah: A Historical Perspective," *Utah Historical Quarterly* 57 (Fall 1989): 302–3; "Forgotten Chapters of History: Evolution of the Livestock Business," n.d., n.p., typescript, MS d 3823, CHL.

38. *Autobiography of Parley P. Pratt*, edited by Parley P. Pratt (Salt Lake City: Deseret Book Company, 1938), 329–31.
39. *Journal History of the Church of Jesus Christ of Latter-day Saints*, November 15, 16, 1847 (henceforth *JH*).
40. J. D. B. DeBow, *Seventh Census 1850 Statistical View, Compendium* (Washington: Beverly Tucker, 1854).
41. Arrington, *Great Basin Kingdom,* 90–92; "History of Grazing," chapter 3, "The Westward Movement," 33–35.
42. Daniel H. Wells, September 10, 1861, *JD* 9:63.
43. Brigham Young, "An Epistle to the Saints in G. S. L. Valley," September 9, 1847, typescript in *JH,* September 9, 1847.
44. Willard Snow to Erastus Snow, October 6, 1847, typescript in *JH,* October 6, 1847; Arrington, *Great Basin Kingdom,* 48. Farmer, *On Zion's Mount*, 88, notes this was a periodic occurrence through the 1890s.
45. Rogers, *Council of Fifty*, 10, 146–47.
46. On branding, see Richard H. Cracroft, "The Heraldry of the Range: Utah Cattle Brands," *Utah Historical Quarterly* 32 (Summer 1964): 219; Levi S. Peterson, "The Development of Utah Livestock Law, 1848–1896," *Utah Historical Quarterly* 32 (Summer 1964), 206–7. For the pound, see "Poundkeeper's Office Records, 1853–1854," MS 1496, CHL. Brigham Young's correspondence contains many disputes over fencing; for example, Mary A. Johnson to Brigham Young, May 16, 1860, Springville, Brigham Young office files: General Correspondence, Incoming, 1840–1877, General Letters, 1840–1877, Ho-J, 1860, CHL.
47. Arrington, *Great Basin Kingdom*, 58; *A Mormon Chronicle: The Diaries of John D. Lee, 1848–1876*, edited by Robert Glass Cleland and Juanita Brooks (San Marino, CA: The Huntington Library, 1955), 1: 81–82, 86.
48. Rogers, *Council of Fifty*, 144–45.
49. Earl M. Christensen and Myrtis A. Hutchinson, "Historical Observations on the Ecology of Rush and Tooele Valleys, Utah," *Proceedings of the Utah Academy of Sciences, Arts, and Letters* 42 (1965): 90–95.
50. Wayne K. Hinton, "The Birth and Infancy of the National Forests in Southern Utah: Settlement to 1910," Delivered as Faculty Honor Lecture, Southern Utah State College, May 16, 1885.
51. Tom Chaffin, *Pathfinder: John Charles Frémont and the Course of American Empire* (New York: Hill and Wang, 2002), 261.
52. Allen D. Roberts, "History of Antelope Island in the Great Salt Lake, Utah," Wallace N. Cooper 2 & Associates, n.d., 3–4. On the mysterious Stump, see 23–24.
53. Perpetual Emigrating Fund Company general files: Minutebook and ledger, 1850–1880, September 14, 1850, CHL.
54. The fund was authorized in fall 1849; Arrington, *Great Basin Kingdom,* 99–106. Garr was a bonded herdsman. He deposited a bond to

compensate the owners who contracted with him for potential losses. The county courts licensed such herdsmen. For the legislation, see *JH*, September 14, 1850; Levi S. Peterson, "The Development of Utah Livestock Law," 201–3. See also Brigham D. Madsen, ed., *Exploring the Great Salt Lake: The Stansbury Expedition of 1849–50* (Salt Lake City: University of Utah Press, 1989), 321.

55. For a brief description of the island and the ranch, see John Hudson's April 11, 1850 entry, *A Forty-Niner in Utah with the Stansbury Exploration of Great Salt Lake: Letters and Journal of John Hudson, 1848–1850*, edited by Brigham D. Madsen (Salt Lake City: Tanner Trust Fund, 1981), 128; Madsen, *Exploring the Great Salt Lake*, 325–26, 370–71. Garr built an adobe ranch house that is considered to be the oldest Euroamerican structure still on its own foundation in Utah. For Young's cattle on Antelope, see, for example, the ninety-six cows, sixty-six calves, seventeen oxen, forty-eight steers, forty-four heifers, and three bulls Briant Stringham counted on June 26, 1855; Livestock record, 1853–1857, Brigham Young office files: Brigham Young Financial Files, 1841–1877, Miscellaneous Account Books, 1851–1865, CHL.
56. Briant Stringham to D. H. Wells, Garr Island, June 6, 1855, Brigham Young office files: General Correspondence, Incoming, 1840–1877, General Letters, 1840–1877, Sk-St, 1855, CHL.
57. "History of Grazing," chapter 3, "The Westward Movement," 49.
58. For example, John Brown was a traveling agent at Kanesville. He bought fifty yoke of oxen in Missouri. See *Autobiography of Pioneer John Brown, 1820–1896*, arranged and published by John Zimmerman Brown (Salt Lake City: Stevens and Wallis, 1941), January 1851, 119.
59. Brigham Young, April 9, 1852, *JD*, 1:53; see also "The Son of God Will Come," *Deseret News*, April 17, 1852, 2, emphasis in original.
60. Brigham Young, May 8, 1853, *JD*, 1:110.
61. Brigham Young, October 6, 1860, *JD*, 8:317.
62. Arrington, *Great Basin Kingdom*, 133–36.
63. Ibid., 188–92.
64. Roberts, "History of Antelope Island," 35.
65. On the gold seekers, see Arrington, *Great Basin Kingdom*, 68–69; Brigham D. Madsen, *Gold Rush Sojourners in Great Salt Lake City, 1849 and 1850* (Salt Lake City: University of Utah Press, 1983); Beckstead, *Cowboying*, 8.
66. "Message of Governor Brigham Young," *Millennial Star*, 16, 214.
67. Heber C. Kimball, February 29, 1856, in *Millennial Star* 18 (June 21, 1856), 396.
68. Young and Sparks, *Cattle in the Cold Desert*, 91.
69. Jeffrey Lockwood, "The Death of the Super Hopper," *High Country News*, February 3, 2003.
70. Andrew Moffitt to Briant Stringham, December 30, 1855, Brigham

Young office files: General Correspondence, Incoming, 1840–1877, General Letters, 1840–1877, Lo-M, 1856, CHL; Arrington, *Great Basin Kingdom*, 149–52.

71. William H. Ashby, quoted in Roberts, 43.
72. Daddy Stump was reportedly killed by Indians in Cache Valley; Roberts, 4, 23–24.
73. Rogers, *The Council of Fifty*, 143.
74. *On the Mormon Frontier: The Diary of Hosea Stout*, vol. 2, 1848–1861, edited by Juanita Brooks (Salt Lake City: University of Utah Press, 1964), 547–48, 572–77, 610–13, 616–19, 650–52, 673–74.
75. Brigham Young, June 5, 1853, *JD*, 1:252.
76. See "History of Grazing," chapter 3, "The Westward Movement," 41; Arrington, *Great Basin Kingdom*, 81.
77. Of the first 3,100 head of cattle, 2,213 were oxen; Don D. Walker, "The Cattle Industry of Utah, 1850–1900: An Historical Profile," *Utah Historical Quarterly* 32 (Summer 1964): 182. In 1860, some 27 percent of cattle were oxen; by 1880 only 3 percent; 184.
78. Arrington, *Great Basin Kingdom*, 206–11; William G. Hartley, "Down-and-Back Wagon Trains: Travelers on the Mormon Trail in 1861," *Overland Journal* 11 (Winter 1993): 23–34.
79. See, for example, the bishop of Pleasant Grove who contributed four teams of the five hundred requested in the territory in 1863; *Autobiography of Pioneer John Brown*, 247.
80. For example, Daniel H. Wells sent 127 cattle and 29 horses and mules taken from the "enemy"; Wells to Brigham Young, November 2, 1857, Bridger, Brigham Young office files: 1857 November 1–16, CHL; David L. Bigler and Will Bagley, *The Mormon Rebellion: America's First Civil War, 1857–1858* (Norman: University of Oklahoma, 2012), 211–13, 217.
81. Bigler and Bagley, *Mormon Rebellion.* Mormon and army ranchers clashed, leading to a sergeant's murder; the so-called Spencer-Pike incident was finally brought to trial nearly thirty years later, but the accused was acquitted. See Hal Schindler, "Is That You, Pike? Feud between Settlers, Frontier Army Erupts and Simmers for Three Decades," Utah History to Go, accessed May 27, 2016, historytogo.utah.gov/salt_lake_tribune/in_another_time/070295.html.
82. Will Bagley, *Blood of the Prophets: Brigham Young and the Massacre at Mountain Meadows* (Norman: University of Oklahoma Press, 2002). For arguments over grass, see 101–3; for the accusations of a poisoned ox (or spring) at Corn Creek, 105–11; for the fate of the Fancher and Dukes parties' cattle, 167–72.
83. "History of Grazing," chapter 3, "The Westward Movement," 45–46.
84. See, for example, the description of Echo Canyon and Chalk Creek in Charles Peterson's notes on Eugene Correll, oral history of Thomas E.

Moore of Coalville, September 13, 1940, box 74, folder 6, Charles S. Peterson Papers, MS B 1608, Utah State Historical Society.

85. For accounts that suggest careful planning, see Wayne K. Hinton, "The Birth and Infancy of the National Forests in Southern Utah: Settlement to 1910," Delivered as Faculty Honors Lecture, Southern Utah State College, May 16, 1985, in Charles S. Peterson papers, box 74, folder 10, 3; Arrington, *Great Basin Kingdom*, 88–91. For accounts of more haphazard settlements led by herdsmen, see Farmer, *On Zion's Mount*, 62; and "History of Grazing," chapter 3, "The Westward Movement," 26–27.
86. For example, in Ephraim; see "Comments by Lauritz Nielson on Changes and Some of Their Causes in Ephraim Canyon," typescript, April 14, 1953, box 74, folder 3, Charles S. Peterson Papers.
87. On the Elk Mountain mission, see Alfred N. Billings, "Diary, May–October 1855," MS 419, CHL; Richard Firmage, *History of Grand County* (Salt Lake City: Grand County and the Utah State Historical Society, 1996), 76–90; and Ethan Pettit, "Diaries, 1845–1875," reel 1, no. 2, May 1855–June 1856, MS 419, CHL; for the eventual settlement, see Firmage, 95–110.
88. *Pioneer Pathways*, compiled by Lesson Committee (Salt Lake City: International Society of Daughters of Utah Pioneers, 2003), vol. 6, 382.
89. "Stock," *Deseret News*, March 1, 1865, 8.
90. Orson Hyde, October 7, 1865, in *JD*, 11: 147–54; quote on 149.
91. On City Creek, see George Stewart and John A. Widstoe, "Contribution of Forest Land Resources to the Settlement and Development of the Mormon-Occupied West," *Journal of Forestry* 41 (September 1, 1943): 633–40. On Rush and Tooele Valleys, see Christensen and Hutchinson, 97–103.
92. Ball and Brotherson, 74; Walter P. Cottam and George Stewart, "Plant Succession as a Result of Grazing and of Meadow Desiccation Since Settlement in 1862," University of Utah and Intermountain Forest and Range Experiment Station, reprinted from *Journal of Forestry* xxxviii, no. 8 (August 1940).
93. Walter P. Cottam, "The Impact of Man on the Flora of the Bonneville Basin," University of Utah, February 20, 1961. See also Ball and Brotherson, 73; Flores, "Agriculture, Mountain Ecology, and the Land Ethic," 169.
94. Brigham Young, October 9, 1865, *JD*, 11:142.
95. George A. Smith, April 11, 1867, *JD*, 11:363.
96. Brigham Young, July 24, 1877, *JD*, 19: 60–65; quote on 62.
97. Woodruff's 1847 quote is cited in Richard H. Jackson, "Mormon Perception and Settlement," 324; for the 1877 speech see Wilford Woodruff, September 16, 1877, *JD*, 19: 224.
98. Alexander and Allen, *Mormons and Gentiles*, 23–24; Farmer, *On Zion's*

Mount, 126–38; Flores, “Zion in Eden,” 326; Richard H. Jackson, “Myth and Reality: Environmental Perceptions of the Utah Pioneers,” *The Rocky Mountain Social Science Journal* 9 (1972): 34–36; Richard H. Jackson, “Mormon Perception and Settlement,” 317–34.

99. Don D. Walker, “The Cattle Industry of Utah,” 183, 190.

100. On sheep/cattle differences, see C. E. Favre to C. C. Anderson, August 20, 1940, box 74, folder 7, Charles S. Peterson Papers; and Young and Sparks, *Cattle in the Cold Desert*, 63–69. On wool, see Arrington, *Great Basin Kingdom*, 247; on Seely, see Noble Warrum, *Utah Since Statehood: Historical and Biographical*, vol. 3 (Chicago and Salt Lake: S. J. Clarke Publishing Company, 1919), 572–75.

101. *Agriculture of the United States in 1860* (Washington: Government Printing Office. 1864); Don D. Walker, “Cattle Industry,” 189–30.

102. On longhorns, see C. C. Anderson, interview with J. M. Titus, Castle Valley, August 1940, box 74, folder 4, Charles S. Peterson Papers; Young and Sparks, *Cattle in the Cold Desert*, 71–76; and Don D. Walker, “Cattle Industry,” 185; on Herefords and other newer breeds, 193–94.

103. Roberts, 9–10; Clay Shelley, Utah Division of Natural Resources, interview by author, September 15, 2015. Shelley is the Utah Division of Natural Resources curator of the Fielding Garr Ranch Historic Site.

104. For example, Pleasant Grove’s United Order began in 1874 but “after a number of years it was considered to be an impractical venture and was discontinued”; *Autobiography of Pioneer John Brown*, 340. On Orderville, see Leonard J. Arrington, *Orderville, Utah: A Pioneer Experiment in Economic Organization* (Logan: Utah State Agricultural College, 1954).

105. Arrington, *Great Basin Kingdom*, 323–37, 363–64; “History of Grazing,” chapter 3, “The Westward Movement,” 33.

106. See Paul Nelson’s contrast between “good land” in the Salt Lake Valley and “bad land” on the Colorado Plateau in chapter 6, *Wrecks of Human Ambition: A History of Utah’s Canyon Country to 1936* (Salt Lake City: University of Utah Press, 2014), 135–76.

107. David E. Miller, *Hole-in-the-Rock: An Epic in the Colonization of the Great American West* (Salt Lake City: University of Utah Press, 1959), 53, 122.

108. Charles S. Peterson, “San Juan in Controversy: American Livestock Frontier vs. Mormon Cattle Pool,” in Thomas G. Alexander, ed., *Essays on the American West, 1972–1973* (Provo: Brigham Young University Press, 1974), 59–61. On such tactics elsewhere, see Paul Starrs, *Let the Cowboy Ride: Cattle Ranching in the American West* (Baltimore and London: Johns Hopkins University Press, 1998), 50–54.

109. Bluff settlers reported that one of the motivations for establishing the town was as a “buffer settlement ... against encroachment by stockmen from Colorado and the Southwest.” Miller, *Hole-in-the-Rock*, 5.

110. Charles S. Peterson, “San Juan in Controversy,” 45–68; box 78, folders

1 and 2, Charles S. Peterson Papers. On Scorup, see Neal Lambert, "Al Scorup: Cattleman of the Canyons" *Utah Historical Quarterly* 32 (Summer 1964): 301–20; on the Redds, see Leonard J. Arrington, *Utah's Audacious Stockman: Charlie Redd* (Logan: Utah State University Press, 1995); Charles S. Peterson, *Look to the Mountains: Southeastern Utah and the La Sal National Forest* (Provo: Brigham Young University Press, 1975), 93, 230, 240.

111. "History of Grazing," chapter 3, "The Westward Movement," 55–56.
112. *Mormon Chronicle*, 1: 82–86, 97.
113. For bears in Ephraim Canyon, see J. W. Humphrey, "My Recollections of the Manti Forest," typescript, June 1953, box 74, folder 3, Charles S. Peterson Papers. Steve Siporin, "A Bear and a Bandit," *Utah Historical Quarterly* 83 (Spring 2015): 98–114, explores the story of "Old Ephraim," a grizzly bear killed for preying on sheep in Cache Valley in 1921 and remembered as the last grizzly in Utah. On predator control, see Thomas R. Dunlap, "Values for Varmints: Predator Control and Environmental Ideas, 1920–1939," *Pacific Historical Review* 53 (May 1984): 141–61.
114. Don D. Walker, "The Cattle Industry of Utah," 186–89.
115. "Government Forestry Work in Utah," United States Forest Service, USDA Miscellaneous Publication no. 99, March 1931, 1, box 74, folder 3, Charles S. Peterson Papers.
116. Flores, "Zion in Eden," 332.
117. Walter Pace Cottam, "Is Utah Sahara Bound?" *Bulletin of the University of Utah* 37, no. 11 (Salt Lake City: Extension Division, University of Utah, 1947).
118. For examples of blame on "outside sheep," see "Statements Taken from a Conversation between A. M. Robertson, Moab, Utah and A. J. Wagstaff," typescript, October 27, 1935, box 76, folder 13, Charles S. Peterson Papers; J. W. Humphrey, "Memorandum for Files," Ephraim, Utah, March 14, 1941, box 74, folder 3, Charles S. Peterson Papers.
119. Richard V. Francaviglia, *The Mormon Landscape: Existence, Creation, and Perception of a Unique Image in the American West* (New York: AMS Press, 1978), 5–6, 105; Charles S. Peterson, "Smallholding Land Patterns," 5.

"The People Cannot Conquer the River"

1. Donald Worster, *Rivers of Empire: Water, Aridity, and the Growth of the American West*, 1st edition (New York: Pantheon Books, 1985), 5, 11.
2. Donald J. Pisani, *Water and American Government: The Reclamation Bureau, National Water Policy, and the West, 1902–1935* (Berkeley: University of California Press, 2002), 287.
3. Ibid., 289–92.
4. Ibid. 292.
5. Worster states that "one of the most remarkable aspects of Mormon

irrigation was the speed of its conquest over the desert." Worster, *Rivers of Empire*, 74, 77. Pisani recounts Mormon success at irrigation in Salt Lake City and that "the same pattern was repeated in other Mormon communities." Donald J. Pisani, *To Reclaim a Divided West: Water, Law, and Public Policy, 1848–1902* (Albuquerque: University of New Mexico Press, 1992), 78. See Worster, *Rivers of Empire*, 64, where he also states that early western irrigation works were "so primitive in fact that they commonly had to be rebuilt after heavy floods" and that the "vast majority of them failed." He sees 1902 as a watershed moment, however, in which the Newlands Reclamation Act spawned an "era of florescence." It is this point that marks the central tension between his interpretation and Pisani's.

6. Juanita Brooks, "The Water's In!" *Harper's Magazine* 182 (1941): 611.
7. Aaron McArthur, *St. Thomas, Nevada: A History Uncovered* (Reno: University of Nevada Press, 2013).
8. Ibid., 44–45.
9. William Cronon, *Nature's Metropolis: Chicago and the Great West* (New York: Norton, 1991), xix, 56.
10. Martin Reuss and Stephen H. Cutcliff, *The Illusory Boundary: Environment and Technology in History* (Charlottesville: University of Virginia Press, 2010), 2.
11. Daniel R. Headrick, Technology: A World History (New York: Oxford University Press, 2009), 17–24, 34.
12. Worster, *Rivers of Empire*, 50.
13. Ibid., 64.
14. Robert L. Thayer, *Lifeplace: Bioregional Thought and Practice* (Berkeley: University of California Press, 2003), 3.
15. Robert G. Bailey, *Description of the Ecoregions of the United States* (Ogden, UT: Forest Service, U.S. Department of Agriculture, 1978), 54–56.
16. Ibid.
17. Leonard J. Arrington, "The Mormon Cotton Mission in Southern Utah," *Pacific Historical Review* 25, no. 3 (1956): 230.
18. George A. Smith, "History of the Settling of Southern Utah," 223, in *Journal History of the Church of Jesus Christ of Latter-day Saints*, CR 100 137, October 17, 1861, LDS Church History Library, Salt Lake City.
19. G. Lynn Bowler, *Zion on the Muddy: The Story of the Moapa (Logandale, Nevada) Stake* (Springville, UT: Art City Publishing, 2004), 18.
20. Ibid., 18.
21. W. Paul Reeve, *Making Space on the Western Frontier: Mormons, Miners, and Southern Paiutes* (Urbana: University of Illinois Press, 2006), 44–45.
22. Citizens of St. Thomas, Overton, and St. Joseph to Governor and Legislature of Nevada, December 21, 1870; reprinted in Virginia Lani Tobiasson and Georgia May Bagshaw Hall, *Muddy Valley Reflections: 145 Years of Settlement* (Mesa, AZ: Cox Printing, 2010), 16–17.

23. Audrey M. Godfrey, "Colonizing the Muddy River Valley: A New Perspective," *Journal of Mormon History* 22, no. 2 (1996).
24. Ibid., 125–26.
25. Andrew Karl Larson, *Erastus Snow; the Life of a Missionary and Pioneer for the Early Mormon Church, University of Utah Publications in the American West*, vol. 5, (Salt Lake City: University of Utah Press, 1971), 545.
26. Ibid.
27. Merle Frehner, "Memories of Matilda Reber Frehner," written by her children and grandchildren, compiled by Relda Whitney Leavitt, Vivian Frehner, and Frances Frehner Cox, UNLV Special Collections, Las Vegas, Nevada, 5.
28. Isabel Whitney and Ethel Bunker, "Memories of Matilda Reber Frehner," 2.
29. Ibid.
30. Merle Frehner, "Memories of Matilda Reber Frehner," 2.
31. Frances Frehner Cox, "Memories of Matilda Reber Frehner," 1.
32. Juanita Brooks, "The Water's In!," *Harper's Magazine* 182 (1941): 608.
33. Ibid.
34. Ibid.
35. Ibid.
36. Eugene H. Perkins, *The First Mormon Perkins Families: Progenitors and Utah Pioneers of 1847–1852: A Contemporary History of the Ute Perkins Line* (Privately printed, 2008), 72.
37. Andrew Karl Larson, *The Red Hills of November: A Pioneer Biography of Utah's Cotton Town* (Deseret News Press, 1957).
38. Donald Worster, *Dust Bowl: The Southern Plains in the 1930s* (1979; Oxford: Oxford University Press, 2004).
39. Ibid., 62.
40. Ibid.
41. Conevery Bolton Valencius, *The Health of the Country: How Settlers Understood Themselves and Their Land* (New York: Basic Books, 2002), 3.
42. Larson, *The Red Hills of November*, 57–58.
43. Ibid.
44. Andrew Karl Larson, *Erastus Snow*, 364.
45. Larson, *The Red Hills of November*, 58.
46. Erastus Snow to Edward Hunter, March 20, 1864, cited in ibid., 59.
47. Ibid.
48. Ibid.
49. Ibid.
50. Ibid.
51. Lynn A. Rosenvall, "Defunct Mormon Settlements, 1830–1930," in *The Mormon Role in the Settlement of the West*, edited by Richard H. Jackson (Provo, UT: Brigham Young University Press, 1978), 52.
52. Ibid., 60. "These 43 defunct settlements represent 8.7 percent of the 497 communities founded in the United States during the period 1847–1900."

53. Ibid., 73n20. See also data on maximum and minimum stream flows for rivers throughout the Colorado Plateau.
54. Charles S. Peterson, *Take Up Your Mission: Mormon Colonizing Along the Little Colorado River, 1870–1900* (Tucson: University of Arizona Press, 1973), 181. Todd M. Compton, "The Big Washout: The 1862 Flood in Santa Clara," *Utah Historical Quarterly* 77 (Spring 2009): 125.
55. Larson, *Erastus Snow*, 452.
56. Compton, "The Big Washout: The 1862 Flood in Santa Clara," 111–12.
57. Daniel Bonelli to Brigham Young, January 19, 1862, box 28, folder 17, Brigham Young Collection.
58. Daniel Bonelli, Brigham Young office files: General Letters, 1840–1877, a-Bo, 1862, LDS Church History Library.
59. Willard Carroll, manuscript history of Kanab Stake, as cited in Andrew Karl Larson, *I Was Called to Dixie: The Virgin River Basin: Unique Experiences in Mormon Pioneering* (Deseret News Press, 1961), 363–64.
60. Joseph I. Earl, diary, in author's possession. A copy of the diary is located at The Family History Center, Salt Lake City, Utah.
61. Larson, *I Was Called to Dixie*, 367.
62. W. Paul Reeve, "'A Little Oasis in the Desert': Community Building in Hurricane, Utah, 1860–1920," *Utah Historical Quarterly* 62 (Summer 1994).
63. LDS Stake Minutes, St. George, Microfilm CR mh 7836, 1882; as cited in Dorothy Dawn Frehner Thurston, *A River and a Road* (self-published, 1994), 8–9.
64. Ibid., 11.
65. Larson, *I Was Called to Dixie*, 368.
66. Hattie Leavitt Black, "History of Bunkerville, Nevada," typescript; as cited in Thurston, *A River and a Road*, 8.
67. Peterson, *Take up Your Mission*.
68. Ibid., 185.
69. Isabel Whitney and Ethel Bunker, "Memories of Matilda Reber Frehner," 6.
70. Merle Frehner, "Memories of Matilda Reber Frehner," 2.
71. Ibid., 3.
72. Rosella Gubler Frehner, "Memories of Matilda Reber Frehner," 1.
73. Ibid.
74. J. W. Crosby, Letter to the Editor, *Deseret News*, July 31, 1884, accessed January 10, 2017, digitalnewspapers.org/.
75. Ibid.
76. McArthur, *St. Thomas*, 78.
77. Ibid., 79.
78. Ibid., 74, 78.
79. Ibid., 91.

80. Ibid., 92.
81. Ibid., 88–89.
82. *Flood Hazard Analyses, Las Vegas Wash and Tributaries, Clark County, Nevada: Special Report, History of Flooding, Clark County, Nevada 1905–75* (Reno, NV: Soil Conservation Service, 1977), 2.
83. *Las Vegas Age*, vol. 12, no. 41, October 7, 1916; cited in *Flood Hazard Analyses*, 11.
84. *Las Vegas Age*, vol. 17, no. 35, August 27, 1921; cited in *Flood Hazard Analyses*, 14.
85. McArthur, *St. Thomas*, 99.
86. Marc Reisner, *Cadillac Desert: The American West and Its Disappearing Water* (New York: Penguin Books, 1986), 126.
87. Ibid., 126.
88. Worster, *Under Western Skies*, 68.
89. Ibid., 73.
90. McArthur, *St. Thomas*, 109.
91. Ibid., 114.
92. Vivian Frehner, "Memories of Matilda Reber Frehner," 8.
93. Isabel Whitney and Ethel Bunker, "Memories of Matilda Reber Frehner," 6.
94. Vivian Frehner, "Memories of Matilda Reber Frehner," 9.
95. Ibid.
96. McArthur, *St. Thomas*, 115.
97. USGS, "Flooding and Streamflow in Utah during Water Year 2005," accessed July 1, 2016, webcache.googleusercontent.com/search?q=-cache:toaStD4bjOIJ: https://pubs.usgs.gov/fs/2006/3085/PDF/FS2006-3085.pdf+&cd=2&hl=en&ct=clnk&gl=us.
98. Jared Farmer, *Glen Canyon Dammed: Inventing Lake Powell and the Canyon Country* (Tucson: University of Arizona Press, 1999), xxvi.
99. Ibid.

"There Are Millions of Acres in Our State"

1. Thomas Jefferson, *Notes on the State of Virginia*, edited by William Peden (Chapel Hill: University of North Carolina Press, 1954), 164—65.
2. *Journal of Discourses*, vol. 5 (Liverpool: Latter-day Saints' Book Depot, 1858), 164 (hereafter referred to as *JD*); *108th Semi-Annual Conference of the Church of Jesus Christ of Latter-day Saints* (Salt Lake City: Church of Jesus Christ of Latter-day Saints, 1937), 66; *93rd Semi-Annual Conference of the Church of Jesus Christ of Latter-day Saints* (Salt Lake City: Church of Jesus Christ of Latter-day Saints, 1923), 7.
3. Excerpts from the Autobiography of William Farrington Cahoon, in Stella Shurtleff and Brent Farrington Cahoon, eds. *Reynolds Cahoon and*

His Stalwart Sons (Salt Lake City: Paragon Press, 1960), accessed July 26, 2016, boap.org/LDS/Early-Saints/WFCahoon.html; Nancy Naomi Alexander Tracy, "Autobiography," Harold B. Lee Library, Brigham Young University, Provo, Utah, accessed July 26, 2016, boap.org/LDS/Early-Saints/NTracy.html; Edward Phillips, Autobiography, accessed July 26, 2016, boap.org/LDS/Early-Saints/EPhillips.html; *JD* 11:165; William B. Smart, comp. and ed., *Mormonism's Last Colonizer: The Journals of William H. Smart* (Logan: Utah State University Press, 2008), 327 (October 21, 1900).

4. *JD* 12:288; *JD* 11:165. For a discussion of Mormon beliefs regarding divine modification of the climate, see Richard H. Jackson, "The Mormon Experience: The Plains as Sinai, the Great Salt Lake as the Dead Sea, and the Great Basin as Desert-cum-Promised Land," *Journal of Historical Geography* 18 (January 1992): 52–54.
5. *80th Semi-Annual Conference of the Church of Jesus Christ of Latter-day Saints* (Salt Lake City: Deseret News, 1909), 24; Heber M. Hall, "The Impact of Man on the Vegetation and Soil of the Upper Valley Allotment, Garfield County, Utah" (master's thesis, University of Utah, 1954), appendix, 125.
6. Donald H. Dyal, "The Agrarian Values of Mormonism: A Touch of the Mountain Sod" (PhD diss., Texas A&M University, 1980), examines the doctrinal roots and historical trajectory of Mormon agrarianism. Rebecca Andersen, "Between Mountain and Lake: An Urban Mormon Country" (PhD diss., Arizona State University, 2015), traces the development in scholarly literature of "Mormon Country's traditional agrarian communal image" (2) while arguing that Mormon country also possesses strong urban and suburban traditions. She briefly discusses church leaders' concerns regarding the urban tilt of many Mormons early in the twentieth century. Ethan R. Yorgason, *Transformation of the Mormon Culture Region* (Urbana: University of Illinois Press, 2003), traces shifts in Mormonism as the Latter-day Saints moved beyond a communal, agrarian identity. For sociological studies of the rural Mormon village, see Howard M. Bahr, *Saints Observed: Studies of Mormon Village Life, 1850–2005* (Salt Lake City: University of Utah Press, 2014); and Bahr, ed., *Four Classic Mormon Village Studies* (Salt Lake City: University of Utah Press, 2014). Studies of the Mormon landscape include Thomas Carter, *Building Zion: The Material World of Mormon Settlement* (Minneapolis: University of Minnesota Press, 2014); Chad F. Emmett, "The Evolving Mormon Landscape of the 20th Century," *Geography, Culture and Change in the Mormon West: 1845–2003*, edited by Richard H. Jackson and Mark W. Jackson, 48–65 (Jacksonville, AL: National Council for Geographic Education, 2003); Richard Francaviglia, *The Mormon Landscape: Existence, Creation, and Perception of a Unique Image in the American West* (New

York: AMS Press, 1978); Richard H. Jackson, "Mormon Wests: The Creation and Evolution of an American Region," *Western Places, American Myths: How We Think About the West*, edited by Gary J. Havsladen, 135–65 (Reno and Las Vegas: University of Nevada Press, 2003); D. W. Meinig, "The Mormon Culture Region: Strategies and Patterns in the Geography of the American West, 1847–1964," *Annals of the Association of American Geographers* 55 (June 1965), 191–220; and Charles S. Peterson, "Imprint of Agricultural Systems on the Utah Landscape," in *The Mormon Role in the Settlement of the West*, edited by Richard H. Jackson, 21–106 (Provo: Brigham Young University Press, 1978).

7. *69th Semi-Annual Conference of the Church of Jesus Christ of Latter-day Saints* (Salt Lake City: Deseret News Company, 1898, 5.
8. Richard D. Poll, et al., eds., *Utah's History* (Provo: Brigham Young University Press, 1978), 690; *80th Semi-Annual Conference of the Church of Jesus Christ of Latter-day Saints* (Salt Lake City: Deseret News, 1910), 70.
9. *74th Semi-Annual Conference of the Church of Jesus Christ of Latter-day Saints* (Salt Lake City: Deseret News, 1903), 7–9; *80th Semi-Annual Conference*, 22; *81st Semi-Annual Conference*, 70–71; Ibid., 35. For a discussion of the Enlarged Homestead Act, see Stanford J. Layton, *To No Privileged Class: The Rationalization of Homesteading and Rural Life in the Early Twentieth-Century American West* (Provo: Charles Redd Center for Western Studies, 1988), 21–35.
10. U.S. Census Bureau, "Urban and Rural Population: 1900–1990" (October 1995), accessed July 27, 2016, census.gov/population/censusdata/urpop0090.txt. On the country life and the back-to-the-land movements, see Layton, 37–59, and Dona Brown, *Back to the Land: The Enduring Dream of Self Sufficiency in Modern America* (Madison: University of Wisconsin Press, 2011).
11. *81st Semi-Annual Conference*, 35–36.
12. U.S. Bureau of the Census, *United States Census of Agriculture, 1925*, part 3, *The Western States* (Washington, DC: Government Printing Office, 1927), 315–41. Many who were not Mormons also established new farms in the Beehive State over the decade. See Marshall E. Bowen, "The Russian Molokans of Park Valley," *Utah Historical Quarterly* 83 (Summer 2015): 160–79; Robert Alan Goldberg, *Back to the Soil: The Jewish Farmers of Clarion, Utah, and Their World* (Salt Lake City: University of Utah Press, 1986); Huchel, 399.
13. William B. Smart, *Mormonism's Last Colonizer: The Life and Times of William H. Smart* (Logan: Utah State University Press, 2008), 316; Andrew Jenson, *Encyclopedic History of the Church of Jesus Christ of Latter-day Saints* (Salt Lake City: Church of Jesus Christ of Latter-day Saints, 1941), 57–58.
14. Roger Walker "The Delta Project: Utah's Successful Carey Act Project,"

4, accessed July 27, 2016, waterhistory.org/histories/delta/delta.pdf; Leonard J. Arrington, "Taming the Turbulent Sevier: A Story of Mormon Desert Conquest," *Western Humanities Review* 5 (Autumn 1951): 400–403; Matthew C. Godfrey, *Religion, Politics and Sugar: The Mormon Church, The Federal Government and the Utah-Idaho Sugar Company, 1907–1921* (Logan: Utah State University Press, 2007), 73–75; Charles S. Peterson and Brian Q. Cannon, *The Awkward State of Utah: Coming of Age in the Nation, 1896–1945* (Salt Lake City: University of Utah Press, 2015), 68.

15. W. Paul Reeve, "A Little Oasis in the Desert: Community Building in Hurricane, Utah, 1860–1930" (master's thesis, BYU, 1994), 56–64; Smart, *Mormonism's Last Colonizer*, 158; Brian Quayle Cannon, "Remaking the Agrarian Dream: The New Deal's Rural Resettlement Program in Utah" (master's thesis, Utah State University, 1986), 158.
16. John Bennion, "The Chimerical Desert," *Brigham Young University Studies* 32 (Summer 1992): 31; Cannon, "Remaking the Agrarian Dream," 177; Sharon Haddock, "Homestead Ruins Tell Story of Benmore," *Deseret News*, June 3, 2009; *Box Elder News*, June 29, 1905.
17. Smart, *Mormonism's Last Colonizer*, 221, 237; Daughters of the Utah Pioneers, *Builders of Utah: A Centennial History of Uintah County, 1872 to 1947* (Springville: Art City Publishing, 1947), 155; Marshall E. Bowen, *Utah People in the Nevada Desert: Homestead and Community on a Twentieth-Century Farmers' Frontier* (Logan: Utah State University Press, 1994), 66; J. Carlos Lambert, *The Metropolis Reclamation Project*, bulletin no. 107, University of Nevada Experiment Station (Carson City: Nevada State Printing Office, 1925), 8.
18. Peterson, "Imprint of Agricultural Systems on the Utah Landscape," 100, 102; U.S. Census Bureau, *Report on the Statistics of Agriculture in the United States at the Eleventh Census: 1890* (Washington, DC: GPO, 1895), 108; U.S. Census Bureau, *United States Census of Agriculture, 1925: Reports for States*, part 3, *The Western States* (Washington, DC: GPO, 1927), 316, 320.
19. U.S. Census Bureau, *Fourteenth Census of the United States Taken in the Year 1920*, vol. unknown, *Agriculture* (Washington, DC: GPO, 1922), 484–85; U.S. Census Bureau, *United States Census of Agriculture, 1925*, 320.
20. U.S. Census Bureau, *United States Census of Agriculture, 1935*, vol. 3, *General Report*, (Washington, DC: GPO, 1936), 18; Conrad Taeuber, *Statement on Farm Population Trends*, presented before the Senate Committee on Education and Labor, May 6, 1940 (Washington, DC: Bureau of Agricultural Economics, 1940), 3–5. During the 1920s Utah's population declined sharply in several rural counties, reflecting the slump in agriculture and mining. Losses were greatest in Piute County (-29.4 percent), Juab County (-12.8 percent), Duchesne County (-9.1 percent),

Sanpete County (-8.5 percent), Box Elder County (-5.2 percent), and Emery County (-5 percent). See U.S. Census Bureau, *Fifteenth Census of the United States, 1930*, vol. 1, *Population* (Washington, DC: GPO, 1931), 1099–1104.

21. *93rd Semi-Annual Conference of the Church of Jesus Christ of Latter-day Saints* (Salt Lake City: Church of Jesus Christ of Latter-day Saints, 1923), 146–47, 158; *97th Semi-Annual Conference of the Church of Jesus Christ of Latter-day Saints* (Salt Lake City: Church of Jesus Christ of Latter-day Saints, 1927), 162.
22. Merrill Kay Ridd, "Influences of Soil and Water Conditions on Agricultural Development in the Delta Area, Utah" (PhD diss., Northwestern University, 1963), 117–18; *Millard County Chronicle*, June 28, 1928; J. Howard Maughan, "A Resume of Community Settlement in Washington County, Utah" (1935), 6–7, land-use folder, Agriculture Planning Board reports, series 1165, Utah State Archives, Salt Lake City.
23. *93rd Semi-Annual Conference*, 147; Sara Gregg, *Managing the Mountains: Land Use Planning, the New Deal and the Creation of a Federal Landscape in Appalachia* (New Haven: Yale University Press, 2010), 92, 98.
24. Thomas L. Martin, "A Depleted Soil Means a Depleted Citizenship," *Improvement Era*, December 1938, 118–21.
25. Economic Research Service, *The Land Utilization Program 1934 to 1964*, Agricultural Report no. 85 (Washington, DC: U.S. Department of Agriculture, n.d.), 4–13; Jess Gilbert, *Planning Democracy: Agrarian Intellectuals and the Intended New Deal* (New Haven: Yale University Press, 2015), 107.
26. *Deseret News*, June 5, 1934.
27. *Logan Herald Journal*, May 4, 1949; *Deseret News*, May 31, 1943, July 12, 1957; *Salt Lake Tribune*, June 1, 1943; *Roosevelt Standard*, September 27, 1934; Leonard J. Arrington and Davis Bitton, *The Mormon Experience: A History of the Latter-day Saints* (New York: Knopf, 1979), 315; William R. Palmer, "The Pioneering Mormon," *Improvement Era*, August 1942.
28. A. F. Bracken, "Utah Report on the Extent and Character of Desirable Adjustment in Rural Land-Use and Settlement Areas," 1934, 1, photocopy in author's files and in box 122, Charles S. Peterson Papers, MS B 1608, Utah State Historical Society, Salt Lake City; Aaron F. Bracken, "Land Utilization," 1935, 12, box 122, folder 3, Peterson Papers. See also Aaron F. Bracken, "State Report on Land-Use Study for Utah," March 20, 1935, box 122, folder 9, Peterson Papers.
29. Howard Maughan, "Continuation of Study of the Extent of Desirable Major Land-Use Adjustments and Areas Suitable for Settlement," 1936, box 122, folder 6, Peterson Papers.
30. Maughan, "Continuation of Study," 7, 14, 16, 24; Bracken, "Land Utilization," 5.

31. Bracken, "Utah Report," 17–19, 23–25.
32. Bracken, "Land Utilization," 6; Maughan, "Continuation of Study," 1, 21, 33–35, 37–38.
33. Bracken, "Land Utilization," 13–15, 24–25, Maughan, "Continuation of Study," 47–48.
34. Brian Q. Cannon, "Remaking the Agrarian Dream: The New Deal's Rural Resettlement Program in Utah" (master's thesis, Utah State University, 1986), 136–93; Economic Research Service, *Land Utilization Program*, v.
35. Resolution of James L. Neilson and Bert Burraston accompanying William H. King to Rexford G. Tugwell, January 21, 1936, box 490, records group 96, National Archives, College Park, MD; Francis Mortensen, interview by author, August 21, 1985, notes in author's possession; *122nd Semi-Annual Conference of the Church of Jesus Christ of Latter-day Saints* (Salt Lake City: Church of Jesus Christ of Latter-day Saints, 1952), 33.
36. 2 Nephi 1:9; Mabel W. Nielsen and Audrie C. Ford, *Johns Valley: The Way We Saw it* (Springville, UT: Art City Publishing Co., 1971), 70, 103; Interview with Reed Beebe by author, August 10, 1985, notes in author's possession.
37. The abandoned land included farms in Hamlin Valley and in the Escalante Valley between Beryl and Lund, Pahvant Valley farms at the nether end of canals near Abraham and Sugarville, and the Beaver Bottoms north of Milford.
38. U.S. Census Bureau, *United States Census of Agriculture: 1954*, vol. 1, *Counties and State Economic Areas*, part 31, *Utah and Nevada* (Washington, DC: GPO, 1946), 2; U.S. Census Bureau, *Sixteenth Census of the United States: 1940, Population*, vol. 1, *Number of Inhabitants* (Washington, DC: GPO, 1942), 1080–84; Not all of the land removed from cultivation was abandoned, though. Between 1933 and 1937 Utah farmers signed 4,752 corn contracts, 18,378 wheat contracts, and 18,639 sugar beet contracts with the New Deal's Agricultural Adjustment Administration to fallow some of their fields. Peterson and Cannon, *Awkward State of Utah*, 299–300.
39. Bowen, *Utah People*, 93; Smart, *Mormonism's Last Colonizer: Journals*, July 7, 1919; Smart, *Mormonism's Last Colonizer: Life and Times*, 300.
40. *111th Semi-Annual Conference of the Church of Jesus Christ of Latter-day Saints* (Salt Lake City: Church of Jesus Christ of Latter-day Saints, 1940), 9.
41. *108th Semi-Annual Conference*, 125; *106th Semi-Annual Conference of the Church of Jesus Christ of Latter-day Saints* (Salt Lake City: Church of Jesus Christ of Latter-day Saints, 1936), 59; *108th Semi-Annual Conference*, 91; C. Orval Stott, "The Agricultural Advisory Committee of the Church Welfare Program," *Improvement Era*, October 1940, 529; C. Orval Stott, "Agricultural Opportunities," *Improvement Era*, November 1940, 650,

678; *The Basin City War: A Place to Grow, 1958–1991* (Basin City, WA: Basin City Ward History Committee, 1991); Brian Q. Cannon, *Reopening the Frontier: Homesteading in the Modern West* (Lawrence: University Press of Kansas, 2009).

42. U.S. Census Bureau, *Historical Statistics of the United States, Colonial Times to 1970*, part 1 (Washington, DC: GPO, 1975), 459; U.S. Census Bureau, *U.S. Census of Agriculture:* 1959, vol. 1, part 44, *Utah* (WDC: GPO, 1961), 3.

43. *106th Semi-Annual Conference*, 59; *104th Semi-Annual Conference of the Church of Jesus Christ of Latter-day Saints* (Salt Lake City: Church of Jesus Christ of Latter-day Saints, 1933), 32; *108th Semi-Annual Conference*, 64, 66; *106th Semi-Annual Conference*, 159–60; *120th Semi-Annual Conference of the Church of Jesus Christ of Latter-day Saints* (Salt Lake City: Church of Jesus Christ of Latter-day Saints, 1945), 63.

44. *123rd Semi-Annual Conference of the Church of Jesus Christ of Latter-day Saints* (Salt Lake City: Church of Jesus Christ of Latter-day Saints, 1952), 70; *124th Semi-Annual Conference of the Church of Jesus Christ of Latter-day Saints* (Salt Lake City: Church of Jesus Christ of Latter-day Saints, 1953), 121; Address before the Pennsylvania Millers' and Feed Dealers' Association, Reading, PA, September 25, 1956, Benson Agricultural Speeches, Microfilm no. 3, Ezra Taft Benson Papers, MS 8462, LDS Church History Library, Salt Lake City, Utah.

"The Prophet Said to Plant a Garden"

1. Michelle Obama, *American Grown: The Story of the White House Kitchen Garden and Gardens across America* (New York: Crown, 2012), 9, 28–29; *National Gardening Association Special Report: Garden to Table; A 5-Year Look at Food Gardening in America* (2014). The Carters had a small herb garden and the Clintons grew tomatoes in potted plants, but the Obamas' garden was the first in-ground "proper" garden in sixty-five years.

2. Terrie Lynn Bittner, "Plant a Garden," *LDS Blogs*, March 20, 2009, accessed November 2, 2015, http://ldsblogs.com/1841/plant-a-garden.

3. Plat of the City of Zion, circa early June–June 25, 1833; in Gerrit J. Dirkmaat et al., eds., *Joseph Smith Papers, Documents*, vol. 3: February 1833–March 1834 (Salt Lake City, UT: Church Historian's Press, 2014), 121–30. Though the June 1833 plat of Zion is irregular in the orientation of blocks and the number of lots per block, the half-acre size and alternating pattern are clarified in the August 1833 plat for Kirtland, Ohio. See Plat of Kirtland, Ohio, not before August 2, 1833, in *Joseph Smith Papers, Documents*, vol. 3, 208–21.

4. Richard V. Francaviglia, *The Mormon Landscape: Existence, Creation, and Perception of a Unique Image in the American West* (New York: AMS Press, 1978), 81.

5. Elizabeth Cumming to Sarah Cummings Wallace, June 17, 1858, in Ray R. Canning and Beverly Beeton, eds., *The Genteel Gentile: Letters of Elizabeth Cumming, 1857–1858* (Salt Lake City: Tanner Trust, 1977), 77.
6. Mark Twain, *Roughing It* (Hartford, CT: American Publishing Co., 1872), 109.
7. John Muir, *Steep Trails* (Boston: Houghton Mifflin, 1918), 106–7.
8. Wallace Stegner, *Mormon Country* (Lincoln, NE: Bison Books, 1981), 21.
9. Joseph Young, April 8, 1857, in *Journal of Discourses*, 6:241–42. The *Journal of Discourses* may be an accurate reflection of what was actually said by an individual, but it is also possible that an individual's words were embellished or enhanced before publication.
10. Brigham Young, April 28, 1872, in *Journal of Discourses*, 15:20.
11. Orson Pratt, May 18, 1877, in *Journal of Discourses*, 19:32.
12. Thomas Jefferson, *Notes on the State of Virginia* (London: John Stockdale, 1787), 274. For an in-depth study of how agrarianism shaped American identity, see Conevery Bolton Valencius, *The Health of the Country: How American Settlers Understood Themselves and Their Land* (New York: Basic Books, 2002), 191–208.
13. Emily Anne Brooksby Wheeler, "The Solitary Place Shall Be Glad for Them: Understanding and Treating Mormon Pioneer Gardens as Cultural Landscapes" (master's thesis, Utah State University, 2011), 37–38.
14. Donald H. Dyal, "Mormon Pursuit of the Agrarian Ideal," *Agricultural History* 63 (Fall 1989), 19.
15. U.S. Census, 1910, 1920, 1925, 1930, 1935, 1940, 1945, 1950, 1954, 1959, 1964, 1969, 1974, United States Department of Agriculture Census of Agriculture Historical Archive, Albert R. Mann Library, Cornell University, agcensus.mannlib.cornell.edu/AgCensus.
16. John A. Widtsoe, *Report of the Semi-Annual Conference of the Church of Jesus Christ of Latter-day Saints*, 6–8, 49, October 1944 (Salt Lake City: Church of Jesus Christ of Latter-day Saints) (hereafter *Conference Report*).
17. Heber J. Grant, *Conference Report*, April 1923, 7–9.
18. J. Reuben Clark, *Conference Report*, October 1940, 10.
19. "Applying the Principles of Church Welfare," April 1979. The importance of agriculture and home food production in Kimball's memory is illustrated in his memoir *One Silent Sleepless Night* (Salt Lake City: Bookcraft, 1975).
20. Caroline E. Miner and Edward L. Kimball, *Camilla* (Salt Lake City: Deseret Book, 1980), 102.
21. Carma Wadley, "A Young, Vigorous Lifestyle," *Church News*, January 6, 1979, 10, quoting Arthur Haycock.
22. Miner and Kimball, *Camilla*, 154.
23. Edward L. Kimball, *Lengthen Your Stride: The Presidency of Spencer W. Kimball*, working draft, *Spencer W. Kimball CD Library* (Provo, UT: BYU Studies, 2006), chapter 10, page 2, note 8.

24. For an overview of environmental concerns in this period, see Phillip Shabecoff, *A Fierce Green Fire: The American Environmental Movement* (New York: Hill and Wang, 1993), 111–202.
25. For example, the nonprofit organization LDS Earth Stewardship, which advocates for environmental issues, carefully avoids identifying itself with environmentalism in order to increase credibility among practicing Latter-day Saints.
26. Paul R. Ehrlich, *The Population Bomb* (New York: Ballantine, 1968); see also Shabecoff, *A Fierce Green Fire*, 94–97.
27. "Two Apostles of Control," *Life*, April 17, 1970, 33.
28. Doctrine and Covenants 104:17.
29. *Conference Report*, April 1974, 178; see also Bernard P. Brockbank, "Today Millions Are Waiting," General Conference, April 1975, https://www.lds.org/general-conference/1975/04/today-millions-are-waiting; and Spencer W. Kimball, "Voices of the Past, of the Present, and of the Future," General Conference, April 1971, https://www.lds.org/general-conference/1971/04/voices-of-the-past-of-the-present-of-the-future.
30. Phillip F. Low, "Realities of the Population Explosion," *Ensign*, May 1971, https://www.lds.org/ensign/1971/05/realities-of-the-population-explosion.
31. Matthew Bowman, *The Mormon People: The Making of an American Faith* (New York: Random House, 2012), 207–8.
32. Jules B. Billard, "The Revolution in American Agriculture," *National Geographic*, February 1970: 147–85, copy in Spencer W. Kimball Collection, Church History Library, Salt Lake City.
33. Dale L. Schurter, "Are We Bringing a Curse on Our Land?" *Plain Truth*, March 1971: 33–35, copy in Spencer W. Kimball Collection, Church History Library, Salt Lake City.
34. Spencer W. Kimball, "Guidelines to Carry Forth the Work of God in Cleanliness," General Conference, April 1974, https://www.lds.org/general-conference/1974/04/guidelines-to-carry-forth-the-work-of-god-in-cleanliness?lang=eng.
35. J. Reuben Clark, address, April 6, 1936, quoted in Henry D. Taylor, "The Church Welfare Plan" (n.p., 1984), copy at Church History Library, Salt Lake City.
36. "An Important Message on Relief," *Deseret News*, April 7, 1836.
37. Dyal, "Mormon Pursuit of the Agrarian Ideal," 31–32.
38. See, for example, *Wel-Fair* (Chatsworth, CA: Chatsworth LDS Stake, 1974), copy in Church History Library, Salt Lake City.
39. Mark Fiege, *The Republic of Nature: An Environmental History of the United States* (Seattle: University of Washington Press, 2012), 361.

40. Raymond F. Hopkins and Donald J. Puchala, "Perspectives on the International Relations of Food," *International Organization* 32, no. 3 (Summer 1978), 583–86.
41. *Conference Report*, April 1974, 176.
42. Ibid., 184–85.
43. "First Presidency Issues Statement on Beautification," *Church News,* published by *Deseret News*, September 28, 1974, 11. This was not the first church-targeted cleanup campaign. As noted previously, J. Reuben Clark spoke disapprovingly of unsightly yards, and in 1967 in northern Utah's Cache Valley, Utah State University Extension Services sent a letter to local LDS ecclesiastical leaders, urging them to work together with their congregations to clean up "weeds, clutter, and debris" in front and backyards. A. Stark, memo on "ward respectability," May 17, 1967, copy in Church History Library.
44. Kimball, *Lengthen Your Stride*, chapter 10, page 3, *Spencer W. Kimball CD Library*.
45. "God Will Not Be Mocked," General Conference, October 1974, https://www.lds.org/general-conference/1974/10/god-will-not-be-mocked. In 1978, historian Richard Francaviglia warned of the cultural heritage the Mormon West stood to lose if fences and structures were torn down. Richard V. Francaviglia, "The Passing Mormon Village," *Landscape* 22, no. 2 (Spring 1978), 40–47.
46. "Why Call Me Lord, Lord, and Do Not the Things Which I Say?" General Conference, April 1975, https://www.lds.org/general-conference/1975/04/why-call-me-lord-lord-and-do-not-the-things-which-i-say.
47. Brigham Young, June 10, 1860, in *Journal of Discourses*, 8:79. Similarly, Joseph Smith gave a revelation in 1832 declaring, "There remaineth a scorge and a Judgment to be poured out upon the children of Zion for shall the children of the kingdom pollute my holy land verily Verily I say unto you na[y]." Revelation, September 22–23, 1832 [D&C 84], in Matthew C. Godfrey, et al., *Joseph Smith Papers, Documents,* vol. 2: July 1831–January 1833 (Salt Lake City: Church Historian's Press, 2013), 298.
48. "Stand by Your Guns," in Edward Kimball collection, box 4, folder 2, BYU. This was a talk prepared for the World's Fair Expo in Spokane, Washington, on July 24, 1974, but Kimball ultimately revised his remarks and spoke more on the theme of pioneers. Kimball, *Lengthen Your Stride*, chapter 10, page 9, note 44, *Spencer W. Kimball CD Library*.
49. "Why Call Me Lord, Lord."
50. He did not mention it at the October 1980 conference, or in his final General Conference address in October 1982, which was read by his secretary, Arthur Haycock. Kimball did not speak in October 1981, the only time since his call as an apostle in 1943 that he missed General Conference. He was suffering from internal bleeding in the stomach.

Kimball, *Lengthen Your Stride*, chapter 38, page 4, *Spencer W. Kimball CD Library*.

51. "The Stone Cut without Hands," General Conference, April 1976, https://www.lds.org/general-conference/1976/04/the-stone-cut-without-hands.
52. Ibid.
53. "The Lord Expects His Saints to Follow the Commandments," General Conference, April 1977, https://www.lds.org/general-conference/1977/04/the-lord-expects-his-saints-to-follow-the-commandments.
54. *Conference Report*, April 1974, 179. In the same time period, Kimball also signaled that there was a limit to the relative importance of things like fixing up old barns. At a devotional address at Brigham Young University in September 1978, he made clear that there were weightier issues, like personal appearance and one's personal status before God. "Barns and gardens and woodsheds matter," he told the students, "but people matter more… . The appearance of that eternal soul with all of its outward manifestations surely takes precedence with us over the important matter of newly painted homes and barns and carefully repaired fences."
55. "Home Vegetable Gardening Statistics," *Mother Earth News*, September–October 1986.
56. See Laura Lawson, *City Bountiful: A Century of Community Gardening in America* (Berkeley: University of California Press, 2005), 213–18.
57. See Joel Garreau, *Edge City: Life on the New Frontier* (New York: Basic Books, 1992), 3–12; Glen M. Leonard, *History of Davis County* (Salt Lake City: Utah State Historical Society, 1999), 367–71; Tom Lewis, *Divided Highways: Building the Interstate Highways, Transforming American Life* (Ithaca, NY: Cornell University Press, 2013), xiii–xxvii. For analysis of the racial aspects of suburbanization, see William J. Collins and Robert A. Margo, "Race and Home Ownership from the End of the Civil War to the Present," *American Economic Review* 101, no. 3 (2011): 355–59.
58. Spencer W. Kimball Collection, box 99, folder 5, Church History Library, Salt Lake City. For instance, he noted "flower pots in Scandinavia" in one set of lecture notes.
59. "Family Preparedness," General Conference, April 1976, https://www.lds.org/general-conference/1976/04/family-preparedness.
60. "A Report and a Challenge," General Conference, October 1976, https://www.lds.org/general-conference/1976/10/a-report-and-a-challenge.
61. *Conference Report*, October 1974, 172–73.
62. "The Lord Expects His Saints to Follow the Commandments," General Conference, April 1977, https://www.lds.org/general-conference/1977/04/the-lord-expects-his-saints-to-follow-the-commandments.
63. "The True Way of Life and Salvation," General Conference, April 1978, https://www.lds.org/general-conference/1978/04/the-true-way-of-life-and-salvation.

64. "We Are on the Lord's Errand," General Conference, April 1981, https://www.lds.org/general-conference/1981/04/we-are-on-the-lords-errand. Kimball also spoke of the spiritual significance of home food preservation: "In the National Geographic magazine last month, we clipped a picture of a woman bringing bottled and canned fruit to her storage room, which was full of the products of her labors and was neat and tidy. That's the way the Lord planned that we should prepare and eat our vegetables." "The Stone Cut without Hands," April 1976.
65. "The Fruit of Our Welfare Services Labors," General Conference, October 1978, https://www.lds.org/general-conference/1978/10/the-fruit-of-our-welfare-services-labors.
66. "The Stone Cut without Hands," April 1976.
67. Ibid.
68. "Family Preparedness," April 1976.
69. *Conference Report*, October 1974, 172–73.
70. "Welfare Services: The Gospel in Action," General Conference, October 1977, https://www.lds.org/general-conference/1977/10/welfare-services-the-gospel-in-action.
71. "Listen to the Prophets," General Conference, April 1978, https://www.lds.org/general-conference/1978/04/listen-to-the-prophets. A few years earlier, he had discounted the importance of gardening making economic sense, saying in April 1974, "I remember when the sisters used to say, 'Well, but we could buy it at the store a lot cheaper than we can put it up.' But that isn't quite the answer." Then, positing a scenario where the grocery supply chain broke down, he concluded, "People would get awfully hungry after two weeks were over." *Conference Report*, April 1974, 185.
72. Barbara W. Winder, "Draw Near unto Me through Obedience," *Ensign*, November 1985, https://www.lds.org/general-conference/1985/10/draw-near-unto-me-through-obedience.
73. James E. Faust, "The Responsibility for Welfare Rests with Me and My Family," *Ensign*, May 1986, https://www.lds.org/general-conference/1986/04/the-responsibility-for-welfare-rests-with-me-and-my-family.
74. For examples of gardening as an activity of dutiful Saints, see Joe J. Christensen, "Greed, Selfishness, and Overindulgence," General Conference, April 1999, https://www.lds.org/general-conference/1999/04/greed-selfishness-and-overindulgence; Aileen H. Clyde, "What Is Relief Society For?" General Conference, October 1995, https://www.lds.org/general-conference/1995/10/what-is-relief-society-for; and Earl C. Tingey, "For the Strength of Youth," General Conference, April 2004, https://www.lds.org/general-conference/2004/04/for-the-strength-of-youth?lang=eng. For gardens as symbols, see D. Todd Christofferson, "As Many as I Love, I Rebuke and Chasten," General Conference, April 2011,

https://www.lds.org/general-conference/2011/04/as-many-as-i-love-i-rebuke-and-chasten; and James E. Faust, "Of Seeds and Soils," General Conference, October 1999, https://www.lds.org/general-conference/1999/10/of-seeds-and-soils.

75. Mary Jane McAllister Davis, "The Prophet Said to Plant a Garden," *Children's Songbook* (Salt Lake City: Church of Jesus Christ of Latter-day Saints, 1989), 237. It was published first in the *Friend* (the church's children's magazine) in 1982. The children's curriculum, both manuals and music, has a strong environmental component, including much on the Creation.
76. "White House Says Prepare for the Upcoming Hurricane Season," message board, June 2, 2011, survivalistboards.com, accessed 30 October 2015, https://www.survivalistboards.com/showthread.php?s=fdfbd93e7a9d96c89a32dc0fdee113c9&t=172823.
77. Craig D. Galli, "Study Guide: LDS Perspectives on Environmental Stewardship," *Mormon Chronicles*, April 30, 2010, accessed November 5, 2015, http://mormon-chronicles.blogspot.com/2010/04/lds-perspectives-on-environmental.html; Richard D. Stratton, *Kindness to Animals and Caring for the Earth: Selections from the Sermons and Writings of Latter-day Saint Church Leaders* (Portland, OR: Inkwater, 2004).
78. Alison Moore Smith, "The Prophet Said to Plant a Garden. Right," *Mormon Momma: Crossing the Plains in the Modern World*, May 6, 2013, accessed November 5, 2015, http://mormonmomma.com/the-prophet-said-to-plant-a-garden-right; Amy, "Let's Do These Things because They Are Right," *Gotta Wanna Needa Getta Prepared*, December 5, 2010, accessed November 5, 2015, http://gottawannaneedagettaprepared.blogspot.com/2010/12/lets-do-these-things-because-they-are.html; Michelle, "A Bounteous Harvest," *Abundant Recompense*, September 14, 2014, accessed May 25, 2016, http://nateandmich.blogspot.com/2014/09/a-bounteous-harvest.html.
79. "The Foundations of Righteousness," General Conference, October 1977, https://www.lds.org/general-conference/1977/10/the-foundations-of-righteousness.

"For the Strength of the Hills"

The title quotation is from Felicia D. Hemans, "For the Strength of the Hills," in *Hymns of the Church of Jesus Christ of Latter-day Saints* (Salt Lake City: Deseret Book Company, 1985), 35.

1. Jared Farmer, *On Zion's Mount: Mormons, Indians, and the American Landscape* (Cambridge: Harvard University Press, 2010), 142–74. See also Jared Farmer, "Restoring Greatness to Utah," accessed December 28, 2017, jaredfarmer.net/e-books/.

2. I am indebted to Genevieve Atwood, former head of the Utah Geological Survey, for this description.
3. Richard Burton, *The City of the Saints and across the Rocky Mountains to California*, edited by Fawn Brodie (New York: Alfred A. Knopf, 1963), 217–18.
4. William Cronon, "The Trouble with Wilderness; or, Getting Back to the Wrong Nature," in *Uncommon Ground: Rethinking the Human Place in Nature*, edited by William Cronon (New York: W. W. Norton & Company, 1996), 80.
5. Henri Lefebvre, *The Production of Space*, translated by Donald Nicholson-Smith (Malden, MA: Blackwell Publishing, 1991), 83–85.
6. "The Wasatch Fault," Utah Geological Survey, Public Information Series 40, 1996, 2–3, files.geology.utah.gov/online/pdf/pi-40.pdf; Genevieve Atwood, in discussion with the author, June 26, 2013.
7. Howard Stansbury, *Exploration and Survey of the Valley of the Great Salt Lake of Utah* (Philadelphia: Lippincott, Gramabo and Co., 1852), 105.
8. Genevieve Atwood, "Lake Bonneville," *Utah History Encyclopedia*, accessed June 21, 2013, heritage.utah.gov/history/uhg-place-lake-bonneville.
9. Grove Karl Gilbert, *Lake Bonneville* (Washington: Government Printing Office, 1890), 94, 106.
10. William F. Case, "Why Is the Wasatch Front 'Blessed with the Abundant Sand, Gravel, and Rock that Were So Useful for the 'Olympian' Interstate 15 Project?" *Utah Geological Survey Notes* 33 (August 2001): 10.
11. Atwood, "Lake Bonneville."
12. Journal History (hereafter JH), July 22, 1847.
13. JH, July 24, 1847.
14. JH, July 26, 1847.
15. George A. Smith, "Historical Discourse" *Journal of Discourses* (hereafter *JD*) 13:85. Joseph F. Smith told a similar story in an address given December 3, 1882, suggesting this was an established event within the developing Mormon historical narrative. Joseph F. Smith, "Interest in the Work of God—Faith in the Destiny of the People," *JD* 24:156. See also Ronald W. Walker, "'A Banner is Unfurled': Mormonism's Ensign Peak," *Dialogue: A Journal of Mormon Thought* 26 (Winter 1993): 74. In his travel account, Richard F. Burton likewise repeated this story, suggesting that Joseph appeared to Young while he stood on Ensign Peak contemplating the valley. Richard Burton, *The City of the Saints*, 217–18.
16. JH, July 21, 1849.
17. Kristen Smart Rogers, "William Henry Smart: Uinta Basin Pioneer Leader," *Utah Historical Quarterly* 45 (Winter 1977): 74.
18. Brigham Young, "Management of the Canyons—Paying Debts—Keeping Stores—Material for the Temple," *JD* 1:218.
19. In "The Natural World and the Establishment of Zion, 1831–1833,"

included in this volume, Matthew C. Godfrey explains that early Mormons believed that God required them to redeem the land through ordered and planned agricultural communities, industry, and infrastructure. See also Dan Flores, "Agriculture, Mountain Ecology, and the Land Ethic: Phases of the Environmental History of Utah," in *Working the Range: Essays on the History of Western Land Management and Environment*, edited by John Wunder (Westport, CT: Greenwood Press, 1985), 164.

20. Brigham Young, "The Order of Enoch," *JD* 15:222.
21. Daniel Wells, "Building Up of the Kingdom of God—Home Manufactures," *JD* 9:60–61.
22. Robert Jones Phillips, "Utah Sand and Gravel Industry" (master's thesis, University of Utah, 1956), 36–37.
23. Brigham Young noted in an 1874 address that he was "very much interested of late with regard to the studies and researches of the geologists who have been investigating the geological character of the Rocky Mountain country. Professor Marsh, of Yale College, with a class of his students, has spent, I think, four summers in succession in the practical study of geology in these mountain regions." Marsh and his students had discovered the fossil remains of no less than "fourteen different kinds of horses, varying in height from three to nine feet," evidence Young supplied to substantiate the presence of horses in the Book of Mormon. "Cease to Bring In and Build Up Babylon," *JD* 17:45–46.
24. Brigham Young, "Management of Canyons," *JD* 1:219.
25. Ibid., 1:220.
26. Thomas Alexander, "Stewardship and Enterprise: The LDS Church and the Wasatch Oasis Environment," 345–46. For statements regarding the earth's spiritual creation, see D&C 29:34 (which Alexander quotes in his article); Pearl of Great Price, Moses 3:4, "And every plant of the field before it was in the earth, and every herb of the field before it grew. For I, the Lord God, created all things, of which I have spoken, spiritually, before they were naturally upon the face of the earth." See also Orson Pratt, "Pre-existence, in Spiritual Form, of Man, the Lower Animals and the Earth," *JD* 21:200, jod.mrm.org/21/200.
27. An example of Brigham Young's dislike for prospecting can be found in an address dated June 17, 1877, in which he states, "This chain of mountains has been followed from the north to the south, and its various spurs have been prospected, and what do they find? Just enough to allure them, and to finally lead them from the faith, and at last to make them miserable and poor." Brigham Young, "Trying to Be Saints," *JD* 19:36, jod.mrm.org/19/36.
28. "Pavement History" September 30, 2008, accessed August 14, 2015, pavementinteractive.org/article/pavement-history; Shirley F. Colby, "Development of the Sand and Gravel Industry," Information Circular,

United States Department of the Interior, Bureau of Mines, I. C. 7203, March 1942, 1.

29. Frederick W. Boone Jr., *Sands of Time: A History of Sand and Gravel Operations in Port Jefferson and Nearby Harbors* (Port Jefferson, NY: Three Village Historical Society, 1998), 3.
30. Robert W. Lesley, *History of the Portland Cement Industry in the United States* (Chicago: International Trade Press Inc., 1924), 4, 34, 39, 40.
31. Shirley F. Colby, "Development of the Sand and Gravel Industry," 2. See also National Sand and Gravel Association, "Case Histories: Rehabilitation of Worked-Out Sand and Gravel Deposits" (Washington: The Association, 1961). A significant number of gravel pits profiled in this report date to the 1920s.
32. Floyd W. Parsons, "Introduction," in *History of the Portland Cement Industry in the United States*, by Robert W. Lesley (Chicago: International Trade Press Inc., 1924), 10.
33. Herbert P. Gillette, *Concrete Construction, Methods and Cost* (New York: The Myron C. Clark Publishing Co., 1908), 14.
34. Tim Cooper, *Laying the Foundations: A History and Archaeology of the Trent Valley Sand and Gravel Industry* (York, England: Council for British Archaeology, 2008), 42–43, 45; W. J. Kotsrean, *Prices of Sand and Gravel*, War Industries Board Price Bulletin No. 39 (Washington: Government Printing Office, 1919), 4.
35. G. K. Gilbert, *Pleistocene Lake Bonneville, Ancestral Great Salt Lake, as Described in the Notebooks of G. K. Gilbert, 1875–1880*, edited by Charles B. Hunt, Brigham Young University Geology Studies, volume 29, part 1 (Provo, UT: Brigham Young University, Department of Geology, 1982), 27–28; G. K. Gilbert, *Lake Bonneville* (Washington: United States Geological Survey, 1890), 348–49.
36. See Agriculturalist Arthur Young's description of limekilns in France in the 1780s in *Travels during the Years 1787, 1788, and 1789*, vol. 2, 2nd edition (Bury St. Edumnds: J. Rackham, 1794), 118.
37. Milton R. Hunter, *Beneath Ben Lomond's Peak: A History of Weber County, 1824–1900* (Salt Lake City: The Deseret News Press, 1944), 363–64.
38. Miles P. Romney, "Utah's Cinderella Minerals: The Nonmetalics," *Utah Historical Quarterly* 31 (Summer 1963): 227.
39. "Native Material Being Used in Hotel Utah," *Salt Lake Tribune*, March 22, 1910.
40. Ezra C. Knowlton, *History of Utah Sand and Gravel Products Corporation, 1920–1958* (Salt Lake City: Utah Sand and Gravel Products Corporation, 1959), 10–11.
41. Knowlton, *History of Utah Sand and Gravel Products Corporation*, 2.
42. Ibid., 2, 10–11, 91–92.
43. Linda Sillitoe, *A History of Salt Lake County* (Salt Lake City: Utah State Historical Society and Salt Lake County Commission, 1996), 135.

44. "LDS Church Forms World Relief Setup," *Salt Lake Telegram*, March 22, 1937; "LDS Church Announces Underground Garage Plans," *Salt Lake Telegram*, November 3, 1939; "Speakers Pay Homage at Ryberg Rites," *Salt Lake Telegram*, February 14, 1950.
45. "Gravel Pits and Hillside Holes," *Deseret Evening News*, November 11, 1908.
46. "Ensign Peak," JH, August 8, 1908, https://dcms.lds.org/delivery/DeliveryManagerServlet?dps_pid=IE311231.
47. "Good Forest Suggestions," *The Deseret News*, September 16, 1908.
48. "Boy Scouts Hold Celebration on Ensign Peak," *Deseret News*, July 27, 1916, recorded in JH, July 26, 1916, https://dcms.lds.org/delivery/DeliveryManagerServlet?dps_pid=IE484087.
49. Ronald Walker wrote several excellent articles about Ensign Peak that concern the banner or flag pioneers waved at its top. See Ronald W. Walker, "A Gauge of the Times: Ensign Peak in the Twentieth Century," *Utah Historical Quarterly* 62 (Winter 1994): 4–25; Ronald W. Walker, "'A Banner is Unfurled.'"
50. "Hundreds to Make Pilgrimage to Crown of Historic Mountain Peak," *Deseret News* July 19, 1919. Recorded in JH, July 19, 1919, dcms.lds.org/delivery/DeliveryManagerServlet?dps_pid=IE313445; "Naming of Ensign Peak to Be Celebrated," *Deseret News,* July 24, 1920. Recorded in JH, July 24, 1920, https://dcms.lds.org/delivery/DeliveryManagerServlet?dps_pid=IE467514; "Ensign Stake Holds Sunset Rites on Peak," *Salt Lake Tribune*, July 27, 1938, recorded in JH, July 26, 1938.
51. "Stolen Ensign Peak Marker Is Recovered after 30 Years," *Church News*, October 17, 1992; Ronald W. Walker, "A Gauge of the Times," 17–18.
52. Newspaper article from JH, July 17, 1934.
53. Knowlton, *History of Utah Sand and Gravel Products Corporation*, 17.
54. Ibid., 12, 50.
55. Ibid., 50–52, 70–73.
56. See *Improvement Era* 57, no. 2 (February 1954): 34; *Improvement Era* 69, no. 3 (March 1966): 57.
57. *Census of Population: 1960*, vol. 1, part A, 46–12; *American Fork Citizen*, February 10, 1955; *American Fork Citizen*, January 20, 1955; *American Fork Citizen*, February 25, 1965.
58. Ora Bundy, compiler, *After Victory: Plans for Utah and the Wasatch Front*, part 4 (Salt Lake City: Utah State Department of Publicity and Industrial Development, June 1943), 7–8.
59. Ezra Knowlton, *History of Highway Development in Utah*, 553.
60. Utah Highway Progress Newsletter, vol. 2, no. 15 (March 1959), box 1, series 21106–7, State Road Commission Utah Highway Progress Newsletter and Utah Highways and Byways Newsletters, Utah State Historical Society, Salt Lake City, Utah.
61. "Neighborly Gesture," newspaper article, recorded in JH, June 11, 1952.
62. "Road Up Ensign Peak," *Salt Lake Tribune*, recorded in JH, October

21, 1959; “Development Proposed for Ensign Peak Area,” *Deseret News*, December 28, 1961.

63. “Stolen Ensign Peak Marker Is Recovered after 30 Years,” *Church News*, October 17, 1992.
64. Geoff Biesinger, letter to the editor, *Deseret News*, November 29, 1993.
65. “‘This is the Place Monument’ Top Utah Tourist Attraction,” *Deseret News*, January 16, 1952, recorded in JH, January 16, 1952.
66. “Development Plan Adopted for ‘This is the Place’ Park,” *Deseret News*, May 28, 1958, recorded in JH, May 28, 1958.
67. “Make It Really ‘The Place,’” *Deseret News*, June 15, 1965, recorded in JH, June 15, 1965.
68. “Monument High Tribute to Mormon Heritage,” newspaper editorial, recorded in JH, July 18, 1953.
69. Gordon Springer, “Put a Statue atop Ensign Peak,” *Deseret News*, May 29, 1978.
70. Rita Keetch, “Huge Gravel Pit,” *Salt Lake Tribune*, April 12, 1971.
71. “Why Utah should adopt controls on strip-mining,” *The Deseret News*, December 6, 1974.
72. Richard Van Horn, “Restore Pits,” *The Salt Lake Tribune*, May 1, 1971.
73. “Beck Street Reclamation Framework and Foothill Area Plan” (Salt Lake City and North Salt Lake City, adopted by Salt Lake City, 1999), 16, 77, 79, 81–82, 87, slcdocs.com/Planning/MasterPlansMaps/beck.pdf.
74. “Traffic Tops the List of Concerns for Candidates,” *Deseret News*, November 1, 1995; “Mountain Rezone Plan Is the Pits for Bountiful,” *Deseret News*, December 13, 1995; “It’s Time to Make a Decision about Bountiful Highway,” *Deseret News*, January 8, 1996.
75. “Beck Street’s View Is the Pits, Group Says,” *Deseret News*, December 12, 1996.
76. Genevieve Atwood, telephone interview, June 26, 2013.
77. John Bowman and Eric Jergensen, *Geoantiquity: An Earth Images Foundation Production*, produced by Doug Prose and Diane LaMacchia, 2006.
78. “Beck Street Reclamation Framework and Foothill Area Plan,” 11.
79. “S.L. May Swap to Get Ensign Peak Land,” *Deseret News*, December 24, 1980.
80. R. Scott Lloyd, “Park at Ensign Peak Dedicated,” *Church News*, August 3, 1996.
81. R. Scott Lloyd, “Unfurling of Truth’s Banner on ‘Zion’s Hill’ Commemorated by Hundreds at Annual Hike,” *Church News*, July 31, 1993; R. Scott Lloyd, “Hikers Commemorate Rising of an ‘Ensign to the Nations,’ *Church News*, August 1, 1992.
82. R. Scott Lloyd, “Park at Ensign Peak Dedicated,” *Church News*, August 3, 1996.
83. R. Scott Lloyd, “New Garden Graces Ensign Peak,” *Church News*, August 2, 1997.

84. "Beck Street Reclamation Framework and Foothill Area Plan," 81.
85. Ibid., 66.
86. Zeth C. Myers, "Things of Beauty," *The Salt Lake Tribune*, September 12, 1998.
87. Brandon Lommis, "Future Is Still Uncertain for S.L.'s 'Gravel Mountain,'" *Salt Lake Tribune*, August 17, 1998.
88. Alan Edwards, "Bunyan-Like Staircase May Be Part of Future Beck Street Hillside," *Deseret News*, April 28, 1997.
89. "Beck Street Reclamation Framework and Foothill Area Plan," 4.
90. In Woodruff's journal entry for July 24, 1847, he writes, "Thoughts of pleasant meditation ran in rapid succession through our minds at the anticipation that not many years hence the House of God would be established ... while the valleys would be converted into orchards, vineyards, fields, etc., planted with cities." Wilford Woodruff, *History of His Life and Labors: As Recorded in His Daily Journals*, edited by Matthias F. Cowley (Salt Lake City: The Deseret News, 1909), 313.

EPILOGUE

1. Willis Jenkins, *The Future of Ethics: Sustainability, Social Justice, and Religious Creativity* (Georgetown: Georgetown University Press,), 5. Emphasis added.
2. For a history of recent religious environmental activism in the United States, see Stephen Ellingson, *To Care for the Creation: The Emergence of the Religious Environmental Movement* (Chicago: University of Chicago Press, 2016).
3. See, for example, Mark Stoll's *Inherit the Holy Mountain: Religion and the Rise of American Environmentalism* (Oxford: Oxford University Press, 2015).
4. E. O. Wilson, *The Creation: An Appeal to Save Life on Earth* (New York: Norton, 2006), 8.
5. See lds.org/topics/environmental-stewardship-and-conservation?lang=eng&old=true and mormonnewsroom.org/article/environmental-stewardship-conservation.
6. For an overview of this scholarship, see my essay "Toward a Greener Faith: A Review of Recent Mormon Environmental Scholarship," *Mormon Studies Review* 3 (2016): 85–103.

APPENDIX

This essay was originally a talk given by Elder Marcus B. Nash on April 12, 2013, at the University of Utah's 18th Annual Wallace Stegner Center Symposium. Nash serves as a general authority–seventy for the Church of Jesus Christ of Latter-day Saints and was assigned by the LDS Church to represent the church in the symposium, which was on faith and the environment. The opinions expressed in this essay are his own.

1. Bible Dictionary, "Scripture."
2. Ibid., "Canon."
3. By way of explanation, the scriptural canon of the LDS Church includes both ancient and modern scripture. These scriptures include the Old and New Testaments, accepted by the LDS Church as a collection of inspired writings and revelations by prophets of antiquity in the "Old World"; the Book of Mormon, accepted by the LDS Church as a collection of inspired writings and revelations by prophets of antiquity in the "New World,"; the Doctrine and Covenants, accepted by the LDS Church as a collection of inspired writings and revelations given in modern times through Joseph Smith; and the Pearl of Great Price, also accepted by the LDS Church as a collection of ancient writings and revelations revealed to Joseph Smith in modern times. The LDS Church believes that the canon of scripture is not closed.This essay weaves both ancient and modern scripture together in setting forth what the author understands to be the doctrine of the LDS Church pertaining to the environment.
4. Psalms 8:5.
5. Moses 1:10. See also Hebrews 1:2.
6. Moses 1:31–33, 39.
7. Acts 17:28–29
8. Abraham 3:24–25.
9. Abraham 3:26.
10. 2 Nephi 2:16.
11. Moses 2:31.
12. See generally, the First Presidency and Quorum of the Twelve Apostles of the Church of Jesus Christ of Latter-day Saints, *The Family: A Proclamation to the World* (Salt Lake City: The Church of Jesus Christ of Latter-day Saints, 1995).
13. Doctrine and Covenants 49:16–17. See also Genesis 1:27–28.
14. 1 Nephi 17:36.
15. Moses 3:4–5.
16. Moses 3:9.
17. Moses 3:19.
18. See D&C 77:2.
19. Joseph Fielding Smith, *Doctrines of Salvation: Sermons and Writings of Joseph Fielding Smith,* 3 vols., edited by Bruce R. McConkie (Salt Lake City: Bookcraft, 1960), 1:84.
20. D&C 29:23–25.
21. D&C 49:19–21.
22. D&C 104:13–14.
23. Genesis 1:26.

24. D&C 49:19–21; see also D&C 78:6.
25. D&C 104:17–18.
26. D&C 59:16–20.
27. Moses 7:28.
28. Mormon 8:29–31.
29. Ezra Taft Benson, "Problems Affecting the Domestic Tranquility of Citizens of the United States of America," *Vital Speeches* 42 (February 1 1976): 240.
30. Quoted in Spencer Garvey, "What It Means to Be Green," *New Era* 22 (July 1992): 21, online at www.lds.org.
31. Moses 7:48–49.
32. D&C 82:19.
33. Neal A. Maxwell, *A Wonderful Flood of Light* (Salt Lake City: Bookcraft, 1990), 103.
34. Ezra Taft Benson, "Born of God," *Ensign* 19 (July 1989): 4, online at www.lds.org.
35. Brigham Young, in *Journal of Discourses,* 26 vols. (Liverpool: F. D. Richards, 1855–86), 11:18, December 11, 1864.
36. Ibid., 8:80, June 10, 1860.
37. Ibid., 9:370, August 31, 1862.
38. Ibid., 11:136, August 1–10, 1865.
39. James E. Faust, "Be Healers," *Clark Memorandum* (Spring 2003), 3.
40. "Church-Affiliated Ranch Balances Agriculture and Conservation in Central Florida," April 1, 2016, http://www.mormonnewsroom.org/article/church-ranch-balances-agriculture-conservation-central-florida (accessed September 25, 2017).
41. D&C 88:45, 47.
42. The Book of Mormon prophet Alma agreed with this concept; trying to convince an unbeliever about God's existence, he said: "All things denote there is a God, Yea, even the earth and all things that are upon the face of it … do witness that there is a Supreme Creator" (Alma 30:44).
43. See D&C 88:18–20.

BIBLIOGRAPHY

Abbey, Edward. *Desert Solitaire: A Season in the Wilderness*. New York: McGraw-Hill, 1968.

Albrecht, Stan L. and Tim B. Heaton. "Secularization, Higher Education, and Religiosity." *Review of Religious Research* 26 (September 1984): 43–58.

Alexander, Thomas G. "Brigham Young and the Transformation of Utah Wilderness, 1847–58." *Journal of Mormon History* 41 (January 2015).

———. "Conflict and Fraud: Utah Public Land Surveys in the 1850s, the Subsequent Investigation, and Problems with the Land Disposal System." *Utah Historical Quarterly* 80 (Spring 2012).

———. "Cooperation, Conflict, and Compromise: Women, Men, and the Environment in Salt Lake City, 1890–1930." *BYU Studies* 35 (1995): 6–39.

———. "Irrigating the Mormon Heartland: The Operation of the Irrigation Companies in Wasatch Oasis Communities, 1847–1880." *Agricultural History* 76 (Spring 2002): 172–87.

———. "Mormon Prophets and the Environment: Creation, Sin, the Fall, Redemption, and the Millennium." In *Dreams, Myths & Reality: Utah and the American West*, edited by William Thomas Allison and Susan J. Matt. Salt Lake City: Signature Books, 2008. 85–103.

———. "Red Rock and Grey Stone: Reed Smoot, the Establishment of Zion and Bryce National Parks, and the Rebuilding of Downtown Washington, D.C." *Pacific Historical Review* 72 (Spring 2003): 1–38.

———. *The Rise of Multiple-Use Management in the Mountain West: A History of Region 4 of the Forest Service*. Washington, DC, 1987.

———. "Senator Reed Smoot and Western Land Policy, 1905–1920." *Arizona and the West* 13 (Autumn 1971): 245–64.

———. "Stewardship and Enterprises: The LDS Church and the Wasatch Oasis Environment, 1847–1930." *Western Historical Quarterly* 25 (Autumn 1994): 340–64.

———. "Sylvester Q. Cannon and the Revival of Environmental Consciousness in the Mormon Community." *Environmental History* 3 (October 1998): 488–507.

———. "Wilford Woodruff, Intellectual Progress, and the Growth of an Amateur Scientific and Technological Tradition in Early Territorial Utah." *Utah Historical* Quarterly 59 (Spring 1991): 164–88.

Alexander, Thomas G. and James B. Allen. *Mormons and Gentiles: A History of Salt Lake City*. Boulder: Pruett Publishing Company, 1984.

Andersen, Rebecca. "Between Mountain and Lake: An Urban Mormon Country." PhD diss., Arizona State University, 2015.

Arrington, Leonard J. *Beet Sugar in the West: A History of the Utah-Idaho* Sugar Company, 1891–1966. Seattle: University of Washington Press, 1966.

———. *Brigham Young: American Moses*. New York: Knopf, 1985.

———. *Great Basin Kingdom: An Economic History of the Latter-day Saints*. Cambridge: Harvard University Press, 1958.

———. "The Mormon Cotton Mission in Southern Utah." *Pacific Historical Review* 25 (August 1956): 221–38.

———. "Property among the Mormons." *Rural Sociology* 16 (1951): 339–52.

———. "Taming the Turbulent Sevier: A Story of Mormon Desert Conquest." *Western Humanities Review* 5 (Autumn 1951).

Arrington, Leonard J., Dean L. May, and Feramorz Fox. *Building the City of God: Community and Cooperation among the Mormons*. Salt Lake City: Deseret Book, 1976.

Ashurst-McGee, Mark. "Zion Rising: Joseph Smith's Early Social and Political Thought." Ph.D. diss., Arizona State University, 2008.

Baer, Jr., Richard A. "Higher Education, the Church, and Environmental Values." *Natural Resources Journal* 17 (July 1977): 477–91.

———. "The Church and Man's Relationship to His Natural Environment." *Quaker Life* 12 (January 1970): 420–21.

Bagley, Will. *The Pioneer Camp of the Saints: The 1846 and 1847 Mormon Trail Journals of Thomas Bullock*. Spokane: Arthur H. Clark, 1997.

Bahr, Howard M. *Four Classic Mormon Village Studies*. Salt Lake City: University of Utah Press, 2014.

———. *Saints Observed: Studies of Mormon Village Life, 1850–2005*. Salt Lake City: University of Utah Press, 2014.

Bate, Kerry William. "John Steele: Medicine Man, Magician, Mormon Patriarch." *Utah Historical Quarterly* 62 (Winter 1994): 71–90.

Baugh, Alexander L. "John C. Frémont's 1843–44 Western Expedition and Its Influence on Mormon Settlement in Utah." *Utah Historical Quarterly* 83 (Fall 2015): 254–69.

Beckstead, James H. *Cowboying: A Tough Job in a Hard Land*. Salt Lake City: University of Utah Press, 1991.

Bennett, Richard E. *We'll Find the Place: The Mormon Exodus, 1846–1848*. Norman: University of Oklahoma, 2009.

Bennion, John. "Water Law on the Eve of Statehood: Israel Bennion and a Conflict in Vernon, 1893–1986." *Utah Historical Quarterly* 82 (Fall 2014): 289–305.

Bennion, Lowell C. "Ben." "Mormondom's Deseret Homeland." In *Homelands: A Geography of Culture and Place across America*, edited by Richard L. Nostrand and Lawrence E. Estaville. Baltimore: The Johns Hopkins University Press, 2001.

Berkhofer, Robert F., Jr. *A Behavioral Approach to Historical Analysis*. New York: Free Press, 1969.

Berry, Wendell. "Religion and the Environment." In *American Environmentalism: Readings in Conservation History*, edited by Roderick Frazier Nash. 3rd edition. New York: McGraw Hill Humanities/Social Sciences, 1990.

Bitton, Davis, and Linda P. Wilcox. "Pestiferous Ironclads: The Grasshopper Problem in Pioneer Utah." *Utah Historical Quarterly* 46 (Fall 1978): 336–55.

Bolton, Herbert E. "The Mormons in the Opening of the Great West." *Utah Genealogical and Historical Magazine* 44 (1926): 40–72.

Bowler, G. Lynn. *Zion on the Muddy: The Story of the Moapa (Logandale, Nevada) Stake*. Springville, UT: Art City Publishing, 2004.

Brehm, Joan M. and Brian W. Eisenhauer. "Environmental Concern in the Mormon Culture Region." *Society and Natural Resources* 19 (May–June 2006): 393–410.

Brooks, Juanita, ed. *Journal of the Southern Indian Mission: Diary of Thomas D. Brown*. Logan: Utah State University Press, 1972.

Bushman, Richard L. *The Refinement of America: Persons, Houses, Cities*. New York: Alfred A. Knopf, 1992.

Cannon, Brian Q. "Remaking the Agrarian Dream: The New Deal's Rural Resettlement Program in Utah." Master's thesis, Utah State University, 1986.

———. *Reopening the Frontier: Homesteading in the Modern West*. Lawrence: University Press of Kansas, 2009.

Carmack, Noel. "Running the Line: James Henry Martineau's Survey in Northern Utah, 1860–1862." *Utah Historical Quarterly* 68 (Fall 2000): 292–312.

Carson, Rachel. *Silent Spring*. Boston: Houghton Mifflin Company, 2002.

Carter, Thomas. *Building Zion: The Material World of Mormon Settlement*. Minneapolis: University of Minnesota Press, 2015.

Cheney, Brock. *Plain but Wholesome: Foodways of the Mormon Pioneers*. Salt Lake City: University of Utah Press, 2012.

Christensen, Earl M., and Myrtis A. Hutchinson. "Historical Observations on the Ecology of Rush and Tooele Valleys, Utah." *Proceedings of the Utah Academy of Sciences, Arts, and Letters* 42 (1965): 90–95.

Christy, Howard A. "Open Hand and Mailed Fist: Mormon-Indian Relations in Utah, 1847–52." *Utah Historical Quarterly* 46 (Summer 1978): 216–35.

Coates, Peter A. "The Strange Stillness of the Past: Toward an Environmental History of Sound and Noise." *Environmental History* 10 (October 2005): 636–65.

Coleman, Jon. *Vicious: Wolves and Men in America*. New Haven: Yale University Press, 2004.
Compton, Todd M. "The Big Washout: The 1862 Flood in Santa Clara." *Utah Historical Quarterly* 77 (Spring 2009): 108–25.
Cook, Lyndon W. *Joseph Smith and the Law of Consecration*. Provo, Utah: Grandin Book Company, 1985.
Cottam, Walter Pace. "Is Utah Sahara Bound?" *Bulletin of the University of Utah* 37, no. 11. Salt Lake City: Extension Division, University of Utah, 1947.
Cracroft, Richard H. "The Heraldry of the Range: Utah Cattle Brands." *Utah Historical Quarterly* 32 (Summer 1964): 217–31.
Cronon, William. "Landscapes of Abundance and Scarcity." In *The Oxford History of the American West*, edited by Clyde A. Milner II, Carol A. O'Connor, and Martha A. Sandweiss, 603–37. New York: Oxford University Press, 1994.
———. "Modes of Prophecy and Production: Placing Nature in History." *Journal of American History* 76 (March 1990): 1122–31.
———. *Nature's Metropolis: Chicago and the Great West*. New York: Norton, 1992.
Crosby, Alfred W. *The Columbian Exchange: Biological and Cultural Consequences of 1492*. Westport, CT: Praeger, 2003.
———. *Ecological Imperialism: The Biological Expansion of Europe, 900–1900*. New York: Cambridge University Press, 1986.
———. "An Enthusiastic Second." *Journal of American History* 76 (March 1990): 1107–10.
Crunden, Robert M. *Ministers of Reform: The Progressive's Achievement in American Civilization, 1889–1920*. New York: Basic Books, 1982.
Dant, Sara. "Field Notes: Brigham Young's 'All the People' Quote Quandary." *Western Historical Quarterly* 46 (Summer 2015): 219–23.
———. *Losing Eden: An Environmental History of the American West*. Hoboken, NJ: John Wiley & Sons, 2017.
Davies, Jeremy. *The Birth of the Anthropocene*. Oakland, CA: University of California Press, 2016.
Davis, John. *The Landscape of Belief: Encountering the Holy Land in Nineteenth-Century American Art and Culture*. Princeton: Princeton University Press, 1996.
DePillis, Mario S. "Christ Comes to Jackson County: The Mormon City of Zion and Its Consequences." *John Whitmer Historical Association Journal* 23 (2003): 21–44.
DeVoto, Bernard. *The Western Paradox: A Conservation Reader*, edited by Douglas Brinkley and Patricia Nelson Limerick. New Haven: Yale University Press, 2001.
Dunbar, Robert G. *Forging New Rights in Western Waters*. Lincoln: University of Nebraska Press, 1983.
Dunlap, Thomas R. "Values for Varmints: Predator Control and Environmental Ideas, 1920–1939." *Pacific Historical Review* 53 (May 1984): 141–61.

Dyal, Donald H. "The Agrarian Values of Mormonism: A Touch of the Mountain Sod." PhD diss., Texas A&M University, 1980.
Ellingson, Stephen. *To Care for the Creation: The Emergence of the Religious Environmental Movement.* Chicago: University of Chicago Press, 2016.
Ely, Richard T. "Economic Aspects of Mormonism." *Harper's Monthly Magazine* 106 (April 1903): 667–78.
Etulain, Richard W., ed. *Does the Frontier Experience Make America Exceptional?* Boston: Bedford/St. Martin's Press, 1999.
Faragher, John Mack. *Women and Men on the Overland Trail.* New Haven: Yale University Press, 1979.
Farmer, Jared. "Crossroads of the West." *Journal of Mormon History* (Winter 2015): 156–73.
———. *Glen Canyon Dammed: Inventing Lake Powell and the Canyon Country.* Tucson: University of Arizona Press, 1999.
———. *On Zion's Mount: Mormon, Indians, and the American Landscape.* Cambridge: Harvard University Press, 2008.
Fiege, Mark. *Irrigated Eden: The Making of an Agricultural Landscape in the American West.* Seattle: University of Washington Press, 1999.
———. *The Republic of Nature: An Environmental History of the United States.* Seattle: University of Washington Press, 2012.
Flores, Dan L. "Agriculture, Mountain Ecology, and the Land Ethic: Phases of the Environmental History of Utah." In *Working on the Range: Essays on the History of Western Land Management and the Environment,* edited by John R. Wunder. Westport, CT: Greenwood Press, 1985.
———. *The Natural West: Environmental History in the Great Plains and Rocky Mountains.* Norman: University of Oklahoma Press, 2001.
———. "Zion in Eden: Phases of the Environmental History of Utah." *Environmental Review* 7 (Winter 1983): 325–44.
Francaviglia, Richard. *Believing in Place: A Spiritual Geography of the Great Basin.* Reno: University of Nevada Press, 2003.
———. "The City of Zion in the Mountain West." *The Improvement Era* 72 (December 1969): 10–11, 14–17.
———. "Geography and Mormon Identity." In *The Oxford Handbook on Mormonism,* edited by Philip Barlow and Terryl Givens. Oxford University Press, 2015.
———. *Go East, Young Man: Imagining the American West as the Orient.* Logan: Utah State University Press, 2011.
———. "'Like the Hajis of Meccah and Jerusalem': Orientalism and the Mormon Experience." *Leonard J. Arrington Mormon History Lecture Series,* no. 17. Logan: Utah State University Press, 2011.
———. *The Mapmakers of New Zion: A Cartographic History of Mormonism.* Salt Lake City: University of Utah Press, 2015.
———. *Mapping and Imagination in the Great Basin: A Cartographic History.* Reno: University of Nevada Press, 2005.

———. *The Mormon Landscape: Existence, Creation and Perception of a Unique Image in the American West.* New York: AMS Press, 1978.

———. "The Passing Mormon Village." *Landscape* 22, no. 2 (Spring 1978): 40–47.

Frémont, John C. *Report of the Exploring Expedition to the Rocky Mountains in the Year 1842 and to Oregon and North California in the Years 1843–44.* Washington, DC: United States Senate, 1845.

Galli, Craig D. "Study Guide: LDS Perspectives on Environmental Stewardship." *Mormon Chronicles*, April 30, 2010.

Garate, Donald T. *The Zion Tunnel: From Slickrock to Switchback.* Springdale, UT: Zion Natural History Association, May 2, 1989.

Geary, Edward. *The Proper Edge of the Sky: High Plateau Country of Utah.* Salt Lake City: University of Utah Press, 2002.

Gilbert, G. K. "Salt Lake Drainage System." In *Report on the Lands of the Arid Region of the United States, with a More Detailed Account of the Lands of Utah*, edited by John Wesley Powell. Washington, DC: U.S. Government Printing Office, 1878.

Godfrey, Audrey M. "Colonizing the Muddy River Valley: A New Perspective." *Journal of Mormon History* 22 (Fall 1996): 120–42.

Godfrey, Matthew C. *Religion, Politics and Sugar: The Mormon Church, the Federal Government, and the Utah-Idaho Sugar Company, 1907–1921.* Logan: Utah State University Press, 2007.

Gottlieb, Roger S. "Introduction: Religion in an Age of Environmental Crisis." In *This Sacred Earth: Religion, Nature, Environment*, edited by Roger S. Gottlieb. 2nd edition. New York City: Routledge, 2004.

Hall, Marcus. *Earth Repair: A Transatlantic History of Environmental Restoration.* Charlottesville: University of Virginia Press, 2005.

Handley, George. *Home Waters: A Year of Recompense on the Provo River.* Salt Lake City: University of Utah Press, 2011.

———. "Toward a Greener Faith: A Review of Recent Mormon Environmental Scholarship." *Mormon Studies Review* 3 (2016): 85–103.

Handley, George B., Terry B. Ball, and Steven L. Peck. *Stewardship and the Creation: LDS Perspectives on the Environment.* Provo: BYU Religious Studies Center, 2006.

Hardin, Garrett. "The Tragedy of the Commons." *Science* 162 (December 13, 1968): 1243–48.

Hartley, William G. "Mormons, Crickets, and Gulls: A New Look at an Old Story." In *The New Mormon History: Revisionist Essays on the Past*, edited by D. Michael Quinn. Salt Lake City: Signature Books, 1992.

Hebner, William Logan. *Southern Paiute: A Portrait.* Logan: Utah State University Press, 2010.

Hinton, Wayne K. "Getting Along: The Significance of Cooperation in the Development of Zion National Park." *Utah Historical Quarterly* 68 (Fall 2000): 313–31.

Honker, Andrew. "'Been Grazed Almost to Extinction': The Environment,

Human Action, and Utah Flooding, 1900–1940." *Utah Historical Quarterly* 67 (Winter 1999): 23–47.
Hough, Franklin B. *Report upon Forestry*. Washington, DC: U.S. Government Printing Office, 1878–1880, 1882.
Howe, Daniel Walker. *What Hath God Wrought: The Transformation of America, 1815–1848*. New York: Oxford University Press, 2007.
Hunter, Milton R. *Brigham Young: The Colonizer*. Salt Lake City: Deseret News Press, 1940.
Hutchins, Zachary McLeod. *Inventing Eden: Primitivism, Millennialism, and the Making of New England*. Oxford: Oxford University Press, 2014.
Jackson, Richard H. "Geography and Settlement in the Intermountain West: Creating an American Mecca." *Journal of the West* 33 (July 1994): 22–34.
———. "Great Salt Lake and Great Salt Lake City: American Curiosities." *Utah Historical Quarterly* 56 (Spring 1988): 128–47.
———. "The Mormon Experience: The Plains as Sinai, the Great Salt Lake as the Dead Sea, and the Great Basin as Desert-cum-Promised Land." *Journal of Historical Geography* 18 (January 1992): 41–58.
———. "Mormon Perception and Settlement." *Annals of the Association of American Geographers* 68 (1978): 317–34.
———. "Myth and Reality: Environmental Perceptions of the Utah Pioneers." *The Rocky Mountain Social Science Journal* 9 (1972): 33–38.
———. "Righteousness and Environmental Change: The Mormons and the Environment." In *Charles Redd Monographs in Western History: Essays on the American West, 1973–74*, no. 5, edited by Thomas G. Alexander. Provo: Brigham Young University Press, 1975.
———. "Utah's Harsh Lands, Hearth of Greatness." *Utah Historical Quarterly* 49 (Winter 1981): 4–25.
Jackson, Richard H. and R. Henrie. "Perceptions of Sacred Space," *Journal of Cultural Geography* 3 (Spring/Summer 1983): 94–107.
Jenkins, Willis. *The Future of Ethics: Sustainability, Social Justice, and Religious Creativity*. Georgetown: Georgetown University Press, 2013.
Kneipp, L. F. "Utah's Forest Resources: Their Administration, Development and Use." In *Third Report of the State Bureau of Immigration, Labor, and Statistics*, 1915–1916. Salt Lake City: State of Utah, 1917.
Knowlton, Ezra C. *History of Utah Sand and Gravel Products Corporation, 1920–1958*. Salt Lake City: Utah Sand and Gravel Products Corporation, 1959.
Larson, Andrew Karl. *The Red Hills of November: A Pioneer Biography of Utah's Cotton Town*. Deseret News Press, 1957.
Layton, Stanford J. *To No Privileged Class: The Rationalization of Homesteading and Rural Life in the Early Twentieth Century American West*. Provo, UT: Charles Redd Center for Western Studies, 1988.
Madsen, David B., and Brigham D. Madsen. "One Man's Meat is Another Man's Poison: A Revisionist View of the Seagull `Miracle." *Nevada Historical Society Quarterly* 30 (Fall 1987): 165–81.

Manning, Richard. *Grassland: The History, Biology, Politics, and Promise of the American Prairie*. New York: Viking, 1995.

May, Dean. "'A Different Mode of Life': Irrigation and Society in Nineteenth-Century Utah." *Agricultural History* 49 (January 1975): 3–20.

McArthur, Aaron. *St. Thomas, Nevada: A History Uncovered*. Reno: University of Nevada Press, 2013.

Mead, Elwood. *Report of Irrigation Investigations in Utah*. Washington, DC: U.S. Government Printing Office, 1904.

Meinig, D. W. "The Mormon Culture Region: Strategies and Patterns in the Geography of the American West, 1847–1964." *Annals of the Association of American Geographers* 55 (June 1965): 191–219.

Merchant, Carolyn. "Gender and Environmental History." *Journal of American History* 76 (March 1990): 1117–21.

Miller, David E. *Hole-in-the-Rock: An Epic in the Colonization of the Great American West*. Salt Lake City: University of Utah Press, 1959.

Morgan, Dale L. *The State of Deseret*. Logan and Salt Lake City: Utah State University Press and the Utah Historical Society, 1987.

Muir, John. *My First Summer in the Sierra*. Mineola: Dover Publications Inc., 2004.

Nash, Roderick. *The Rights of Nature: A History of Environmental Ethics*. Madison: University of Wisconsin Press, 1989.

———. *Wilderness and the American Mind*, 3rd ed. New Haven: Yale University Press, 1982.

Nelson, Lowry. *The Mormon Village: A Pattern and Technique of Land Settlement*. Salt Lake City: University of Utah Press, 1952.

Nelson, Paul. *Wrecks of Human Ambition: A History of Utah's Canyon Country to 1936*. Salt Lake City: University of Utah Press, 2014.

Nibley, Hugh W. "Brigham Young on the Environment." In *To the Glory of God: Mormon Essays on Great Issues—Environment, Commitment, Love, Peace, Youth, Man*. Edited by Truman G. Madsen and Charles D. Tate Jr. Salt Lake City: Deseret Book Company, 1972.

———. "Subduing the Earth." In *Nibley on the Timely and Timeless: Classic Essays of Hugh W. Nibley*. Religious Studies Monograph Series. Provo, UT: Religious Studies Center, Brigham Young University, 1978.

Nibley, Preston. *Brigham Young: The Man and His Work*. Salt Lake City: Deseret News Press, 1937.

O'Dea, Thomas F. "The Mormon Village." *American Journal of Sociology* 59 (July 1953): 99–101.

Oelschlaeger, Max. *Caring for Creation: An Ecumenical Approach to the Environmental Crisis*. New Haven: Yale University Press, 1994.

Olsen, Steve L. *The Mormon Ideology of Place: Cosmic Symbolism of the City of Zion, 1830–1846*. Provo, UT: Joseph Fielding Smith Institute for Latter-day Saint History and BYU Studies, 2002.

Opie, John. "Renaissance Origins of the Environmental Crisis." *Environmental Review* 11 (Spring 1987): 2–17.

Palmer, William R. "Indian Names in Utah Geography." *Utah Historical Quarterly* 1 (January 1928): 5–26.

Peterson, Charles S. "Imprint of Agricultural Systems on the Utah Landscape." In *The Mormon Role in the Settlement of the West*, edited by Richard H. Jackson. Provo: Brigham Young University Press, 1978.

———. "San Juan in Controversy: American Livestock Frontier vs. Mormon Cattle Pool." In *Essays on the American West, 1972–1973*, edited by Thomas G. Alexander. Provo: Brigham Young University Press, 1974.

———. "Small Holding Land Patterns in Utah and the Problem of Forest Watershed Management." *Forest History* 17 (July 1973): 5–13.

———. *Take up Your Mission: Mormon Colonizing along the Little Colorado River, 1870–1900*. Tucson: University of Arizona Press, 1973.

Peterson, Levi S. "The Development of Utah Livestock Law, 1848–1896." *Utah Historical Quarterly* 32 (Summer 1964): 198–216.

Phillips, Robert Jones. "Utah Sand and Gravel Industry." Master's thesis, University of Utah, 1956.

Pisani, Donald J. *Water and American Government: The Reclamation Bureau, National Water Policy, and the West, 1902–1935*. Berkeley: University of California Press, 2002.

Pomeroy, Earl. "Toward a Reorientation of Western History: Continuity and Environment." *Mississippi Valley Historical Review* 41 (March 1955): 579–600.

Pyne, Stephen J. "Firestick History." *Journal of American History* 76 (March 1990): 1132–41.

Quammen, David. "The Paradox of the Park." *National Geographic* 229 (May 2016): 55–67.

Rath, Richard Cullen. *How Early America Sounded*. Ithaca, New York: Cornell University Press, 2003.

Reeve, W. Paul. *Making Space on the Western Frontier: Mormons, Miners, and Southern Paiutes*. Urbana: University of Illinois Press, 2006.

Reisner, Marc. *Cadillac Desert: The American West and Its Disappearing Water*. New York: Penguin Books, 1986.

Reuss, Martin, and Stephen H. Cutcliff. *The Illusory Boundary: Environment and Technology in History*. Charlottesville: University of Virginia Press, 2010.

Ricks, Joel E. *Forms and Methods of Early Mormon Settlement in Utah and the Surrounding Region, 1847 to 1877*. Logan: Utah State University Press, 1964.

Ridd, Merrill Kay. "Influences of Soil and Water Conditions on Agricultural Development in the Delta Area, Utah." PhD diss., Northwestern University, 1963.

Robinson, Michael J. *Predatory Bureaucracy: The Extermination of Wolves and the Transformation of the West*. Boulder: University Press of Colorado, 2005.

Rogers, Jedediah S. *Roads in the Wilderness: Conflict in Canyon Country*. Salt Lake City: University of Utah Press, 2013.

———, ed. *The Council of Fifty: A Documentary History*. Salt Lake City: Signature Books, 2014.

Rogers, Kristen. "William Henry Smart: Uinta Basin Pioneer Leader." *Utah Historical Quarterly* 45 (Winter 1977): 61–74.

———. "Steward of the Earth." *This People* (Spring 1990): 12–16.

Sahlins, Marshall. *Stone Age Economics*. New York: Routledge, 1972.

Schroeder, Walter A. "The Presettlement Prairie in the Kansas City Region (Jackson County, Missouri)." *Missouri Prairie Journal* 7 (1985).

Schwägerl, Christian. *The Anthropocene: The Human Era and How It Shapes Our Planet*. Santa Fe: Synergetic Press, 2014.

Smaby, Beverly P. "The Mormons and the Indians: Conflicting Ecological Systems in the Great Basin." *American Studies* 16 (Spring 1975): 35–48.

Smart, William B., ed. *Mormonism's Last Colonizer: The Life and Times of William H. Smart*. Logan: Utah State University Press, 2008.

Smith, Craig S. "James E. Talmage and the 1895 Deseret Museum Expedition to Southern Utah." *Utah Historical Quarterly* 84 (Spring 2016): 136–51.

Smith, Joseph Fielding. *Essentials in Church History*. Salt Lake City: Deseret News Press, 1922.

Smythe, William. *The Conquest of Arid America*. New York: The Macmillan Company, 1905.

Sonntag Bradley, Martha. "Creating the Sacred Space of Zion." *Journal of Mormon History* 31 (Spring 2005): 1–30.

Sorensen, Victor. "The Wasters and Destroyers: Community-Sponsored Predator Control in Early Utah Territory." *Utah Historical Quarterly* 62 (Winter 1994): 26–41.

Speth, William W. "Environment, Culture and the Mormon in Early Utah: A Study in Cultural Adaptation." *Yearbook of the Association of Pacific Coast Geographers* 29 (1967): 53–67.

Stansbury, Howard. *Exploration and Survey of the Valley of the Great Salt Lake of Utah, Including a Reconnaissance of a New Route through the Rocky Mountains*. Philadelphia: Lippincott, Grambo, 1852.

Stegner, Wallace. *Beyond the Hundredth Meridian: John Wesley Powell and the Second Opening of the West*. Boston: Houghton Mifflin Company, 1954.

———. *Mormon Country*. Lincoln: University of Nebraska Press, 1942.

———, ed. *This Is Dinosaur: Echo Park Country and Its Magic Rivers*. New York: Knopf, 1955.

Stewart, George and John A. Widtsoe. "Contribution of Forest Land Resources to the Settlement and Development of the Mormon-Occupied West." *Journal of Forestry* 41 (September 1, 1943): 633–40.

Stoffel, Richard W., and Michael J. Evans. *Kaibab Paiute History: The Early Years*. Fredonia, AZ: Kaibab Paiute Tribe, 1978.

Stoll, Mark R. *Inherit the Mountain: Religion and the Rise of American Environmentalism*. Oxford: Oxford University Press, 2015.

Stott, Clifford. *Search for Sanctuary: Brigham Young and the White Mountain Expedition*. Salt Lake City: University of Utah Press, 1984.

Strahorn, Robert E. *To the Rockies and Beyond: Or a Summer on the Union Pacific Railway and Branches*. Omaha: Omaha Republican Print, 1878.

Stratton, Richard D. *Kindness to Animals and Caring for the Earth: Selections from the Sermons and Writings of Latter-day Saint Church Leaders*. Portland, OR: Inkwater, 2004.

Sutter, Paul S. "The World with Us: The State of American Environmental History." *Journal of American History* 100 (June 2013): 94–119.

Tarr, Joel A. "The Material Basis of Urban Environmental History." *Environmental History* 10 (October 2005): 744–46.

Thayer, Robert L. *Lifeplace: Bioregional Thought and Practice, Life Place*. Berkeley: University of California Press, 2003.

Thomas, George. *The Development of Institutions under Irrigation: With Special Reference to Early Utah Conditions*. New York: Macmillan, 1920.

Trimble, Stephen L. *Bargaining for Eden: The Fight for the Last Open Spaces in America*. Berkeley: University of California Press, 2008.

Turner, Frederick. "The Invented Landscape." In *Beyond Preservation: Restoring and Inventing Landscapes*, edited by A. Dwight Baldwin Jr., Judith de Luce, and Carl Pletsch. Minneapolis: University of Minnesota Press, 1994.

Turner, John G. *Brigham Young: Pioneer Prophet*. Cambridge, MA: The Belknap Press of Harvard University Press, 2012.

Umbach, Gregory. "Learning to Shop in Zion: The Consumer Revolution in Great Basin Mormon Culture, 1847–1910." *Journal of Social History* 38 (Fall 2004): 29–61.

Valencius, Conevery Bolton. *The Health of the Country: How Settlers Understood Themselves and Their Land*. New York: Basic Books, 2002.

Van Wagoner, Richard S., ed. *The Complete Discourses of Brigham Young: Volume 1, 1832–1852*. Salt Lake City: The Smith-Pettit Foundation, 2009.

Walker, Don D. "The Cattle Industry of Utah, 1850–1900: An Historical Profile." *Utah Historical Quarterly* 32 (Summer 1964): 182–97.

Webb, Walter. *The Great Plains: A Study in Institutions and Environment*. Waltham: Ginn and Co., 1931.

West, Elliott. *The Contested Plains: Indians, Goldseekers, and the Rush to Colorado*. Lawrence: University of Kansas, 1998.

Wheeler, Emily Anne Brooksby. "The Solitary Place Shall Be Glad for Them: Understanding and Treating Mormon Pioneer Gardens as Cultural Landscapes." Master's thesis, Utah State University, 2011.

White, Lynn, Jr. "The Historical Roots of Our Ecological Crisis." *Science* 155 (March 1967): 1203–7.

White, Richard. "Environmental History, Ecology, and Meaning." *Journal of American History* 76 (March 1990): 1111–16.

Wilford Woodruff's Journal, 9 vols., edited by Scott G. Kenney. Midvale, Utah: Signature Books, 1985.

Wilkinson, Charles. *Fire on the Plateau: Conflict and Endurance in the American Southwest*. Washington, DC: Island Press, 1999.

Williams, Terry Tempest. *Red: Passion and Patience in the Desert*. New York: Pantheon Books, 2001.

Williams, Terry Tempest, William B. Smart, and Gibbs M. Smith, eds. *New Genesis: A Mormon Reader on Land and Community*. Layton: Gibbs Smith, 1999.

Wilson, E. O. *The Creation: An Appeal to Save Life on Earth*. New York: Norton, 2006.

Woodbury, Angus. "A History of Southern Utah and Its National Parks." *Utah Historical Quarterly* 12 (July–October 1944): 111–209.

Worster, Donald. *Dust Bowl: The Southern Plains in the 1930s*. New York: Oxford University Press, 1979, 2004.

———. *Rivers of Empire: Water, Aridity, and the Growth of the American West*. New York: Pantheon Books, 1985.

———. "Seeing beyond Culture." *Journal of American History* 76 (March 1990): 1142–47.

———. "Transformations of the Earth: Toward an Agroecological Perspective in History." *Journal of American History* 76 (March 1990): 1087–1106.

———. "The Vulnerable Earth: Toward a Planetary History." *Environmental Review* 11 (Summer 1987): 8–14.

Wright, John B. *Rocky Mountain Divide: Selling and Saving the West*. Austin: University of Texas Press, 1993.

Yorgason, Ethan R. *Transformation of the Mormon Culture Region*. Urbana: University of Illinois Press, 2003.

CONTRIBUTORS

Thomas G. Alexander is the Lemuel Hardison Redd Jr. professor emeritus of western American history at Brigham Young University. He is the author of numerous books and articles including: *Utah: The Right Place* (2nd edition, 2003); *Mormonism in Transition: A History of the Latter-day Saints, 1890–1930* (3rd edition, 2012); *The Rise of Multiple-Use Management in the Intermountain West: A History of Region 4 of the Forest Service* (1987); and *Things in Heaven and Earth: The Life and Times of Wilford Woodruff, A Mormon Prophet* (2nd edition, 1993).

Rebecca Andersen is a lecturer in history at Utah State University. She recently completed her PhD in American and public history at Arizona State University. Her work focuses on twentieth-century Mormon suburban growth along Utah's Wasatch Front.

Brian Q. Cannon is a professor of history and director of the Charles Redd Center for Western Studies at Brigham Young University. Three books, two edited collections, and over two dozen articles reflect his interests in western American, Mormon, Utah, and rural history. He is the past president of the Agricultural History Society and the Mormon History Association.

Sara Dant is a professor and chair of history at Weber State University, whose work focuses on environmental politics in the United States with a particular emphasis on the creation and development of consensus and bipartisanism. Her latest book is *Losing Eden: An Environmental History of the American West* (Wiley, 2017), and she is the author of several prize-winning articles and chapters on western environmental politics and water.

Brett D. Dowdle holds a PhD in American history from Texas Christian University. He is currently a historian for the LDS Church History Department, working on the Joseph Smith Papers Project.

RICHARD FRANCAVIGLIA, who received a PhD from the University of Oregon in 1970, is a historical geographer with a longstanding interest in the way places change through time, and how that change is depicted in maps. The role of religion in this process is of special interest to him. He began studying the Latter-day Saints in the late 1960s and has authored numerous articles and books about them, including *The Mormon Landscape* (1978) and *The Mapmakers of New Zion* (2015). He is a professor emeritus of history and geography at University of Texas–Arlington and is currently an associated scholar at Willamette University in Salem, Oregon.

BRIAN FREHNER is a historian of energy, environment and the American West. He teaches and writes about these topics at the University of Missouri–Kansas City. He is the author of *Finding Oil: The Nature of Petroleum Geology* and coeditor of *Indians and Energy: Exploitation and Opportunity in the American Southwest.* He is at work on another monograph related to the twentieth-century oil industry and is coediting a volume tentatively titled, Great Plains: An Environmental History. This project is funded by the National Science Foundation.

MATTHEW C. GODFREY is the managing historian of the Joseph Smith Papers and has served as a documentary editor on several volumes of the series. He holds a PhD in American and public history from Washington State University. He is the author of *Religion, Politics, and Sugar: The Mormon Church, the Federal Government, and the Utah-Idaho Sugar Company, 1907–1921* (2007), which was a cowinner of the Mormon History Association's Smith-Pettit Award for Best First Book. He has also published articles in journals such as *Agricultural History*, *Public Historian*, *Pacific Northwest Quarterly*, *BYU Studies Quarterly*, the *Journal of Mormon History*, and various collections of essays.

GEORGE B. HANDLEY is a professor of interdisciplinary humanities at Brigham Young University, whose writing focuses on the intersection between religion, literature, and the environment. A comparative ecocritic of the Americas, he is also known for his work in Mormon ecotheology and his creative writing that blends nature, theology, and family history. Handley is the author of *Home Waters: A Year of Recompenses on the Provo River* (2010), among other works.

JEFF NICHOLS is a professor of history at Westminster College in Salt Lake City, where he codirects the Institute for Mountain Research. A former U.S. Navy officer, he is the author of *Prostitution, Polygamy, and Power: Salt Lake City, 1847–1918* (2002) and coeditor of *Playing with Shadows: Voices of Dissent in the Mormon West* (2011).

BETSY GAINES QUAMMEN recently received her PhD in environmental history from Montana State University, focusing on Mormon settlement and public land conflict. She founded the Tributary Fund, an organization that amplified religious traditions celebrating wildlife and their habitats. Her project sites included Mongolia, Bhutan, and the American West. Betsy worked for the East African Wildlife Society in Kenya, later moving to Montana to focus on ecosystem protection, endangered species, and grazing reform in the Northern Rockies. She earned her MS in environmental studies at the University of Montana and served on the boards of the Sierra Club and American Wildlands.

JEDEDIAH S. ROGERS is co-editor of *Utah Historical Quarterly* and a senior state historian for the Utah Division of State History. Rogers is the author of *Roads in the Wilderness: Conflict in Canyon Country* (2013), winner of the Wallace Stegner Prize in American Environmental or Western History, and editor of *In the President's Office: The Diaries of L. John Nuttall, 1879—1892* (2007) and *The Council of Fifty: A Documentary History* (2014).

NATHAN N. WAITE is an editor with the LDS Church History Department and an associate editorial manager for the Joseph Smith Papers Project. He is coeditor of *A Zion Canyon Reader* (2014) and *Settling the Valley, Proclaiming the Gospel: The General Epistles of the Mormon First Presidency* (2017).

INDEX